MINING HAULAGE

I. C. S.
REFERENCE LIBRARY

A SERIES OF TEXTBOOKS PREPARED FOR THE STUDENTS OF THE INTERNATIONAL CORRESPONDENCE SCHOOLS AND CONTAINING IN PERMANENT FORM THE INSTRUCTION PAPERS, EXAMINATION QUESTIONS, AND KEYS USED IN THEIR VARIOUS COURSES

SCRANTON
INTERNATIONAL TEXTBOOK COMPANY
148

©2008-2010 Periscope Film LLC

All Rights Reserved

www.PeriscopeFilm.com

ISBN #978-1-935700-13-5

Copyright, 1907, by INTERNATIONAL TEXTBOOK COMPANY.

Entered at Stationers' Hall, London.

Hoisting: Copyright, 1906, by INTERNATIONAL TEXTBOOK COMPANY. Entered at Stationers' Hall, London.

Haulage: Copyright, 1906, by INTERNATIONAL TEXTBOOK COMPANY. Entered at Stationers' Hall, London.

Mine Drainage: Copyright, 1906, by INTERNATIONAL TEXTBOOK COMPANY. Entered at Stationers' Hall, London.

All rights reserved.

This book has been digitally watermarked to prevent illegal duplication.

©2008-2010 Periscope Film LLC

All Rights Reserved

www.PeriscopeFilm.com

ISBN #978-1-935700-13-5

CONTENTS

HAULAGE

Mine Cars	54	1
Ore Cars	54	6
Coal Cars	54	12
Mine-Car Details	54	24
Resistance to Mine-Car Haulage	54	37
Mine-Haulage Systems	54	43
Man Power for Transportation	54	44
Animal Haulage	54	47
Locomotive Haulage	55	1
Steam-Locomotive Haulage	55	4
Construction of Steam Mine Locomotives	55	5
Power of a Locomotive	55	10
Hauling Capacity of Locomotives	55	15
Compressed-Air Locomotive Haulage	55	22
Pipe Lines	55	24
Electric Locomotive Haulage	56	1
Wiring for Electric Mine Haulage	56	7
Electric Mine Locomotives	56	19
Haulage Locomotives	56	35
Gathering Locomotives	56	45
Rating and Capacity	56	52
Operation of Electric Mine Locomotives	56	61
Rope Haulage	57	1
Endless-Rope Haulage	57	2
Power for Driving Endless-Rope System	57	4
Combined Endless-Rope and Gravity Plane	57	13
District Haulage	57	13

CONTENTS

Haulage—*Continued*	Section	Page
Overhead Endless-Rope Haulage System	57	22
Track Rollers, and Carrying Sheaves	57	24
Curves	57	29
Endless-Rope Haulage Calculations	57	33
Operation of Endless-Rope System	57	36
Tail-Rope Haulage	57	37
District Tail-Rope Haulage	57	42
Crooked-Entry Haulage	57	45
Power for Operating Tail-Rope	57	52
Surface Tail-Rope System of Haulage	57	55
Signaling	57	56
Tail or Return Sheaves	57	58
Couplings	57	64
Third-Rail Tail-Rope Haulage	57	67
Tail-Rope Haulage Calculations	57	71
Gravity Planes	58	1
Arrangement of Tracks	58	2
Calculations for Gravity Planes	58	26
Engine Planes	58	32
Calculation for an Engine Plane	58	41

HAULAGE
(PART 1)

MINE CARS

1. Mine Cars in General.—Mine cars are used for the purpose of transporting materials from the working place in a mine to a place either in or out of the mine where they are unloaded. There is great variety in their shape and construction, but one of the chief considerations in their design is whether they are to be dumped by hand or by the aid of a tipple, and whether they are to be dumped from the end or side.

Some cars are constructed of wood and iron, others entirely of iron; some are very narrow gauge and others broader gauge; some have a very short wheel base and others a longer wheel base. Wooden mine cars are usually built at the mine, so there is no universal shape or design. As only a few cars can be described here, only those will be described that have proved satisfactory in places where they have been used.

2. Size of Car.—In mines where headroom for passage and loading is limited by the thickness of the material mined, or where the width of the passageways is restricted by the nature of the roof, floor, or the material mined, the cars must conform in size to the heights and widths of the passages in the mine through which they must be moved; where the passageways are not restricted by the thickness and nature of the material mined, the size of the cars is determined by the ease with which they may be handled and loaded.

Copyrighted by International Textbook Company. Entered at Stationers' Hall, London

On good, clean, level tracks, men can push cars weighing 3,000 pounds and carrying a load of $2\frac{1}{2}$ tons, but it is difficult to start and to stop them. To stop and to control them, therefore, brakes are required or the wheels must be skidded by sprags placed between the spokes.

Since mine cars must be moved about by hand, to a certain extent, probably the maximum practicable size is reached when they have a capacity of $2\frac{1}{2}$ to 3 tons. Mine cars with a larger capacity than this are not only difficult to handle, both for men and mules, but they must be made extra strong for the reason that heavy cars, when bumped together, even when empty, are more severely racked and otherwise damaged than light ones. Exceptions are, of course, found to this general rule and occasionally mine cars are found in the anthracite fields of Pennsylvania with a capacity of 140 cubic feet, or $3\frac{1}{2}$ to 4 tons of run-of-mine coal, and even this capacity in some mines could be much larger as far as the underground openings are concerned were it not for the trouble encountered in loading, hauling, raising cars to the surface and in dumping their contents.

In loading coal cars in flat workings, the coal must be shoveled into the cars; and if the coal is low, the loader is compelled to open the door of the car and throw the coal with his shovel to the rear end until the car is about three-fourths loaded, when he closes the door and loads the remainder of the car over the top. In case the car is long, it is extremely hard work to throw run-of-mine coal to the rear of a car, particularly under a car door; it is also hard work to throw coal more than 4 feet high into a car. This would not be such difficult work outdoors where the air is pure, since then the men would not get so short of breath or be so anxious to fill their cars, that they may not lose their turn with the drivers who gather the cars. If the miner does not top his car when mining by the car, he will be docked; if he is mining by weight, the company will not haul out his car unless fully loaded; adding to this the fact that the coal must be cleaned after mining, it will be seen that the men must work fast and hard; and that if the cars are too large, the output will be decreased, or more rooms, more cars, and more miners will be needed to keep up a steady supply of coal.

From a company standpoint, the size of cars is important, for if too small it will require more men, more cars, more hoisting, and more expense generally to produce a large daily tonnage; therefore, in mines where the coal seam is less than 5 feet in thickness, the cars are made long and broad to increase their capacity; and in mines where the coal seam is above 5 feet in thickness, the cars are so proportioned that they can be readily loaded and will hold about $1\frac{1}{2}$ to $2\frac{1}{2}$ tons.

Where the deposits being worked are inclined, the cars are sometimes loaded from chutes; in other cases, from platforms above the cars. In such cases, the car loader has not the trouble of the car loader in the flat seam, since most of the work is done by gravity. At some of the mines of the Messabi Range in the Lake Superior iron-ore district, where the milling system of mining is practiced, standard-gauge railroad cars have been run into the mine and loaded from chutes placed 60 feet apart.

3. Capacity of Mine Cars.—The capacity of a mine car is obtained by the following rule:

Rule.—*Multiply the capacity of the inside of the car box, in cubic feet, by the weight of a cubic foot of the loose material loaded into the car.*

The weight of a cubic foot of the unbroken material is obtained by multiplying the weight of a cubic foot of water (62.5 pounds) by the specific gravity of the material in the solid. As rock and coal, when broken into run-of-mine size, occupy more space than when in the solid, that is, before being broken from the face, the weight of a cubic foot of broken rock or coal is less than a cubic foot of the same material in the solid, and to obtain the weight of a cubic foot of the broken material the weight of a cubic foot of the solid material must be multiplied by the ratio between the space occupied by the broken rock in the solid and the space occupied by the same amount of rock when broken.

Expressed as a formula, this is,
$$C = 62.5 \times \text{Sp. Gr.} \times R \times S$$
in which C = capacity of car, in pounds;
Sp. Gr. = specific gravity of material loaded;
R = ratio between weight of a cubic foot of broken and a cubic foot of solid material;
S = contents of car, in cubic feet.

4. Data for Calculating Capacity of Coal Cars. Before the capacity of a car intended for a certain material can be calculated, it is necessary to know the weight of a cubic foot of material in the broken condition. The following data in regard to anthracite and bituminous coal will, therefore, be useful in this connection.

The specific gravity of anthracite may be assumed to average 1.5; a cubic foot of solid anthracite will, therefore, weigh 62.5 pounds $\times$ 1.5 = 93.75 pounds. Run-of-mine anthracite will occupy about two-thirds more space than anthracite in the solid before it is mined; that is, 1 cubic foot of solid anthracite will occupy $\frac{5}{3}$ cubic foot when broken, and a cubic foot of broken anthracite will therefore weigh $\frac{3}{5}$ as much as a cubic foot of solid; a cubic foot of broken run-of-mine anthracite will, therefore, weigh 93.75 $\times \frac{3}{5}$ = 56.25 pounds; 56 pounds per cubic foot of anthracite is close enough for ordinary calculations. The long ton of 2,240 pounds is used in calculating a ton of anthracite. Hence, a ton of anthracite occupies 2,240 $\div$ 56 = 40 cubic feet.

NOTE.—The ratio between the weights of a cubic foot of solid anthracite and a cubic foot of broken anthracite is sometimes taken as $\frac{3}{4}$ instead of $\frac{3}{5}$.

The specific gravity of bituminous coal varies more widely than anthracite and is from 1.25 to 1.35. In the rule for calculating the capacity of a mine car 1.3 may be assumed. Hence, a cubic foot will weigh 62.5 $\times$ 1.3 = 81.25 pounds. Run-of-mine bituminous occupies approximately two-thirds more space than the same coal in the solid before being mined; hence, a cubic foot of such coal will weigh 81.25 $\times \frac{3}{5}$ = 48.75 pounds. A ton of bituminous coal is 2,000 pounds;

hence, a ton will occupy 2,000 ÷ 48.75 = 40+ cubic feet or the *short* ton of bituminous coal occupies the same space as the *long* ton of anthracite.

A bushel of bituminous coal in Pennsylvania is 76 pounds, and a short ton contains 26.31 bushels. In Iowa, Illinois, Kentucky, and Missouri, the bushel weighs 80 pounds, and there are 25 bushels to the ton. In Montana, the weight of a bushel of coal is 76 pounds.

It is customary to "top off" the load of coal, that is, to build it up above the sides of the car so that the calculated capacity of the box must usually be increased by adding a given number of inches to the height of the box. Anthracite cars are usually built box shaped. Bituminous cars, on the other hand, are commonly built with flaring sides above the tops of the wheels. This must be taken account of in calculating the capacity of a bituminous car.

EXAMPLE 1.—What is the capacity of an anthracite car that is 7 feet long, 3.5 feet wide, and 3 feet high, inside measurements, assuming the topping to be 6 inches above the top of the car?

SOLUTION.—The height of the topping is added to the height of the car box. The capacity of the car is then $7 \times 3.5 \times 3.5 = 85.75$ cu. ft. If, then, the specific gravity of anthracite is 1.5 and it occupies two-thirds more space broken than in the solid, substituting in the formula in Art. **3**,

$$C = 62.5 \times 1.5 \times \tfrac{3}{5} \times 85.75 = 4,823 \text{ lb.} \quad \text{Ans.}$$

EXAMPLE 2.—Calculate the capacity, in cubic feet and tons, of a car for bituminous coal, 8 feet 2 inches long and having the dimensions given on the cross-section shown in Fig. 1.

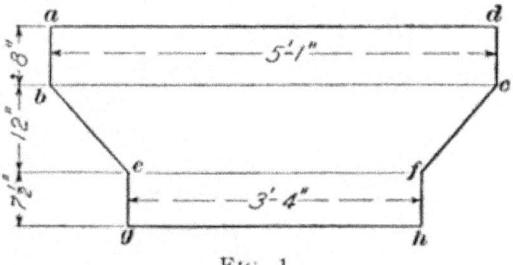

FIG. 1

SOLUTION.—To solve the problem, the cross-section must be divided into three parts, as shown, and each part calculated separately. 8 ft. 2 in. = 98 in.; 5 ft. 1 in. = 61 in.; 3 ft. 4 in. = 40 in.

	SQUARE INCHES
Area of $abcd$ is.............$61 \times 8 =$	488
Area of $bcfe$ is $\dfrac{40+61}{2} \times 12 = 50.5 \times 12 =$	606
Area of $efhg$ is............$40 \times 7\tfrac{1}{2} =$	300
Total area of cross-section........ $=$	1,394

Capacity of car 98 in. long with this cross-section is $1{,}394 \times 98 = 136{,}612$ cu. in. $= \frac{136612}{1728} = 79+$ cu. ft.   Ans.

Assuming 1 T. of bituminous coal $= 40$ cu. ft., $\frac{79}{40} = 1.975$ T.   Ans.

EXAMPLE 3.—If the specific gravity of a bituminous coal is 1.25 and it breaks in mining so that 1 cubic foot of solid coal occupies 1.5 cubic feet when loaded, how many cubic feet must a mine car contain to hold 2 tons?

SOLUTION.—The weight of a cubic foot of the broken coal is $62.5 \times 1.25 \times \frac{2}{3} = 52.08$ lb.   Two tons $= 2{,}000 \times 2 = 4{,}000$ lb. $\frac{4{,}000}{52.08} = 76.80$ cu. ft.   Ans.

EXAMPLES FOR PRACTICE

1. What is the capacity of an anthracite car that is 3 feet 6 inches wide, 6 feet long, and 4 feet high, assuming that it is loaded without topping? Ans. 2.1 T.

2. If the specific gravity of bituminous coal is 1.35 and it breaks in mining so that a cubic foot of broken coal occupies two-thirds more space than a cubic foot whole, how many cubic feet will a car contain to hold $1\frac{1}{2}$ tons of coal? Ans. 59.26 T.

3. How high must a mine car be that is 7 feet long and $3\frac{1}{2}$ feet wide, if it is to hold $2\frac{1}{2}$ tons of anthracite without topping? Ans. 4.08 ft. = 4 ft. 1 in.

4. What is the capacity of a bituminous-coal car with a cross-section similar to that shown in Fig. 1, in which $ad = 5\frac{1}{2}$ feet, $ab = 10$ inches, $ef = 3$ feet 6 inches, $eg = 10$ inches, the distance between bc and ef is 10 inches, the car is 7 feet long, and it is topped in loading 4 inches above the top of the side board? Ans. 2.289 T.

ORE CARS

5. Wooden Ore Cars.—The car shown in Fig. 2 was used with economy and satisfaction at a magnetic iron-ore mine where the cars were hauled about 1 mile on the surface, in trains. Its capacity was about 27 cubic feet or about $2\frac{1}{2}$ tons of magnetite. The frame was made of oak lumber mortised and tenoned. The end sills a and the cross-sills b were tenoned into mortises in the side sills c and held together by $\frac{1}{2}$-inch tie-bars d. The bumpers were made by allowing the side sills to project at each end of the car and placing beside each end an auxiliary block e; this wooden bumper was covered with sheet iron to prevent the wooden blocks being split when the cars came together. The draw-bar *f* was bolted to each cross and end sill so that the strain in starting would be equally distributed over the frame. Originally, the posts were tenoned for mortises in the side sills and in the top rails. But as these numerous mortises and tenons weakened the frame, added expense to the construction and repairs, besides with wet ore and in wet weather

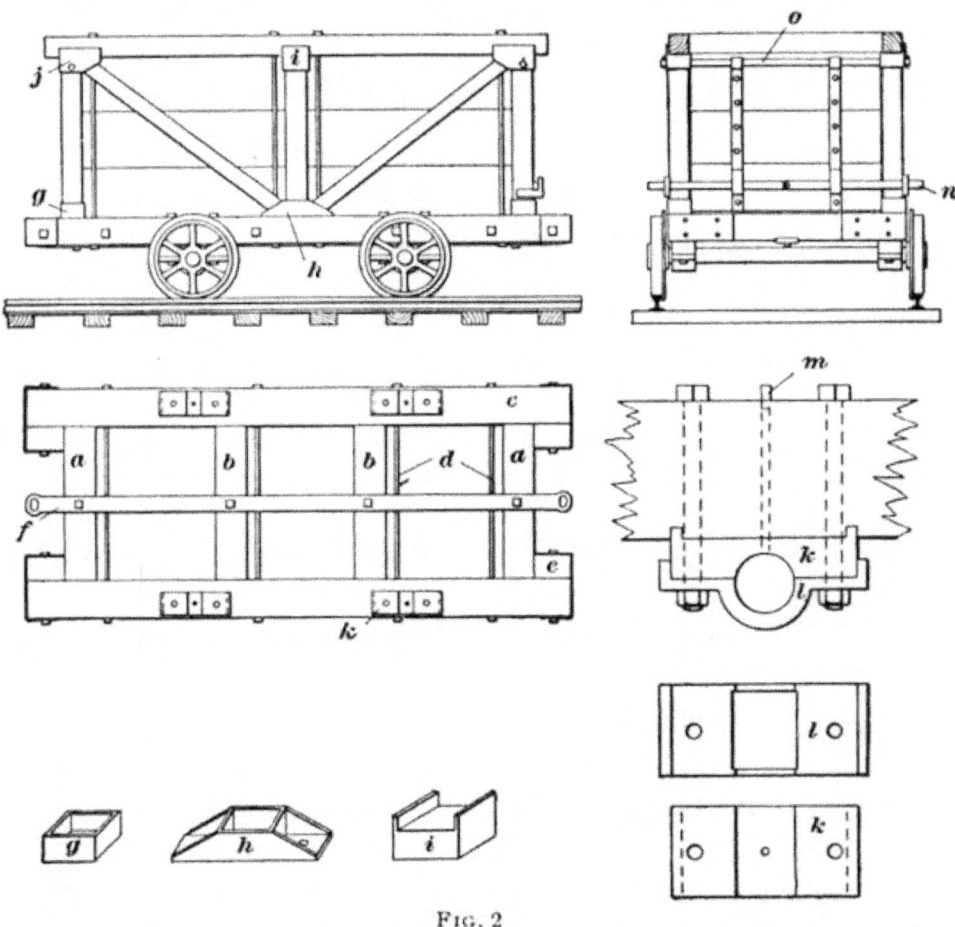

Fig. 2

being instrumental in causing quick decay of the timbers, the cast-iron cups *g*, *h*, *i*, *j* were substituted, considerably decreasing the cost of the car. The wheel base on these cars was 36 inches, 18-inch wheels being shrunk on to 3-inch axles. The latter were turned down to $2\frac{3}{4}$ inches diameter and given a face for 4-inch journal bearings. The journal-box *k* was made of cast iron in two sections, the upper part *k* being babbitted for the axle, and the lower part *l* being cast with a small recess for waste. This inexpensive journal-box was easily babbitted by pouring Babbitt metal through the oil hole of the upper casting. When the car needed oiling, the wooden plug *m* was removed and a little oil poured into the hole. The cars dumped at the end, and the door was provided with a latch *n* and a hinge bar *o*. As ore was shoveled into these cars, the floors and top side rails needed renewal from time to time as they were worn by the jagged corners of the ore. The weight of such a car was about 1,500 pounds, and its cost complete about $60.

6. Iron Ore-Cars.—In the western United States, steel cars are frequently used because it is often difficult to obtain proper lumber for wooden cars, and the sharp rock wears

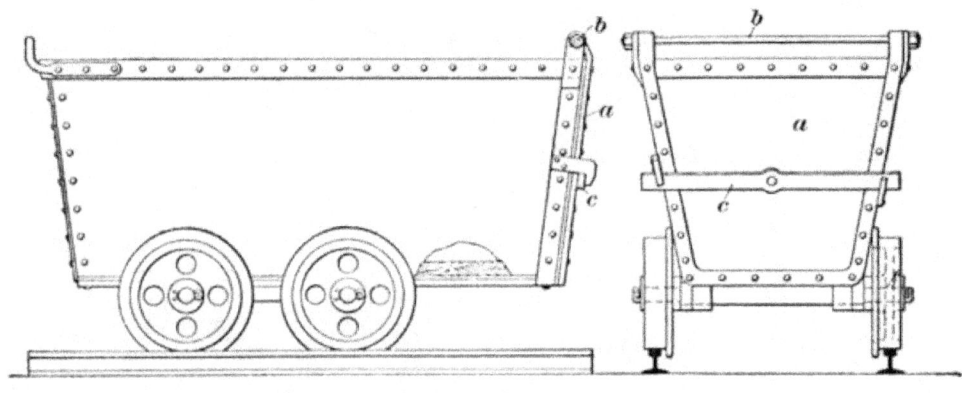

Fig. 3

wood quickly. Another reason is that light-weight cars are required in many ore mines and the size must also be small on account of narrow headings.

Fig. 3 shows an end-dump steel car with the door *a* swinging from a rod *b* at the top of the car, and a latch *c* extending across the car. These cars are made with fixed or self-oiling loose wheels, and stiffness is given them by using angle irons on the corners and flat iron bars along the edges.

The principal wear comes on the bottoms of these cars due to the sharp ore being thrown into the car and to the abrasion when the car is dumped. To provide for this wear, a false bottom of 2-inch oak, maple, or beech planks, or of steel plate is provided, which can be replaced when worn out. A timber false bottom is usually cheaper than the steel and can be more easily replaced; also, when removed, it can be used for fuel under the boiler, while the steel-plate linings are a total loss.

The track gauge for this car varies, but is usually about 24 inches. The length varies from 48 to 72 inches and the height above the rail from 48 to 72 inches. The capacity is from 12 to 32 cubic feet and the weight from 650 to 1,370 pounds, according to the capacity.

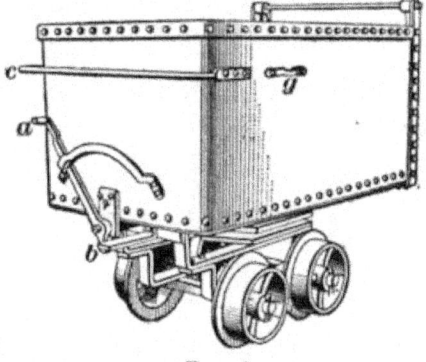

Fig. 4

7. Rotary Dump Cars. Figs. 4 and 5 illustrate a common form of ore car so constructed that it can be dumped either to the right or left, as well as at the end. The body is fastened to the trucks by means of a pivot at e, Fig. 5. When it is desired to dump the car, the operator takes hold of the handle c and then throws the handle a far enough to release the latch b, but this does not rotate the shaft d far enough to release the latch that holds the front end f of the car from swinging open. After this, he lifts the handle c and swings the car to the right or left, as may be desired; then by throwing the handle a clear over, as shown in Fig. 5, the door, or front end, of the car will swing out and the contents be discharged. After this, the car can be swung around and dropped into place. The front end f closes of itself and is secured by returning the lever a to the position shown in Fig. 4. The same style of truck is sometimes employed with wooden cars, but, as a rule, steel is the best material from which to manufacture mine cars for use at ore mines. The little loop g riveted on the side of the car is for inserting

Fig. 5

the latch that holds the car in place while it is being hoisted on the cage. When ore is brought to the surface in skips, it is frequently dumped into cars of this pattern before it is run out to the pocket or stock pile. The car is made from 42 to 60 inches long, 24 to 33 inches wide, 21 to 28 inches deep, and having capacities that range from 12 to 32 cubic feet. The wheels run loose on fixed axles, and the cars are made with gauges to suit the buyer. These and the scoop-box cars, Fig. 6, are top heavy and should have small wheels with wide gauges.

8. The scoop-box car, Fig. 6, has a body of sheet steel with a false bottom. One advantage that this ore car possesses is that no door is needed at the front, as the scoop body forms a sort of chute over which the ore slides. One lever a beneath the car prevents the car body from dumping or rotating except as desired. The lever can be worked by the foot, and since the majority of the ore is in the box rather than the scoop end of the car, a man must raise the car body so that it will tilt on the hinge b and dump. The car is provided with a fifth wheel c so that it may be dumped over the side as well as at the end. The handle d is for rotating the car body and for pulling it back on the truck after dumping. The cars are made from 49 to 67 inches in length, from 27 to 32 inches in width, and from 38 to 40 inches above the top of the rail. They have capacities ranging from 12 to 24 cubic feet, which amounts to from 1 to 2 tons of quartz rock. The gauge for these cars is narrow, from 18 to 24 inches, but can be made **wider**; however, wide gauges throw the body of the car

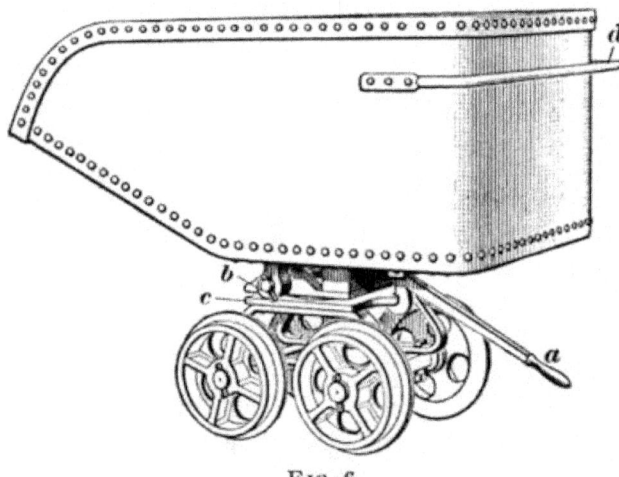

Fig. 6

§ 54　　　　　　　　HAULAGE　　　　　　　　11

higher in order that the scoop can clear the wheels for side dumping. Cars of this description are trammed and dumped by one man, and being without a door have one less part to get out of order.

9. Figs. 7 and 8 show ore cars without doors at either

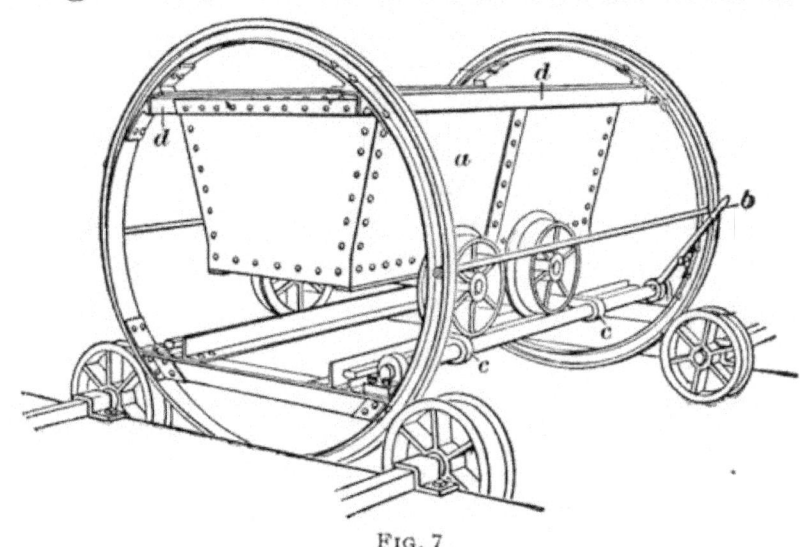

Fig. 7

end, the cars being dumped by a revolving cradle turning them upside down. The advantage derived from the use

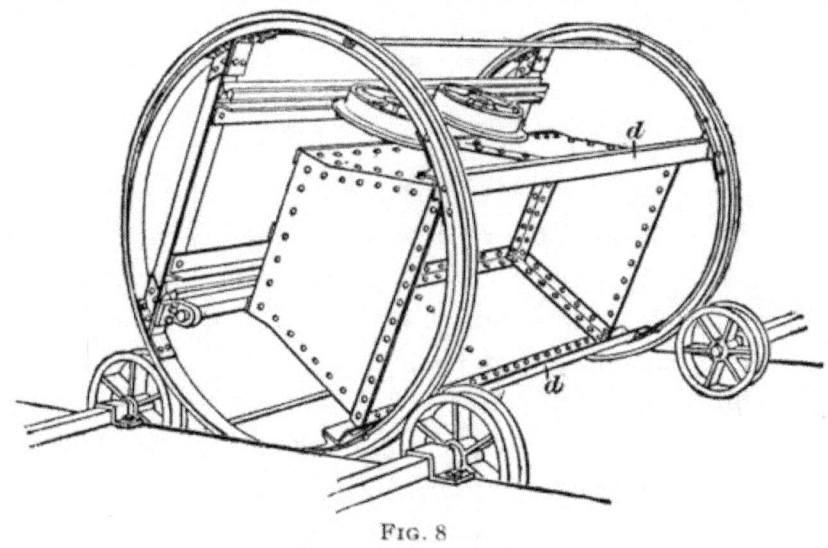

Fig. 8

of such cars is in the first cost and repairs, since their construction is simple. After the car a is in position on the dump, the end lever b is pulled up in such a way as to drop the dogs c, which hold the wheels so that the car is fastened in the dump. The upper edges of the car are held in place by the angle irons d. After the loaded car is in position, the entire dump is revolved into the position shown in Fig. 8. In most cases, the cradle makes a complete revolution, and when it returns to its upright position, a dog or latch (not shown in the figure) catches the rails and holds the cradle in such a position that the portion of the track on which the car rests is continuous with the regular railroad track. After this, the lever b is thrown down and the car run out on to the track. When the car is loaded, the greater part of the weight is above the center of gravity of the cradle, and hence it turns very easily, while after the ore is out, the heavy car wheels and track enable the operator to bring it back into its upright position with very little effort.

Instead of the cradle being revolved automatically or by hand, it may be revolved by ropes or chains encircling it and having the ends attached to the ends of the piston rods that are moved by pistons operated by steam or compressed air. The cradle may also be made large enough to receive a number of cars at once.

COAL CARS

ANTHRACITE MINE CARS

10. Cars for anthracite are usually built of timber and the sides of the box are usually straight. The wheels are generally placed inside the bed frame of the car, but there is a wide variation in the gauge of track, in the wheels and axles, and in the construction of the cars, each of the large coal companies often having a number of different designs of its own. The average capacity is probably from 2 to $2\frac{1}{2}$ tons, though this varies widely. The following are a few of the types of cars used in the anthracite mines of Pennsylvania.

11. Diamond, or Hollenback, Car.—Fig. 9 shows a hopper-shaped car with 27-inch gauge, which is a narrow

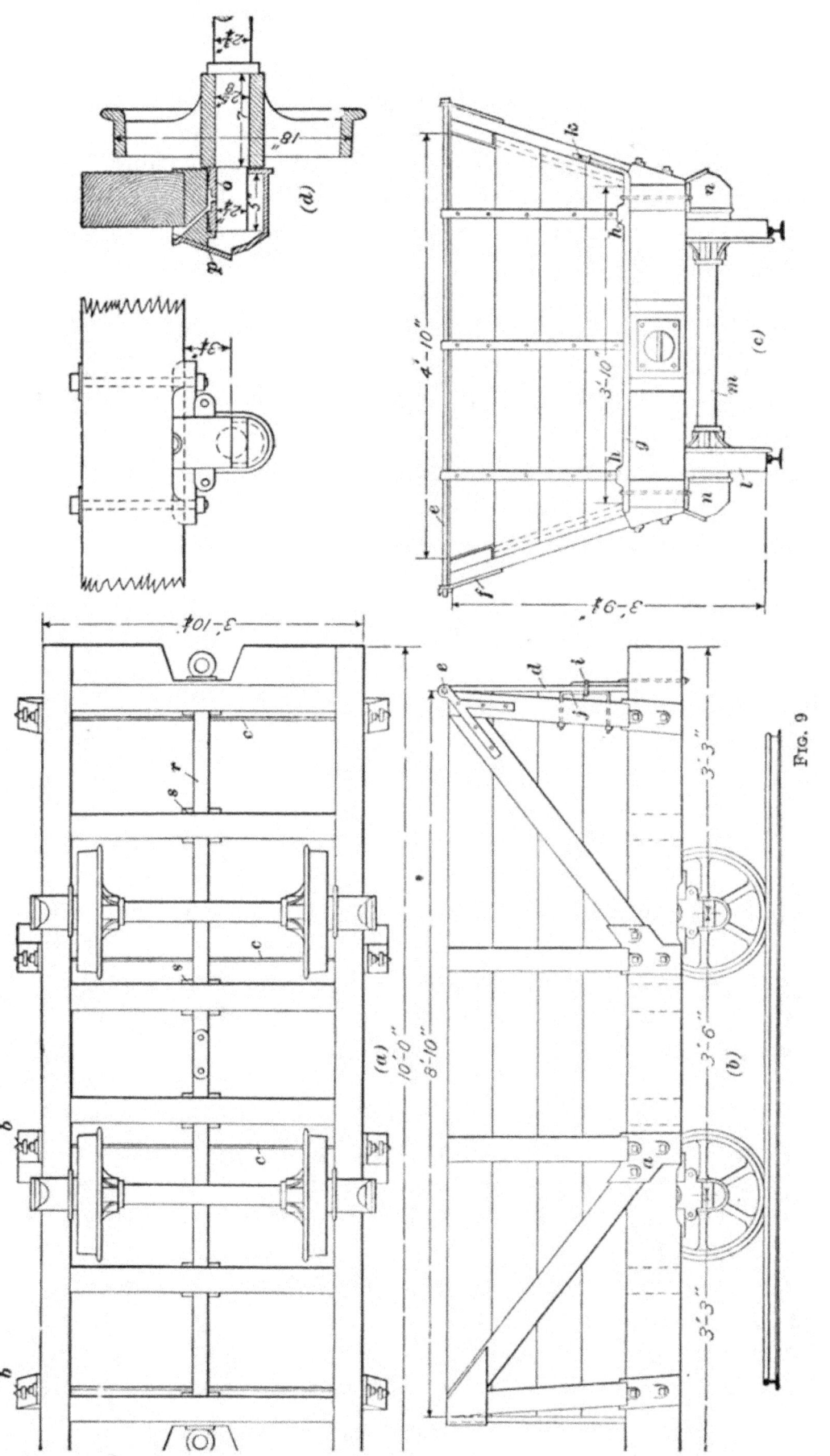

Fig. 9

gauge for anthracite mines. It is given this shape in order to increase its capacity without increasing its length or height. Like most anthracite mine cars, the wheels are placed inside the frame. Fig. 9 (*a*) shows the plan of the car bottom, (*b*) the side elevation, (*c*) the door end, with the general dimensions on each view, and (*d*) a detail of the car box.

The cup castings *a* hold the side posts and both are held in place by bolts *b* and tie-rods *c*, which rods also stiffen the bed frame. The car is an end dumper and the door *d* swings on an iron rod *e* running across the top of the car and supported in suitable irons *f*. The door is locked by a rocking shaft *g* with iron lips *h* that are held upright by a movable ring *i* that slides in the staple *j* and fits over the lever *k* on the end of the shaft *g*. When the ring is raised, the lever is loosened and falls, permitting the car door to swing backwards. The wheels *l* are tight on the axle *m* and the ends of the axle rest in journal-boxes *n*. These journal-boxes are patterned after those used on a standard-gauge railroad car, with journal brasses *o* fitted into the journal-box of the car above the axle. The box is packed with oily waste for lubricating the journal and has a swinging cover *p* for the purpose of inspecting the condition of the brass and packing the waste and oil. The drawbar *r* is in two parts and is riveted together at the center. An attempt is made to divide the pull between the several cross-pieces of the bed frame by means of plates and cotters *s*, but it is difficult to keep an arrangement of this kind rigid.

12. Fig. 10 shows an anthracite car in which the sides are supported by flat iron braces instead of being supported by timber braces set in pedestals or mortised into the sills as has usually been done with anthracite cars. In this way, the width of the car is materially increased, permitting of a decrease in its height and length if desired. The cross-ties *a* of the frame are tenoned, as shown in Fig. 10 (*e*), and the side pieces *b* are held together by tie-rods *c*. The frame is stiffened lengthwise by tie-rods *d*, and the drawbars are

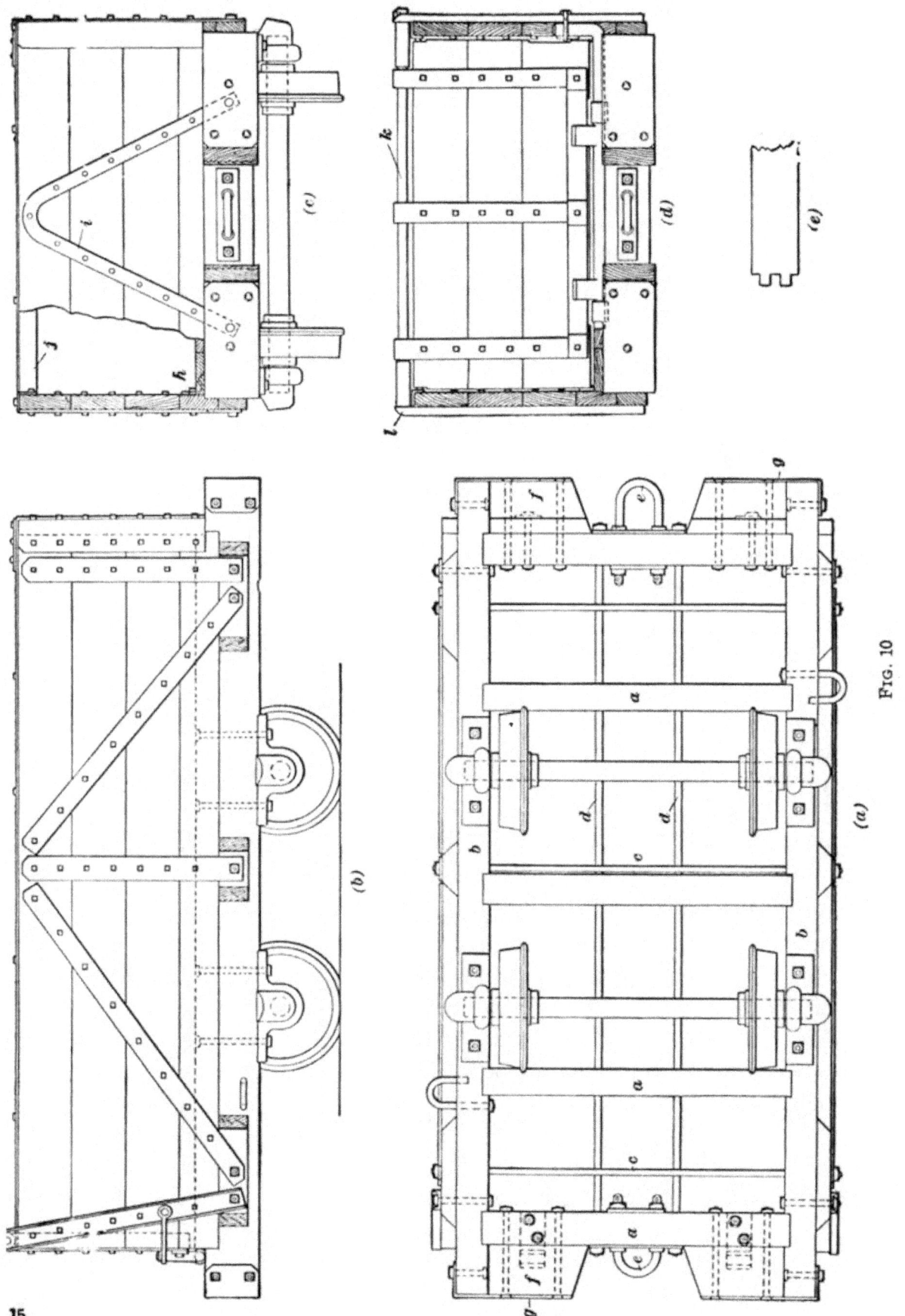

Fig. 10

simply a loop e bolted through the end sills. The bumpers f are covered by sheet iron g, which is held in place by countersunk bolts, as shown. The bottom of the car is of $1\frac{1}{2}$-inch plank and the sides and ends of 2-inch plank. The sides are braced by iron strips 3 inches wide and $\frac{3}{8}$ inch thick bolted to the planking, as shown. The sides of the cars are further braced inside by angle irons h. The back of the car is braced by a **V**-shaped brace i, which bolts to the cross-sill at the bottom and to timber and angle iron j at the top. The top of the car is braced by angle irons laid on top of the side and end timbers and held in place by screws. The door is hung from a cross-bar k supported at the top of the end braces l. The door is held shut by an ordinary lever latch, but the loop holding the latch is bolted through the side timber instead of being held by a ring, as is ordinarily the case.

13. In Fig. 11 is shown a car that is used on slopes as well as on gangways. One of its notable features is the method of fastening the drawbar a to the side sills, Fig. 11 (a). The hooks b, Fig. 11 (b), are for fastening the chains when the car is being hoisted up the slope. The rounded tie-pieces c, Fig. 11 (c) and (d), hold the sides together and fasten the sheet iron that covers the top of the back half of the car to prevent coal from rolling from the car as it is being drawn up the slope. The wheels are loose keyed to axles and are outside the frame. The track gauge is $39\frac{1}{2}$ inches, which is an unusual gauge, an even number of inches being preferable since the gauge outside must be the same as inside the mine and on the slope. The side braces and the upper part of the posts are tenoned into mortises, while the side braces are cut to join flush with the center post, and the posts are fitted into pockets on the sill.

The car is lined with 1-inch boards and these lined with No. 12 B. W. G. sheet iron. The wheel base is 31 inches, which is necessary on account of sharp curves.

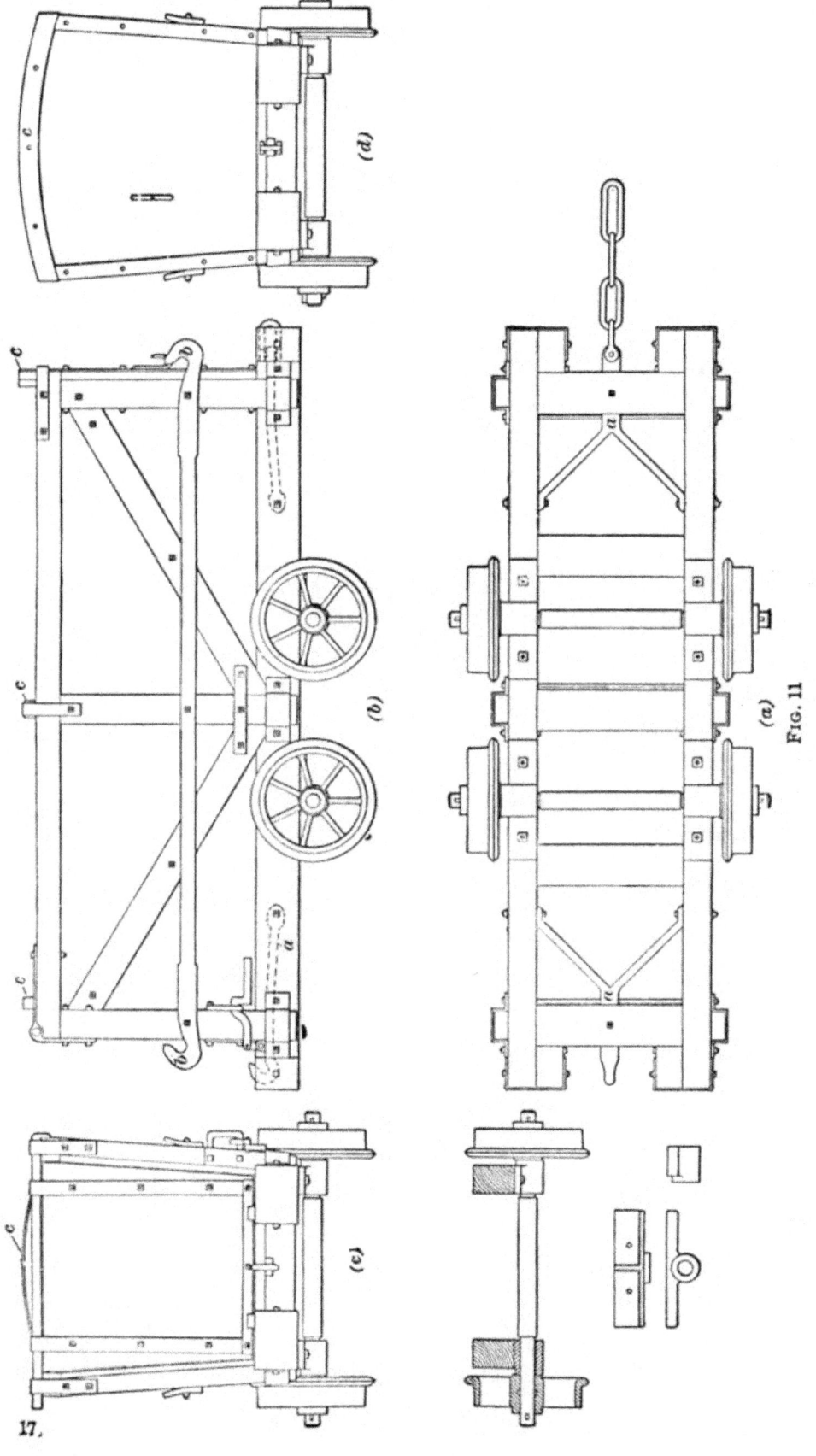

FIG. 11

BITUMINOUS MINE CARS

14. Bituminous-coal cars are built in a great variety of sizes and for widely varying gauges, but they seldom have a capacity of over 100 cubic feet, or about $2\frac{1}{2}$ tons, and are frequently smaller than this, generally averaging from $1\frac{1}{2}$ to 2 tons. They usually have outside wheels and the sides are then flared above the wheels to provide for increased capacity. Both loose and fixed wheels are used. Cars are generally built of wood, although steel cars are used to a limited extent. There is as much variety in details of construction as in anthracite mine cars, so that it is only possible to give a few of the types in common use.

15. Fig. 12 shows a common form of bituminous car, (*a*) being a plan, (*b*) an elevation, (*c*) the door end, and (*d*) the back end. The plank floors *a* and sides *b* are bolted to properly bent car irons *c* that serve as stays to keep the car body from spreading. The car, while having an 18-inch car wheel, stands 4 feet $\frac{5}{8}$ inch above the rail at the door end and 3 feet 6 inches at the back end. Although loose wheels are used, the cars are broad and low so that the center of gravity coming near the axle tends to promote smooth running and steadiness. The car is made wide in order to obtain capacity without increasing the height, and a good feature is the vertical top plank *d*. This car is capable of carrying $2\frac{1}{2}$ tons when heaped, but this load is too much for the stiffness of the lumber. The gauge is rather wide, 40 to 42 inches being most generally used.

16. Fig. 13 shows the car used almost universally in the Flat Top, West Virginia, coal field. Some details, such as latches, diameter of wheels, and brakes differ from those shown. The regulation car of this field has a capacity of 92 cubic feet, and when topped 6 inches has a capacity of about 110 cubic feet; if not strapped across the center, it will bulge at the sides and thus probably bring the capacity up to 113 cubic feet, or approximately 3 tons.

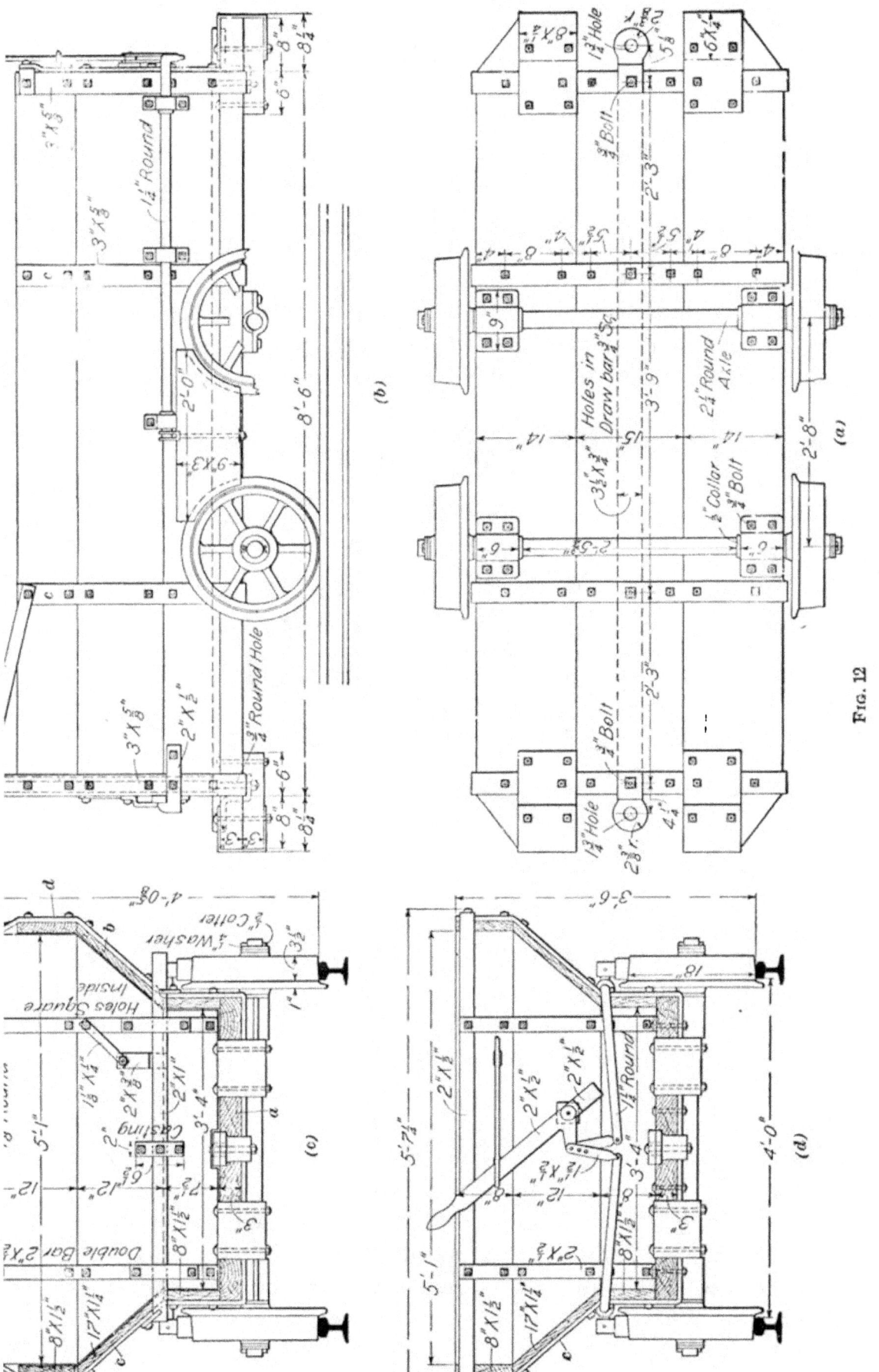

Fig. 12

These cars are built of oak, the bottom planks a being 3 inches thick and the side planks b $1\frac{1}{2}$ inches thick; the doors c are of $1\frac{1}{2}$-inch plank, but 2-inch plank is better; the back-end planks d are $1\frac{1}{2}$ inches thick. The back end is not carried to the top, a space being left to make the loading of the car easier. The cars are serviceable but with the short wheel base, 30 inches, are springy on rough tracks, particularly when there is no tie-strap across the top at the center and they are overloaded. Putting on the brake e also racks the

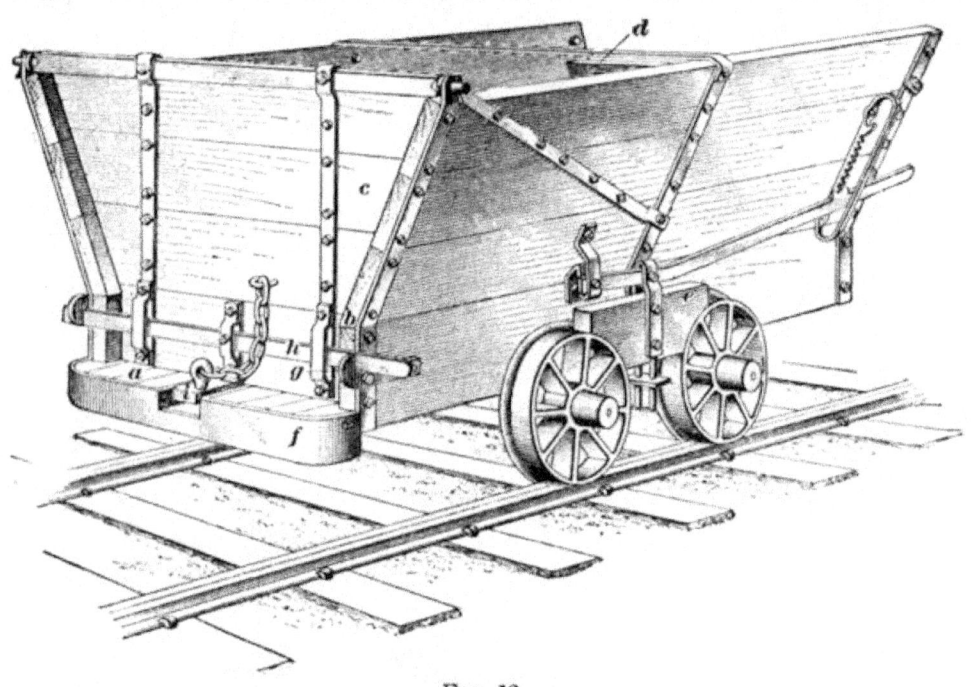

Fig. 13

cars badly if there is no tie-strap on the car. The bumpers f are made by continuing the bottom planks a past the end and reenforcing them by 3-inch planks underneath. The bumper blocks are rounded on the outside end and sheathed with $\frac{1}{4}$-inch sheet iron to prevent the ends becoming broomed. In some cases, the planks are only covered with sheet iron and are not doubled; one car is then liable to climb on the bumper of another and cause a wreck. The reenforcing plank may be carried back and bolted to the car floor to advantage, since the sheet iron often comes loose and the nuts to the bolts come off, allowing the bumpers to become jammed. When business is good, the cars cannot often be spared; hence, often before they are repaired the bumper is wrecked and considerable expense entailed for repairs. The catch g for the latch h on the front end of the car is apt to get out of repair after the car has been in service a short time. The tipple man usually knocks up the catch with a coupling pin; this enlarges the bolt hole in the door, and eventually allows so much play that unless repaired the catch is apt to work loose and allow coal to spill along the track and on the tipple floor. Where end dumps and cradle tips are used, the cars strike the horns with a jar that enlarges the pedestal bolt holes, allowing side motion due to loose pedestals, so that cars are kept on the track merely by the wheel flanges. In case the axles of the self-oiling wheels are worn conical on the car side of the hub, the cars will leave the track on the least provocation, sometimes causing considerable damage to other cars. The drawbar shown in Fig. 12, is a very good type for such cars. The hole for the coupling pin should be $1\frac{3}{4}$ inches in diameter for a $1\frac{1}{2}$-inch coupling pin. The middle cross-bar is forged to a shape that will fit the top side planks, or if not of the proper shape the iron should be heated in the blacksmith's fire and when at a dull-red heat bent over a form. There should be no attempt made to bend this cross-bar to fit the car when the iron is cold or improperly heated as the endeavor will probably break the iron or injure it so that it will eventually break. For the preservation of such cars, cross-bars are necessities and there is no economy in running cars without them, although it is frequently done. Car-iron makers sometimes neglect to forge these cross-bars properly, either giving them too much or too little heat. This can be ascertained by inspection when the irons arrive at the mine previous to their being placed on cars. If too much heat has been given them, the forging will appear burned; if too little heat, the fibers of the iron will be stretched at the bends. The car irons, wheels, and axles can usually be bought from a manufacturer of such irons cheaper than they can be made by the mine blacksmith.

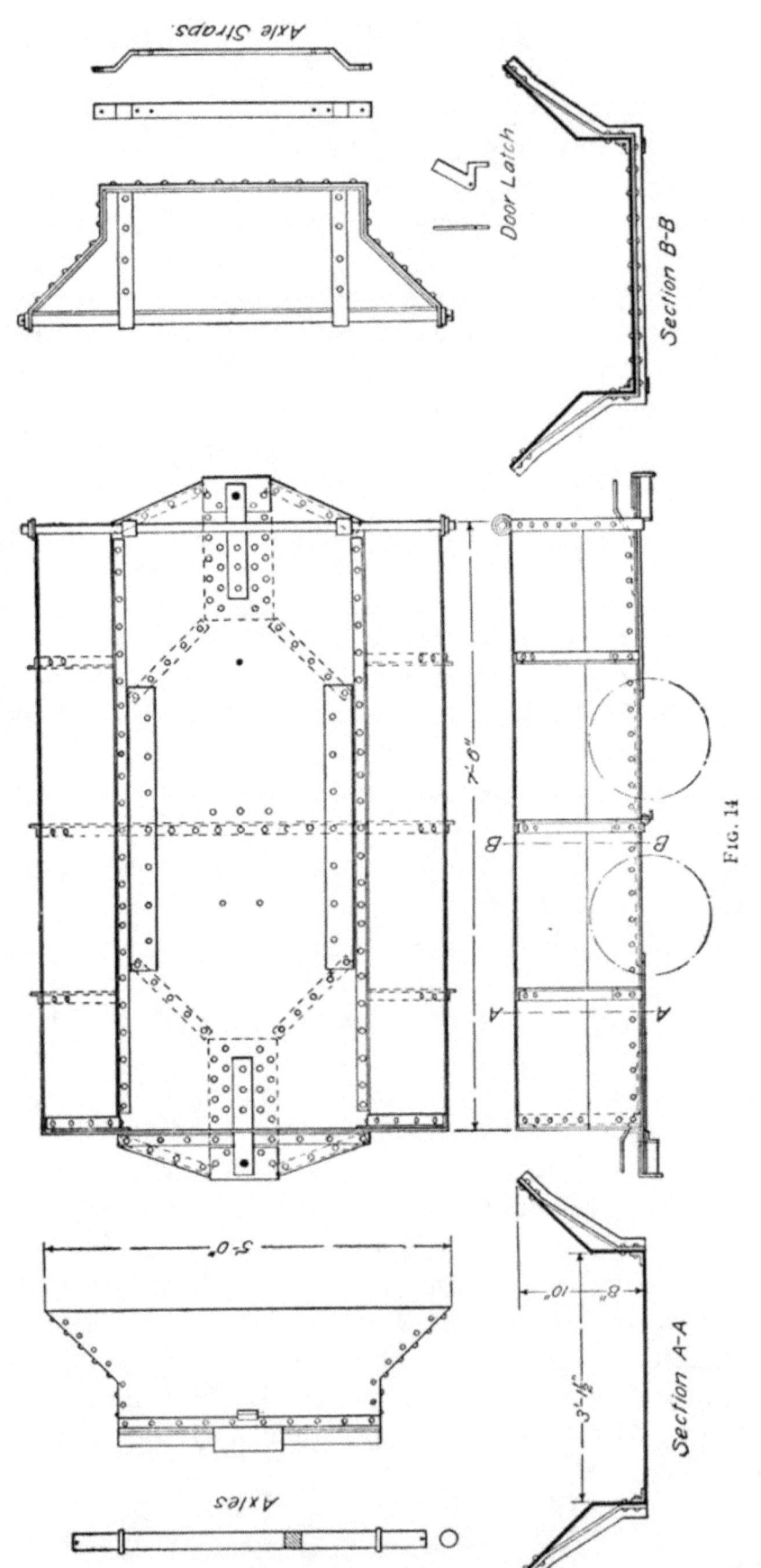

Fig. 14

STEEL COAL CARS

17. Steel cars for use in coal mines have not been generally adopted in America up to the present time on account of their cost and the additional equipment required about a mine to build and repair them, as it requires much more elaborate machine shops to build the steel than the wooden cars. An objection raised to steel cars by many is the difficulty of clearing up a wreck when one occurs in the mine, owing to the lack of facilities and the cramped quarters in which the work must be done.

18. Fig. 14 shows a steel car used at the mines of the Cambria Steel Company, Johnstown, Pennsylvania. This

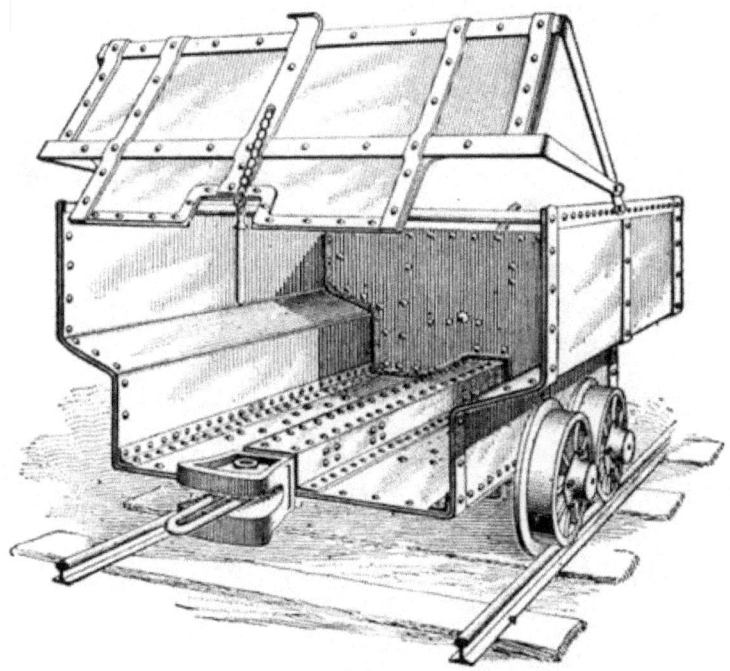

Fig. 15

car is the result of extensive experiments in building steel cars and combines a number of novel features. The drawbar is placed directly on top of the bottom plate of the car and is reenforced on the under side by a $\frac{1}{4}$-inch plate and by angle irons placed diagonally and riveted on the under side to distribute the bumping and pulling stresses, approximately, throughout the whole bottom of the car.

The general dimensions of this car are as follows: Gauge of track, 3 feet 9 inches; thickness of metal, $\frac{3}{16}$ inch; length over all, 8 feet 9 inches; effective width of car, inside, 5 feet; length of car, inside, 7 feet 6 inches; extreme width of car, 5 feet $4\frac{1}{2}$ inches; height of car above top of rail, 2 feet $2\frac{1}{2}$ inches; extreme height of car from top of rail, 2 feet $5\frac{1}{2}$ inches; contents of car, level full, 41.02 cubic feet. It will be noted that the gauge of this car 3 feet 9 inches is unusually wide; this is to give as large a capacity as possible in a short and low car for a thin seam of coal.

19. Fig. 15 shows another form of steel car equipped with spring bumpers arranged as shown in Fig. 16. These

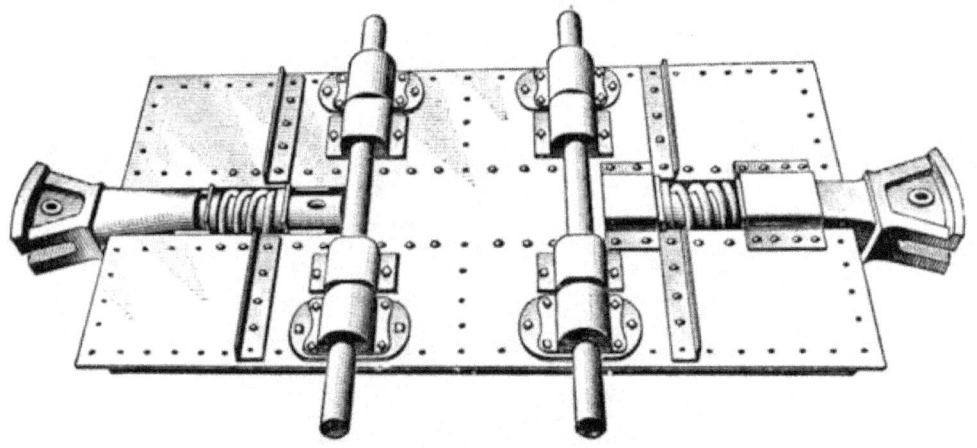

Fig. 16

are said to decrease the strain on the locomotive hauling a trip, especially in starting. These cars dump at the end, and the door is raised as shown in Fig. 15.

MINE-CAR DETAILS

DRAWBARS

20. To prevent the strain of starting a loaded car from coming entirely on the end sills, **drawbars** should transmit the strains to the other sills so that the entire bottom frame receives the shock, as if it were one solid piece. The drawbar shown in the plan, Fig. 17, requires too many cotters to transmit the pull evenly to the various sills, and in this particular car the end sill is made small and reenforced by another sill with no evident purpose except to partly take up

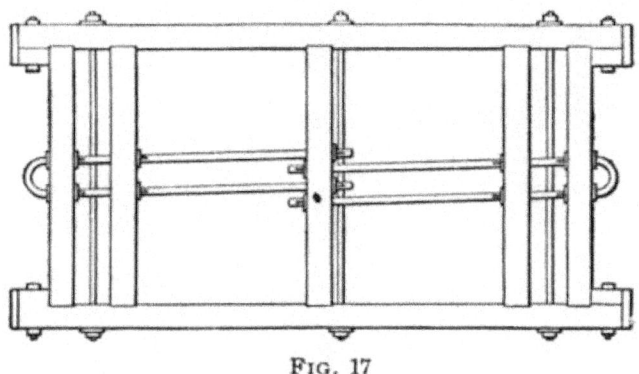

Fig. 17

the pulling strain. Another bad feature is that no provision is made to prevent the drawbars being pushed inwards when the coupling hooks and links jam against the projecting loop.

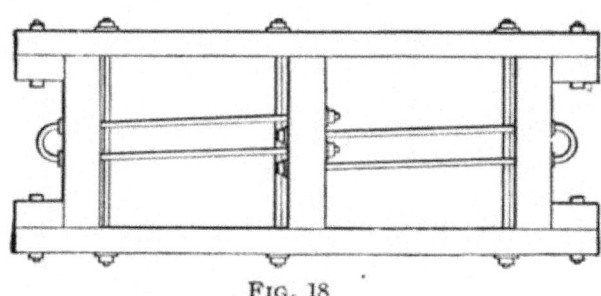

Fig. 18

21. Fig. 18 shows a somewhat similar arrangement, with nuts on the drawbars to take up the strain. This method of attaching the drawbar causes all the strain to come on the

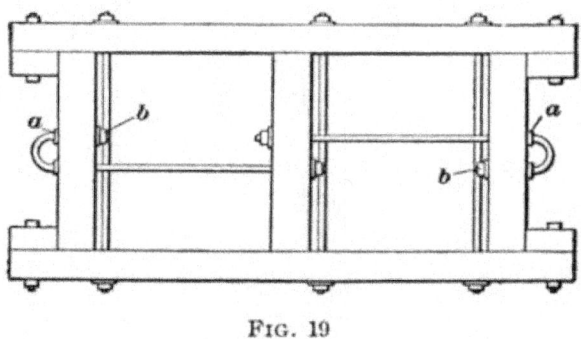

Fig. 19

center sill, which is scarcely better than if it came on the end sill. The arrangement shown in Fig. 19 has some features that are better than those for the other drawbars, but it requires that the loop have shoulders a forged on it and that

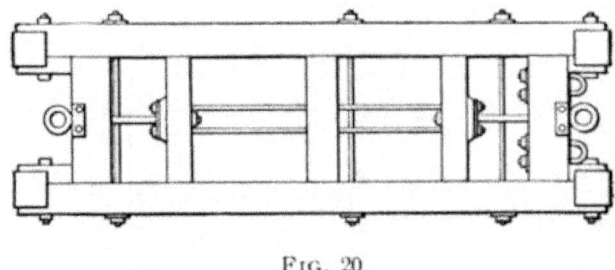

Fig. 20

it be held tight against the cross-sills by nuts b. The strain, however, practically comes on all three sills if the nuts are tight.

22. Figs. 19, 20, 21, and 22 show methods of transmitting

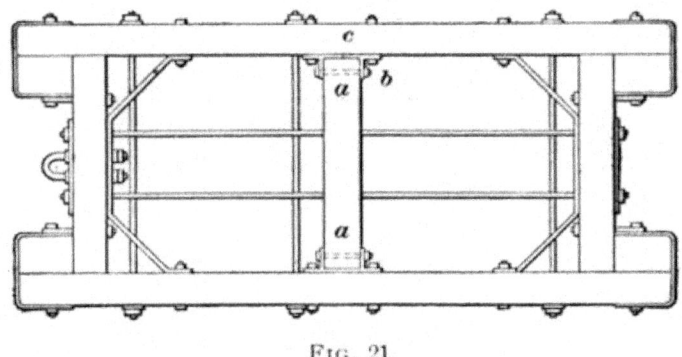

Fig. 21

strains from drawbars to car sills, in addition to those already described.

In Fig. 21, the center cross-sill a is shown resting in a

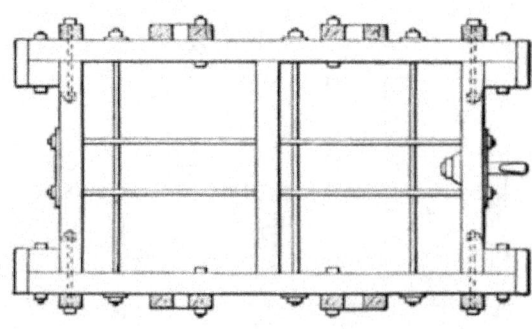

Fig. 22

metal seat b instead of being tenoned in the side sill and kept rigid by long tie-bolts.

MINE-CAR WHEELS AND AXLES

23. General Features.—The best mine-car wheels are made of charcoal pig iron so cast that the rims and flanges will be chilled and thus hardened. Since charcoal pig iron is high priced, mine-car wheels are usually cast from mixtures of mottled and white pig iron and gray pig iron and then chilled. Car wheels made of special steel, such as manganese steel, give excellent service. They are lighter than cast-iron wheels and do not groove on the tread as do cast-iron wheels.

Mine cars will run smoother over rough tracks and run easier when the diameter of the wheels is large. Coal-mine-car wheels vary in diameter from 12 to 20 inches and ore-mine-car wheels from 7 to 18 inches. The diameter of a wheel is measured at the back of the tread and not at the outside of the flanges.

The tread varies from about $2\frac{1}{2}$ to $3\frac{3}{4}$ inches in width, and the flange from $\frac{7}{8}$ to $1\frac{3}{4}$ inches in depth. The dimensions of the hub vary greatly, but depend, in a great measure, on the size of the car axle. The bore of the hub varies from $1\frac{1}{2}$ to $4\frac{1}{2}$ inches.

Usually, the wheels have four to eight spokes of cast iron, but occasionally small sizes are made solid; others have round wrought-iron or steel spokes. Nothing is added to the strength of the wheels by using wrought iron or steel, but the wheels are considerably lighter. The weights of wheels vary from 36 to 141 pounds, depending on the size and the material out of which it is made. An average weight for a 16-inch cast-iron wheel is about 120 pounds and for a 16-inch steel wheel about 80 pounds. There is wide variation in the details of car wheels, chiefly in connection with the hub and the oiling devices, so that only a few of the more common types can be shown here.

24. Inside Versus Outside Wheels.—The wheels on a mine car are called **inside wheels** when they are placed inside the frame, thus giving a broad bottom to the car but increasing its height and making it more difficult to remove or repair the wheels. An **outside wheel** is placed outside the frame of the car, thus decreasing the width of the car bottom, but rendering the wheel easy of access for oiling or repairs. The difference in arrangement between an inside and an outside wheel is shown in Figs. 23 and 24.

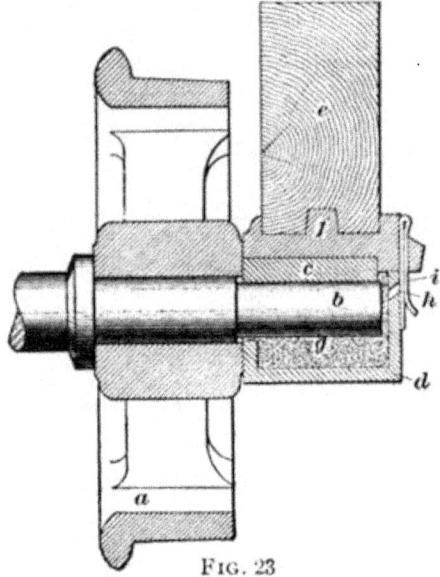

Fig. 23

In Fig. 23, the wheel a is **tight**; that is, it is fixed on the axle and the journal b runs in the bushing c placed in the box d, which is bolted to the side timber e of the car frame and kept from moving sidewise by the tongue f. The lower part of the box is filled with waste g, which is kept soaked with oil, the oil being put in through a hole h kept closed by the stop i.

In Fig. 24, the axle a is held firmly in the seat b bolted to the bed of the car c while the wheel d is loose; that is, it turns on the journal e, which is made smaller than the main axle a. The wheel is kept on the journal by a split pin or cotter driven into the hole f and the wheel is oiled through the hole g, which is usually closed by a plug of wood to keep out the dirt.

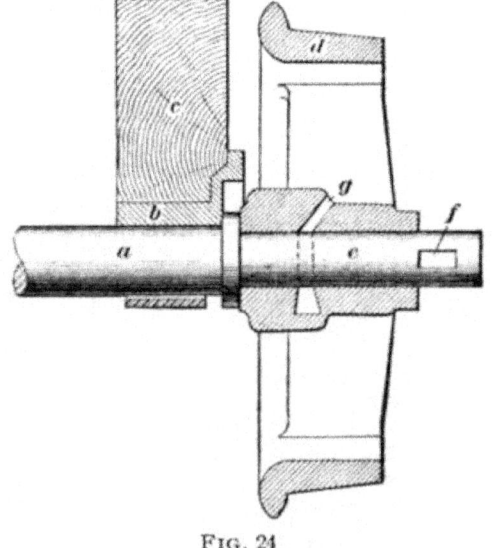

Fig. 24

The wheels on bituminous cars are generally placed outside the body of the car, since such cars usually have no bed frame, while anthracite mine cars usually have inside wheels.

25. Loose Versus Tight Wheels.—There is a wide difference of opinion and practice in regard to loose and tight wheels. Some prefer all tight wheels; others, all loose wheels; and still others, part tight and part loose on a car. Since one wheel must slip in rounding a curve, many think that a loose wheel on each axle assists the running. The objection to this arrangement is that the hub toward the car body wears conical when rounding curves, which causes wobbling, so that cars will leave the tracks unless the wheels are replaced with new ones. There is more or less side motion with loose wheels, which also is a detriment to smooth running, since the wheels will be crowded against one flange and then the other. It makes no difference in this result whether the wheel is underneath or outside the car body.

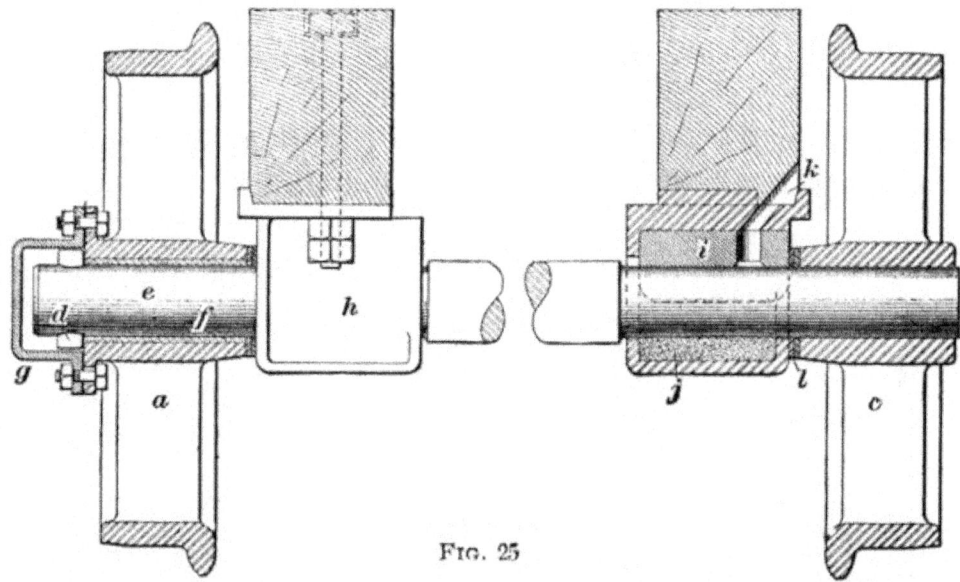

Fig. 25

There is no question but that tight wheels will run smoother than loose wheels, in fact it is difficult to get any two loose-wheeled cars to run alike, whether the axle is fixed or turns in pedestals.

Loose wheels are cheaper, since when one becomes broken it can be replaced quickly and allowance made for any wear on the axle bearing; on the other hand, a broken tight wheel means that the axle and good wheel must be discarded and sent to the wheel makers to have the old wheel pressed off and a new one pressed on the axle.

26. Fig. 25 shows a combination of a loose and a tight wheel on the same axle that has given very satisfactory results in the anthracite region. Both wheels are made of manganese steel, the wheel a being loose on the axle while the wheel c is pressed on to the axle. The loose wheel a is kept on the axle by the $\frac{7}{8}$-inch pin d, while the journal e, which is turned to $2\frac{7}{8}$ inches in diameter, is surrounded by the brass bushing f. The end of the axle is covered by a cast-iron cover g. The axle turns in the cast-steel boxes h, which are alike, the one on the right being shown in section, and in which there are soft cast-iron seats i. The bottom of the box is filled with waste j and oil is poured through the hole k. Between the fixed wheel c and the box are two wrought-iron washers l.

27. Car-Wheel Bushings.—The chief point of wear on a car wheel, aside from the tread, is between the hub and the axle, and to allow for this wear a bushing may be used, which can be replaced from time to time and the life of the wheel thus greatly increased. This bushing may be simply a cast-iron or brass sleeve a, Fig. 26, driven into the hub of

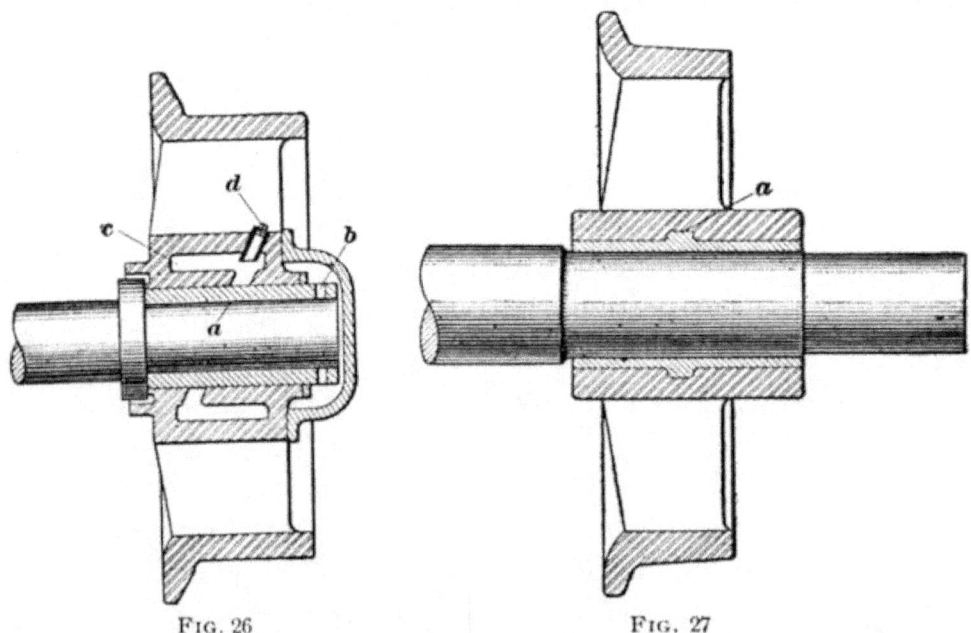

Fig. 26   Fig. 27

the wheel and drilled at b for the cotter. When the bushing becomes worn, it is driven out and replaced by another. The space c is filled with waste and oil run in through an opening that is closed by a screw plug d.

28. In order to facilitate the placing and removal of the bushing, it may be made loose and in parts with one or more lugs a, Fig. 27, cast on the bushing, or lugs a, Fig. 28, may be cast inside the hub to hold the bushing in place.

29. Self-oiling wheels are loose wheels having a cap over the end of the axle, as shown in Fig. 26, to keep the oil from running out. The hub is also frequently made hollow so that a considerable amount of waste and oil can be placed there at one time. The objections to these wheels are that the caps become loose if they are separate from the wheels and fall off; the hubs are not solid and the wheels are therefore more expensive and are more likely to get out of order owing to the increased number of parts. There is a great variety of self-oiling wheels and great ingenuity has been displayed in devising them. A number of these have been shown in connection with the cars described in Figs. 9 (d), 23, 25, and 26.

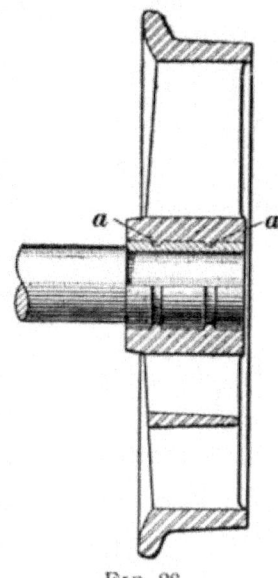

FIG. 28

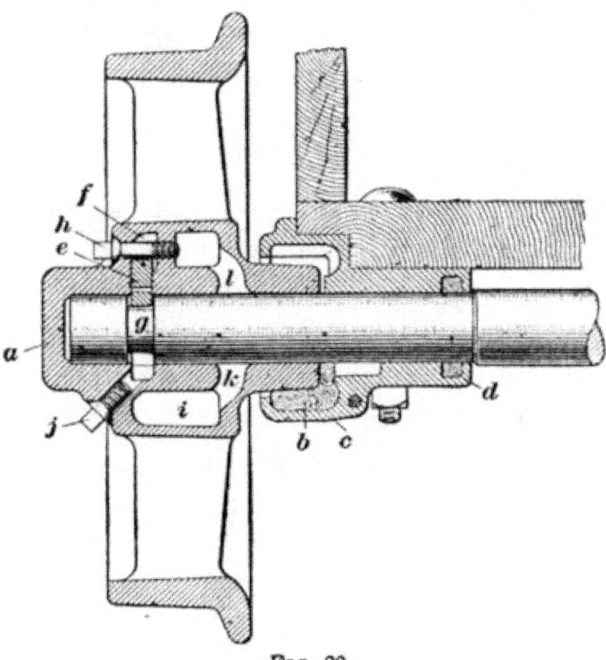

FIG. 29

30. The Faught wheel, shown in Fig. 29, has the outer end extended and cast hollow so as to form a cap a that encloses the end of the axle. The inner end of the hub fits tightly into a dust collar b in the pedestal c. This dust collar is lined with a broad band of hair and wool felt to keep out dust and dirt and to prevent leakage of the oil.

Further protection from dust and leakage is secured by the small piece of felt placed in the groove d near the inner end of the pedestal. The wheel is kept on the axle by a key block e that is socketed in the groove f in the hub and enters a groove g near the end of the axle, being retained in place by the key-block plug h. To take the wheel from the axle, it is turned so that the plug is at the lower side. The plug is then removed and the key block e drops out of the groove g in the axle. Oil is poured into the annular chamber i through the hole, which is kept closed by a screw plug j. The passage for the oil from the hole i into j is not shown. The openings k and l carry the oil from the oil cavity to the axle. The oil cavity i and the passages k and l are filled with felt packing, which prevents the oil being thrown from the axle by centrifugal force when the wheel is turning rapidly.

31. The Fleming car box is shown in Fig. 30. The hub of the wheel is bored out to a groove m into which oil runs from the chamber n containing waste and oil and fed through the opening o, which is covered by an ordinary flap. The journal p bears on a semicircular bearing q that is held in place by the cap r, which is easily removable in order to replace the bearings q. The upper part of the bearing q is beveled so that as the car and axle move, due to inequalities on the road, the car frame will rock on this beveled surface and the axle will not be worn, as is ordinarily the case.

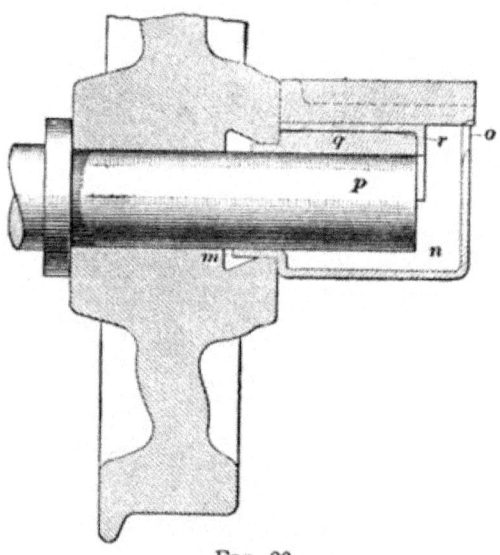

FIG. 30

32. **Mine-car axles** are either of steel or wrought iron, with turned faces for the journal brasses if the wheels are shrunk on the axle, or turned faces for the wheels if the wheels run loose on the axle. The diameter of the axle varies from 2 to 3 inches, according to the size of the car and the weight carried. Axles having wheels shrunk on them are more liable to bend than when the wheels are loose. This is probably due to a certain amount of play taking place in the hub of the loose wheel that permits such a wheel to take up the shock that would otherwise be entirely borne by the axle.

33. Oil for Mine Cars.—Where mine cars are supplied with boxes that can be packed with waste, that is, boxes with covers so that they can be readily inspected, car grease, such as is used on surface railroads, is the proper lubricant. Outside of the anthracite field, very few mine cars are provided with such boxes and consequently oil that will flow must be used. If the cars are used below ground, a mineral oil known as *summer oil* is advisable; that is, an oil that is not too thin or so affected by the heat of the mine that it will run out of the boxes. Where cars are used both above and below ground, it is customary, in temperate climates, to use summer oil in hot weather and zero oil in winter. The latter is a cheap mineral oil that is supposed not to congeal before the temperature reaches zero. Miners frequently call it *black strap* and borrow it for their lights; it smokes so abominably, however, that it is not suitable for burning in anything but torches in the open air.

34. Oil-Saving Device.—The cost of oil for lubricating car wheels amounts to considerable during a year, particularly where there are a large number of cars. The oilers are not as saving in lubricants as they might be and some such device as is shown in Fig. 31 is desirable. This device is a trough made from $\frac{1}{16}$-inch sheet iron bent and placed lengthwise between the empty and loaded tracks. Above the sheet-iron trough are placed cross-pieces of wood 2 in. × 4 in., and on top of these and parallel with the tracks is laid a 12-inch plank for the greaser to walk on while oiling the cars. The sheet-iron drain catches all the oil that drips from the greaser's oil can, also that excess which drips from the boxes or caps of the car wheels. This overflow runs into a receptacle or is sunk into the ground and placed at the lower end of the trough. From here, the oil is taken and strained through two thicknesses of duck.

Fig. 31

With this device and improved car wheels with caps, a great saving in oil may be effected.

MINE-CAR BRAKES

35. When mine cars are used on a pitch, some form of brake is advisable as the speed of the car can be controlled better than by spragging the wheels and the wheels are not so badly worn. A simple block brake similar to that shown in Figs. 12 and 13 is the type generally used. Such a brake may be operated either by a system of levers, as shown in Fig. 12 (b) and (d) or the pressure may be applied to the blocks directly by means of two long levers running alongside and connected across the end by a cross-bar on which the runner presses, similar to that shown in Fig. 32.

36. The Logan brake, made entirely of iron, is shown in Fig. 32. The brake shoes a are of cast iron and are rigidly fastened to the radial arms b with a bolt. These radial arms are of the form shown in the detailed view, Fig. 32 (b), and are made of flat iron 2 inches wide by $\frac{3}{8}$ inch thick. The upright c is a flat piece of iron bent in the shape of a **U**, which connects to the long lever d, which is pivoted at c and runs across the rear of the car, as shown. The radial arms b are slotted at the lower end so as to give

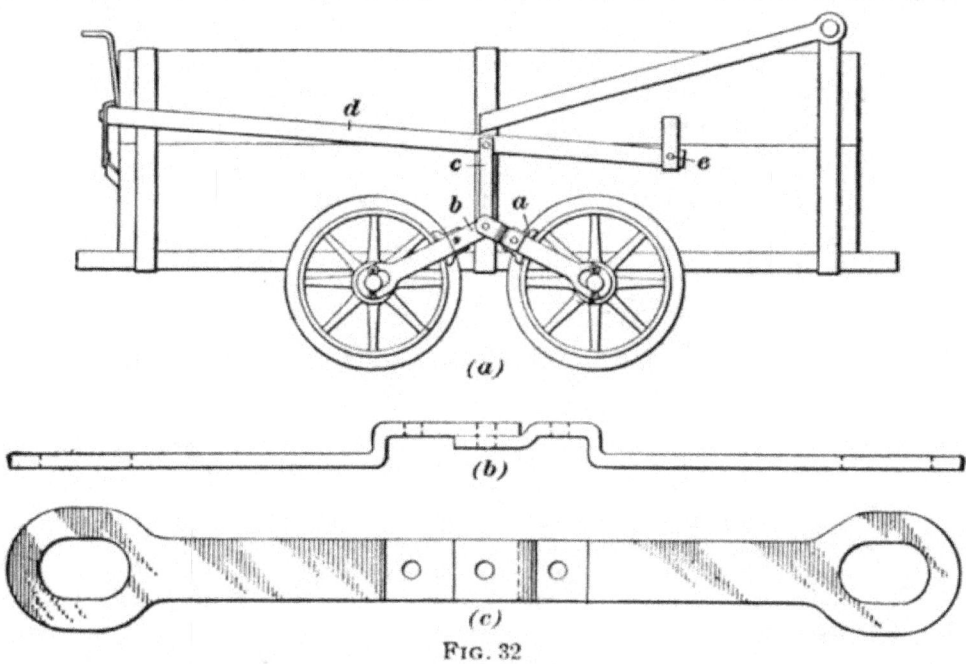

Fig. 32

them a certain amount of play and to allow for wear in the face of the brake shoe.

CAR COUPLINGS

37. Cars are frequently coupled by three links, as shown in Fig. 33, the end links being hooked over hooks on the end of the drawbar. Plain link couplings of this character are, however, apt to be lost and some form of clevis to attach the couplings to the drawbar is preferred.

Fig. 33

38. Clevis Hitchings.—Fig. 34 shows a common form of **double clevis hitching.** These hitchings, as commonly made, are about 20 inches long and weigh about 30 pounds each. When it is desired to have the hitchings permanently attached to a

Fig. 34

car, the pin of one clevis is provided with a small split pin, as shown in Fig. 35. The **hook clevis**, Fig. 36, is exten-

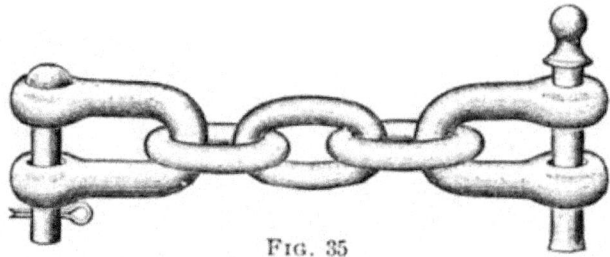

Fig. 35

sively used where a hitching is attached to the drawbar of one car permanently and the hook is attached to a drawbar

Fig. 36

of the next car. Where necessary, the length of a coupling may be shortened by means of a **twisted-link coupling**, as shown in Fig. 37.

Fig. 37

39. Table I shows the weights and capacities of coal cars as ordinarily used.

TABLE I

Approximate Capacity	Weight of Empty Car Pounds	Average Weight of Load Pounds
15 bushels run-of-mine coal	500	1,200 bituminous coal
20 bushels run-of-mine coal	600	1,600 bituminous coal
25 bushels run-of-mine coal	850	2,000 bituminous coal
30 bushels run-of-mine coal	950	2,400 bituminous coal
33 bushels run-of-mine coal	1,050	2,640 bituminous coal
35 bushels run-of-mine coal	1,150	2,800 bituminous coal
40 bushels run-of-mine coal	1,250	3,200 bituminous coal
46 bushels run-of-mine coal	1,400	3,680 bituminous coal
54 bushels run-of-mine coal	1,700	4,320 bituminous coal
2½ long tons coal	2,000	5,600 anthracite coal
3 long tons coal	2,500	6,720 anthracite coal

NOTE.—A bushel is assumed to weigh 80 lb.

RESISTANCE TO MINE-CAR HAULAGE

40. The resistance that must be overcome in moving a mine car is usually divided into two parts known as *frictional resistance* and *grade resistance*.

41. Frictional Resistance.—Frictional resistance is the resistance to moving a car horizontally along the track and depends on a large number of items, such as the condition of the track, the kind, size and condition of the wheels, etc. It is represented in Fig. 38 by the force it is necessary to exert at P to move the weight W represent-

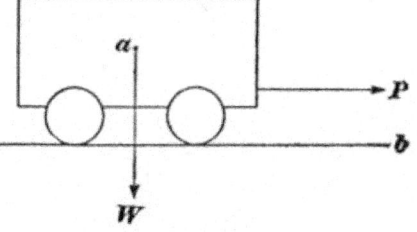

FIG. 38

ing the weight of the car and its load along the level track b. This may be expressed as a percentage of the weight W, but more generally it is expressed as pounds of resistance per ton of weight W.

In this case, the weight W acts through the center of gravity of the load a and perpendicular to the track b.

If P = amount of pull, in pounds;
W = weight of car and load to be moved, in pounds;
f = coefficient of friction;
$$P = fW$$

42. Starting Effort.—The greatest effort is necessary in starting a load from rest, and a rule quite commonly used for estimating the amount of this effort on a clean track is to assume it to be one-fifth of the total load. The amount of this effort is, however, dependent entirely on the condition of the car and the track and may be greater or less than this.

43. Frictional Resistance for Cars in Motion.—The frictional resistance for mine cars is given by the H. K. Porter Company as varying between 20 and 30 pounds per ton for ordinary conditions with fairly good track. Other authorities have given the frictional resistance of coal-mine cars with fixed wheels as varying between 32 and 52 pounds per ton.

If the track is poor and the car wheels scrape along the sides of the car, the frictional resistance will sometimes amount to 100 pounds per ton even when the cars and journals are well lubricated. The resistance will be increased by curves in the track, but this is usually neglected when considering animal haulage, because it is so variable a factor and one that decreases as the radius of the curve increases and as the wheel base of the car and the speed of car decrease.

It is customary to assume a frictional resistance of from 30 to 50 pounds per ton when no definite information is given in a problem that will fix the amount to be assumed, and a very common rule is to take f in the above formula equal to $\frac{1}{40}$.

44. The frictional resistance is sometimes divided into two parts—*track resistance* and *car resistance*—for the purpose of discussion, though in haulage calculation this division is not made and the frictional resistance (sometimes also called *total resistance* or *track resistance*) includes all resistances except that due to gravity and which will be explained later.

45. The **track resistance** depends on the condition of the track and road bed, that is, the size of the rails and the care with which they are laid, the condition as to repairs, curves, the amount of dirt on the track, etc. The track on a main haulage road should have rails weighing at least 40 pounds per yard and laid on ties placed 2 feet between centers. With the track thus laid and the road bed kept clean, the track resistance is quite small; but these conditions do not often prevail in a mine and the track resistance is often a very serious matter. It is necessary to assume, however, for calculation, that a good track is provided, such as is described above.

46. Car resistance depends on the diameter of the wheels, the amount of lubrication, and the kind of axles and journal-boxes used. A large journal bearing reduces the friction, and car wheels of large diameter roll more easily than small wheels over any inequalities in the rails. Wheels that have grooved treads or have worn flat, owing to their having slid along the track, increase the car resistance. If the car resistance is taken at 30 pounds for a 16-inch wheel running on a $2\frac{1}{2}$-inch axle, the leverage exerted will be in the ratio of $\frac{2.5}{16}$. If the wheels are enlarged to 18 inches, the lever arms will be in the ratio of $\frac{2.5}{18}$. The car resistance will therefore be reduced by the use of 18-inch wheels from 30 to $26\frac{2}{3}$ pounds per ton in accordance with the proportion $18 : 16 = 30 : x$, and $x = 26\frac{2}{3}$.

47. Grade Resistance.—Grade resistance is due to gravity and is the resistance that must be overcome in raising the weight through a given height. If a car is on an inclined track, the action of gravity will not be perpendicular to the track, or in the line ad, Fig. 39, as was the case illustrated in Fig. 38, but will be in the direction ac. In this position, the action of gravity holds the car on the track, and at the same time tends to move it down the plane. The force represented by the weight W, Fig. 39, is thus split into two forces represented by the line ad perpendicular to the track and the line ab parallel to the track.

If the length of the line ac represents the amount of the weight W, then the line ad represents the component of the force W acting perpendicular to the track. $ad = ac \cos \theta = W \cos \theta$, and since angle θ = angle a,

$$ad = W \cos a \qquad (1)$$

in which a = angle of inclination of track

The line ab represents the component of the force W acting parallel to the track and tending to move the car down

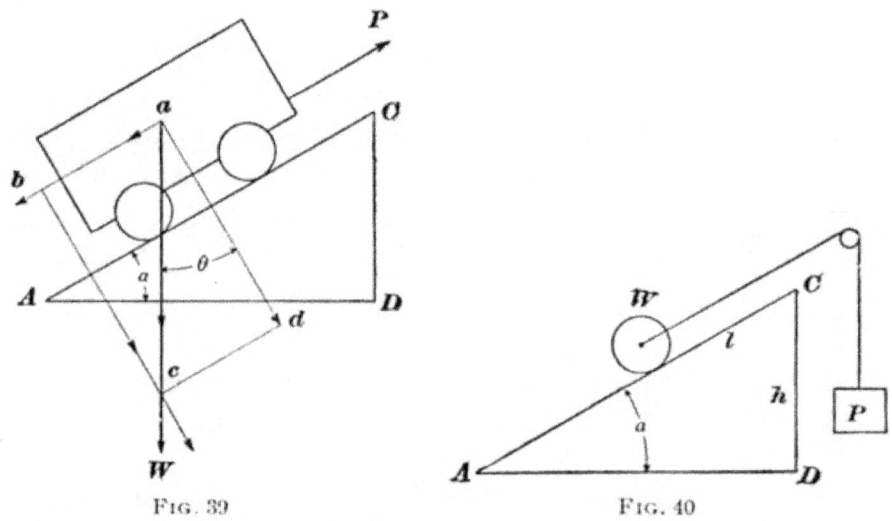

Fig. 39   Fig. 40

the track; this force must be overcome by a force P if the car is to move up the plane. If, then, the force P is represented by the line ab, $P = ab \; (= cd) = ac \sin \theta = W \sin \theta = W \sin a$,

$$P = W \sin a \qquad (2)$$

In Fig. 39, $CD = AC \sin a$. Since the triangles acd and ACD are similar, $\dfrac{P}{W} = \left(\dfrac{cd}{ac}\right) = \dfrac{CD}{AC}$ and $P \times AC = W \times CD$. This means that the work done by the force P in moving the weight W from A to C, not including friction, is the same as in raising the weight W through the height DC or by P falling through CD.

§54   HAULAGE

The resistance to be overcome in moving W up the incline, not including frictional resistance, is called the *grade resistance*, which increases as the inclination of the track increases; it varies in amount directly as the sine of the angle of inclination. If, in Fig. 40, $CD = h$ (height of plane) and $AC = l$ (length of plane), $P : W = h : l$; or $Pl = Wh$ and

$$P = \frac{Wh}{l} = W \sin a$$

$$W = \frac{Pl}{h} = \frac{P}{\sin a}$$

EXAMPLE 1.—Find the weight that will balance a loaded coal car weighing 6,000 pounds on an incline of 18°.

SOLUTION.—
$$p = w \times \sin a = 6{,}000 \times .30902 = 1{,}854+ \text{ lb. Ans.}$$

EXAMPLE 2.—If the length l is 300 feet, the rise h 92.7 feet, and the loaded car weighs 6,000 pounds, what will be the balancing weight?

SOLUTION.—When the length of the plane and vertical height are given, multiply the load by the quotient of the vertical height h divided by the length, thus

$$6{,}000 \times \frac{92.7}{300} = 1{,}854 \text{ lb. Ans.}$$

48. The frictional resistance on an incline decreases as the inclination increases, and since this resistance depends on the force acting perpendicular to the plane it varies as the cosine of the angle of inclination. Hence, for an incline, the formula given in Art. **41**, $P = fW$ becomes

$$P = fW(\cos a)$$

Where the angle of inclination is small, the cosine of the angle is practically 1, and it is not necessary to consider the effect of the inclination on the frictional resistance, and this resistance is calculated by means of the formula in Art. **41**. For the ordinary grades of mine roads this method of calculating gives results that do not differ materially from the theoretical calculations which take account of the inclination; and the difference is on the safe side. For steep inclinations, however, such as are found on engine and gravity planes the inclination should be considered.

49. The total theoretical resistance on a grade for a force pulling up the grade is,
$$P = W \sin a + f W \cos a$$
$$= W(\sin a + f \cos a)$$

50. The grade is frequently expressed in per cent. instead of in degrees, and each per cent. of grade (i. e., each foot of rise in 100 feet measured horizontally) is then considered to represent a grade resistance of 1 per cent. of the weight, or 20 pounds per ton of 2,000 pounds. For instance, a $1\frac{1}{2}$-per-cent. grade will cause $2,000 \times .015 = 30$ pounds gravity resistance per ton to traction up grade, or similar amounts in favor of a load on a down grade. The per cent. of grade is the tangent of the angle of inclination. The tangent and the sine of a small angle are practically equal, hence the results of calculating grade resistance by per cent. and by formula in Art. **49** are practically the same for small angles.

The grades are, when possible, made in favor of the loads; and there is a theoretical grade where it requires the same energy to propel a loaded car on a down grade as an empty car on an up grade. This is, however, seldom realized.

EXAMPLE 1.—Assuming that the cars have loose wheels and that the cars and the track are in first-class condition, and that a mule can exert 150 pounds tractive effort, how many cars weighing 1,500 pounds each can he pull up a 1-per-cent. grade, the total track resistance being taken at 30 pounds per ton?

SOLUTION.—The frictional resistance per car is $1,500 \times .01 = 15$ lb.;
the grade resistance per car is $\dfrac{1,500 \times 30}{2,000} = 22.5$ lb.;
the number of cars hauled is $\dfrac{150}{15 + 22.5} = 4$ cars. Ans.

EXAMPLE 2.—Assume that a car, with its load, weighs 4,500 pounds; what force is necessary to draw it when the total track resistance is 30 pounds per ton and the grade 1.5 per cent.?

SOLUTION.—The frictional resistance per car is $4,500 \times .015 = 67.5$ lb.;
the grade resistance per car is $\dfrac{4,500 \times 30}{2,000} = 67.5$ lb.;
the total resistance is $67.5 + 67.5 = 135$ lb. Ans.

MINE-HAULAGE SYSTEMS

GENERAL CONSIDERATIONS

51. Scope of Subject.—Arrangements must be made at most mines for the transportation of material below ground and on the surface. The problem of getting the material mined from the face of the workings to the dump in the most economical way possible is one of the most important problems with which a mine manager has to deal, and an efficient haulage system is frequently an essential if a mine is to be a commercial success.

52. Choice of Haulage System.—There are three general haulage powers to choose from, *man power*, *animal power*, and *machine power*, and in many cases all three are to be found in one mine. Mechanical haulage is divided into *locomotive haulage* and *rope haulage*. Locomotive haulage is divided into steam locomotives, compressed-air locomotives, and electric locomotives; rope haulage is divided into tail-rope haulage, endless-rope haulage, and haulage on planes, which may be by gravity plane or by engine plane. The kind of power adopted and the particular system of haulage depend, in great measure on local conditions, such as the quantity and kind of material mined, the distance it must be transported, the presence or absence of explosive gases in the mine, and, whether the haulage ways are intake or return airways, the conditions of the track, including such items as grade, curves, length, future extensions, the size of cars and the haulageways, also whether the mine is wet or dry, and a number of other considerations. The cost of installation and operation of the different systems will necessarily be compared with each other, also the cost of future extensions and the probable life of the mine. The number of factors that enter into the choice of a system are so numerous that only very general rules can be given in connection with the comparative advantages and disadvantages of the different systems. Before any particular haulage system is adopted, all the local conditions having a bearing on a successful operation should be investigated. No system of haulage is universal, and each system has peculiarities that are adapted better to one mine than to another.

MAN POWER FOR TRANSPORTATION

53. Previous to the introduction of modern methods of transportation, and even in some out-of-the-way places at the present time it is customary for men and women to carry broken mineral out of the mines in bags and baskets. **Man power for transportation** at modern mines, however, is confined to pushing loaded wheelbarrows or cars from the working places to a point where the material in the car or wheelbarrow can be more readily transported to its destination by some other power and returning the empty wheelbarrow or car from this point to the face. The pushing of cars by men is usually termed **tramming** and is practiced to a much greater extent in ore than in coal mines.

54. Wheelbarrow Transportation.—Wheelbarrows are frequently used in mines to move ore for short distances on stopes and in cross-cuts. They are also used in coal mines when making break-throughs or in other localities where the distance is too great for shoveling. It is the general custom, when opening mines, to wheel the broken material in barrows from the face of the excavation until the distance becomes too great. Men, when wheeling barrows, move on an average at the rate of $2\frac{1}{2}$ miles per hour or 200 feet per minute. Hence, if the lead, or the distance the load is wheeled, be 100 feet, a man will do useful work only on a run-out, the run-back with the empty barrow simply occupying time. A wheelbarrow load is approximately 2 cubic feet and to load and dump it, without considering wheeling it, requires at least 1.25 minutes. To load, wheel, dump, and return to the loading place requires 1.25 + 1 minute when the lead is 100 feet, or, if the lead is 200 feet 1.25 + 2 minutes. A man cannot work steadily all day at this rate, and besides taking occasional rests he is obliged to adjust his wheeling plank and to do other matters that are estimated to amount to one-tenth of his total time. Therefore, if a equals the number of minutes in a working day, $a \times .9 =$ the actual time worked. If b is the number of 100 feet lengths of lead, $1.25 + b =$ the time for a round trip, in minutes. The number of wheelbarrow loads removed per day can be found as follows: Let c equal the number of trips made, then $\dfrac{a \times .9}{1.25 + b} = c$. Since each wheelbarrow load is about 2 cubic feet, fourteen loads will constitute a cubic yard ($\frac{27}{2} = 13.5$). The wheelers' wages added to the wear and tear on the barrow (the latter being assumed as 5 cents per day), when divided by the number of cubic yards removed, will give the cost per cubic yard for loading, wheeling, and emptying the material.

EXAMPLE.—(a) How many cubic yards of earth will a man load, wheel, and empty in a working day of 9 hours, the distance to which the earth is removed being 500 feet? (b) What will be the cost, per cubic yard, when the man's wages are $1.50 per day and the wear on the wheelbarrow is 5 cents per day?

SOLUTION.—(a) $\dfrac{(60 \times 9).9}{1.25 + 5} = \dfrac{486}{6.25}$, or, approximately, seventy-eight trips per day. $78 \div 14 = 5.57$ cu. yd. removed per day.

(b) The cost per cubic yard will be $\dfrac{(150 + 5)}{5.57} = 27.8$ ct. per cu. yd.

Rock, ore, or coal requires more time than earth for loading, so that in loading such material, if 1.6 is used instead of 1.25, the number of cubic yards loaded and the cost per cubic yard can be approximately calculated for wheelbarrows.

55. Tramming.—In ore mining, where cars are in use, the trammers push loaded cars either from the face of a level or from an ore chute to the shaft. In some cases, they dump the contents of the car into ore bins; in other cases, directly into skips; or, if a slope car or a cage is in use, the mine car is pushed on the platform and hoisted out of the mine.

Where buckets are used to carry ore out of a mine, they are either loaded in the shaft or are landed on low platform trucks and pushed to a loading station and back to the shaft. In order to prevent delays, several buckets and trucks are in commission, two at least being in motion while one is being loaded. There is not much difference in the cost of transportation by these systems, which are chosen more from necessity than with a view to economy.

In coal mines, cars are transported by man power only in exceptional cases, such as when opening a mine; in buggy-breast mining occasionally for pushing cars between the face of a room and the entry; and in long-wall mining. The distance the man has to travel in such cases is probably not much over 300 feet. In opening a mine, the work would be carried on too slowly if this distance were greatly exceeded. In buggy-breast mining and in room-and-pillar mining, the rooms are limited in length. In long-wall mining, the face is seldom worked 300 feet from the haulage road, either from the cross or from the main headings.

56. Cost of Loading Mine Cars.—When coal or rock of reasonable size, that is, fit for shoveling, is loaded into a car having 1 cubic yard capacity, the time consumed in loading and dumping will approximate 25 minutes, irrespective of the distance. The average speed on a good tram road would be about 200 feet per minute; thus, the cost of removing per cubic yard of material may be estimated by finding the number of trips that can be made and dividing the wages by the number of trips, as in the following examples.

EXAMPLE 1.—Suppose that there is a run of 1,000 feet, that a working day is 9 hours long, that a trammer's wages are $1.50 per day, that he loads his own car, and that there are $1\frac{1}{4}$ tons of broken rock in a cubic yard; under such conditions: (a) How many cars holding 27 cubic feet (1 cubic yard) can be loaded with rock, trammed, and dumped during the day? (b) What will be the cost per cubic yard? (c) What will be the cost per ton?

SOLUTION.—(a) $\dfrac{2 \times 1,000}{200} = 10$ min. time required for one round trip. $\dfrac{(60 \times 9) \times .9}{25 + 10}$ = approximately 14 trips, or 14 cars holding 1 cu.

yd. each. Ans. (*b*) $\frac{150}{14}$ = 10.7 ct. per cu. yd. of broken rock. Ans. (*c*) Since there are about $1\frac{1}{4}$ T. of broken rock in 1 cu. yd., the cost per ton of broken rock will be about $10.7 \div 1\frac{1}{4}$ = 8.56 ct. Ans.

EXAMPLE 2.—If a mine laborer is paid $1.50 per day of 9 hours and the distance from the face of a buggy breast to the gangway is 300 feet and the buggy holds $\frac{1}{2}$ ton of coal, what is the cost, per ton, of loading and transporting the coal from the face to the bottom of the breast and dumping it, assuming that the laborer occupies nine-tenths of his time loading coal and that he can load the buggy in 15 minutes?

SOLUTION.— $\frac{300 \times 2}{200}$ = 3 min. time consumed in making one round trip. $\frac{(60 \times 9) \cdot .9}{15 + 3}$ = 27 trips = 13.5 T. of coal. $\frac{150}{13.5} = 11\frac{1}{9}$ ct. per T. Ans.

ANIMAL HAULAGE

57. Extent.—The use of animals in mines is generally confined to short hauls on the main haulage roads and for delivering empty and hauling away loaded cars from the working places. No absolute rule can be given for the distance to which mule haulage can be extended economically, but unless the conditions are such as not to warrant the necessary outlay for the installation of a mechanical system, animal haulage should not be extended a greater distance than $\frac{1}{2}$ mile.

Underground haulage is carried on in two stages; viz., gathering the loaded cars to a central station, and hauling them to their destination. Animals have up to this time been generally used for delivering empty cars to the rooms and gathering loaded cars; but mechanical haulage has proved better than animal haulage for the long hauls from the gathering stations to the shaft bottom or drift mouth.

58. Mule Haulage.—The animals employed for underground haulage are usually mules, since they are quick in their movements and generally have intelligence in adapting themselves to the work. They appear to have better eyesight; carry their heads lower; dodge obstacles better and take care of their heels better than horses. They are also tougher, eat less, do not overfeed, and are more to be depended on day after day than horses.

59. Mule Teams.—Mules are used in mines either singly or in teams. When several mules are hitched tandem, they are used on main haulageways to pull trains of empty or loaded cars. String teams, as these are called, are also used in drift mines to pull cars from inside the mine to the tipple, when for any reason it is not considered advisable to employ mechanical haulage. To collect loaded cars from the miners in their rooms and haul empty cars from the partings to them, single mules are used. Probably the average mine mule will weigh 900 pounds, and can exert a pull of one-sixth of its weight at the rate of 2.5 miles per hour, or 220 feet per minute on a level track. This is equivalent to an exertion of 33,000 foot-pounds per minute or 1 horsepower. The maximum effort is exerted in starting a car from a state of rest to motion, and on clean, level rails is about one-fifth of the weight of the car and its contents. The steady effort exerted by the mule after a car is started will depend on the condition of the track and car, also on the track grade. Suppose that a car weighs 2,000 pounds and is loaded with 4,000 pounds, then to start the car on a clean level track will require an exertion of $(6,000 \div 5)$, or 1,200 pounds.

60. Gathering Cars With Mules.—In large mines, a certain number of cars, equal to a trip for the main haulage system, must be gathered at a parting in a given time. Gathering is done by single mules in flat deposits, but in pitching deposits, such as some of the anthracite mines, string-team gathering is practiced. In the former case, the driver hitches his mule to a car at the parting and takes it to the entrance of the miner's room. If the miner has his car loaded, the driver hitches his mule to it and takes it to the loaded side of the parting and then hitching to another empty car takes it to another miner. As it would require too long a time for one mule to collect a sufficient number of cars for a trip, it is customary to use several mules in gathering; and that there may be no delay the drivers take the empties from the parting at the same time and bring in the loaded cars one after the other. Each driver has a certain number of miners to look after, and this he does in rotation; but should one miner fail to have his car loaded in time, the driver skips him and goes on to the next room, thus the miner loses his turn.

Suppose that a trip of ten cars is to be gathered from the working places and taken from the parting to its destination every 40 minutes. This will require that ten empty cars be taken from the parting to the working places and ten loaded cars substituted, and the average time for each car will be but 4 minutes. This is not enough time for one driver and mule, since they cannot enter the room and haul out a loaded car on the run, neither can the changes be made in less than a minute; consequently, 10 minutes should be allowed for a round trip; and as this will be but four cars collected in 40 minutes, three mules and three drivers will be needed for the work, and trips should be arranged, if possible, so that there will be twelve cars at the parting instead of ten.

Three mules in gathering 120 cars from twenty rooms having 60-foot centers will travel about 18 miles each, provided that the last room is 1,760 feet from the parting. Each mule delivers forty empty cars to the rooms and hauls forty empty cars to the parting. The gathering work should be concentrated so that the collecting of loaded cars may be accomplished in a given time in order to prevent delays and keep the men at the surface fully employed during working hours. If, therefore, in this case the distance from the parting to the last working face exceeds 1,760 feet, a parting should be constructed nearer the working places so that the haulage may be concentrated and gathering done on time. The condition that determines the distance of haul from the face to the parting where the trips are made up is the ability of the mules to haul the required quantity of coal to the parting as fast as it is hauled to the shaft bottom. The maximum distance that a mule should travel in gathering should probably not exceed 2,000 feet, and the average distance should be considerably less.

61. Speed of Mule Haulage in Mines.—The normal speed of a mule when hauling loaded cars on an ordinary mine car track is about 2.5 miles per hour. When empty cars are delivered to the miners in a hurry and loaded cars are hauled down grade to the parting under the same conditions, the mule may travel at the rate of 6 miles per hour. This speed tires the animal although it has little to pull. The work of a mine mule under average conditions may be assumed as 5 ton-miles per hour.

62. Safe Grade for Mule Haulage.—While the grade against empties on the main haulways can be 1.5 per cent. the grade on cross-entries should not exceed .5 to 1 per cent., where mules must gather cars in a hurry. If the mules are winded in taking in empties, the loaded cars must necessarily come out slower, so that the advantage gained by quick delivery of empty cars is offset by the loss of time in returning the loaded cars. Often mine mules are injured by winding them and then not giving them time to recover their breath for the return trip. The driver has not so much control over his car and animal that he can stop instantly; and if the mule lags or stumbles, the car will probably run against the mule and injure its legs. A safe down grade for mule haulage should not exceed 3 per cent. and great care will be needed in that case. On such steep grades, while the mule can pull up the ordinary mine car, the brakeman or driver should run the car down independent of the mule. A loaded mine car will slide on rails even with four wheels spragged when the grade is 6 to 8 per cent. depending on the condition of the rails.

63. Cost of Mule Haulage.—A four-mule string team on main haulageways having 1.25-per-cent. grade in favor of the load, can haul between 400 and 500 tons of coal in 10 hours, provided that the cars and tracks are in suitable condition and that the distance is not more than $\frac{1}{2}$ mile. This can be done at less cost per ton than by any system of mechanical haulage; but beyond $\frac{1}{2}$ mile, the output decreases or the cost of haulage increases rapidly. If there are 800 to 1,000 tons of coal to be hauled $\frac{1}{2}$ mile, one four-mule team cannot accomplish the task and an additional four-mule team will increase the cost to such proportions that mechanical haulage will be advisable. Animal haulage, although cheaper in first cost, within certain limits, than other systems of haulage, usually requires more money for maintenance—by which is meant feed, care, harness, shoeing, temporary disabling injuries, deaths due to natural causes and accidents, and for a reserve on which to draw in case of emergency. The cost of gathering cars should not exceed 3 to 5 cents per ton, but it is better to pay 6 cents rather than to kill a mule worth $150 by fast traveling with the idea of economizing. The cost of team mule haulage on the main track should not exceed 3 cents per ton, and when it does mechanical haulage should be substituted. Where two or more mule teams are used for hauling on the main entries, better results will be obtained by double than by single tracks, for when single tracks are used there must be partings for teams to pass and this uses up time, besides the cars may run past the parting before they are stopped, causing both annoyance and loss of time.

64. Animal Versus Mechanical Haulage.—Mechanical haulage makes quicker time than animal haulage and can replace a number of string teams on main haulways and thus reduce the number of partings. The relative economy in the use of mechanical haulage over mule teams depends on the length of haul, since speed is the chief factor in economy where large tonnage is required. There are certain matters, such as filling up the track between the ties and the renewal of ties, that add somewhat to the expense of mule haulage, but, with every objectionable feature weighed against animal haulage, probably it is the cheapest up to the limit of $\frac{1}{2}$ mile.

65. Purchasing Animals.—Large heavy mules are preferable to small light mules, but they should not be thick legged about the hocks and clumsy and should have good feet. The height of the animal is also a consideration; where coal beds are thin, the mule is purchased to fit in the haulageways. While it is difficult to tell the age of mules after they have reached their full growth, it is customary to stipulate that the animals shall be young.

Young mules 4 to 6 years old are easier to break into mine work than others, but even such animals should be purchased with the understanding that they are to be returned if they are not intelligent or active on their feet. Some mules are better adapted to mine work than others; in fact, some are so nervous or dumb that they cannot be taught to act rightly. The stable boss or one thoroughly accustomed to handling mules should break them into mine work, and not a driver alone. After the mule has become accustomed to the work, almost any driver, if careful, can handle the animal successfully.

Mules are subject to lampers and scratches, which should be looked for when the mules are purchased. The former causes indigestion and the latter lameness; and although neither may be severe enough to put the mule out of commission, they are very annoying.

66. Feeding Mules.—An average feed for a mule weighing 1,000 to 1,100 pounds is 12 pounds of grain and 15 pounds of hay. Hay is digested chiefly in the intestines and grain in the stomach, hence if possible a mule should be first watered, then given hay, and lastly grain. If the water is given last, it washes the food into the intestines before it is acted on by the gastric juices in the stomach. If the hay is given after the grain, it carries the grain with it into the intestines. This order of feeding is not always practicable in a mine and it is of advantage to place watering troughs about the mine so that the mules can be watered during the day while at work. As the feed is in the boxes when a mule is put in the stable at night, there should also be water in his water trough so that he can drink at intervals while feeding. Fresh food should never be placed on top of any left over from the previous feeding. A mule should have plenty of water the first thing in the morning, and care should be taken to have the water pure and the troughs clean.

67. Care of Mules.—Care should be taken to see that the mule is properly harnessed and that the harness fits. Particular attention should also be given to the shoeing of

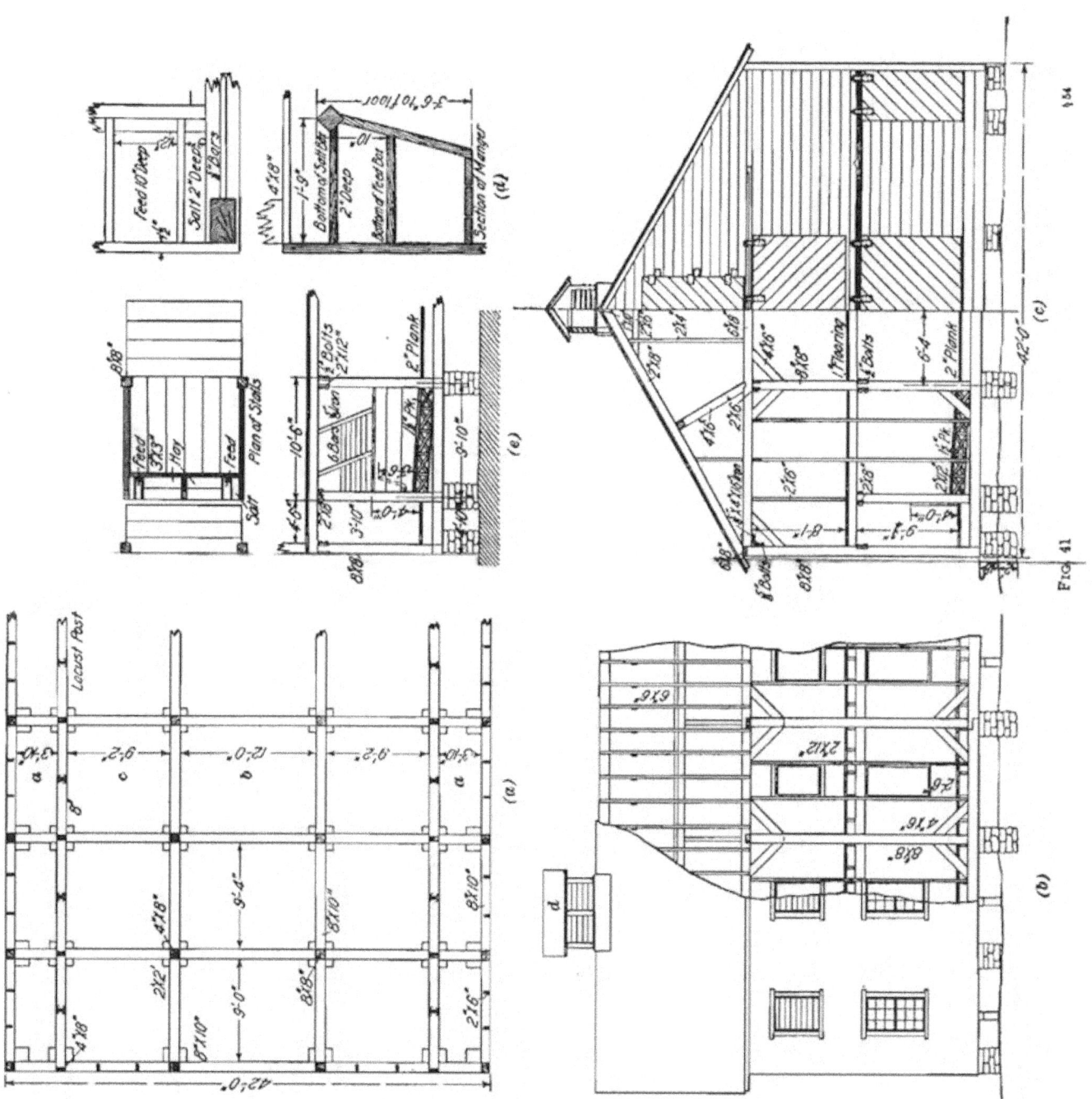

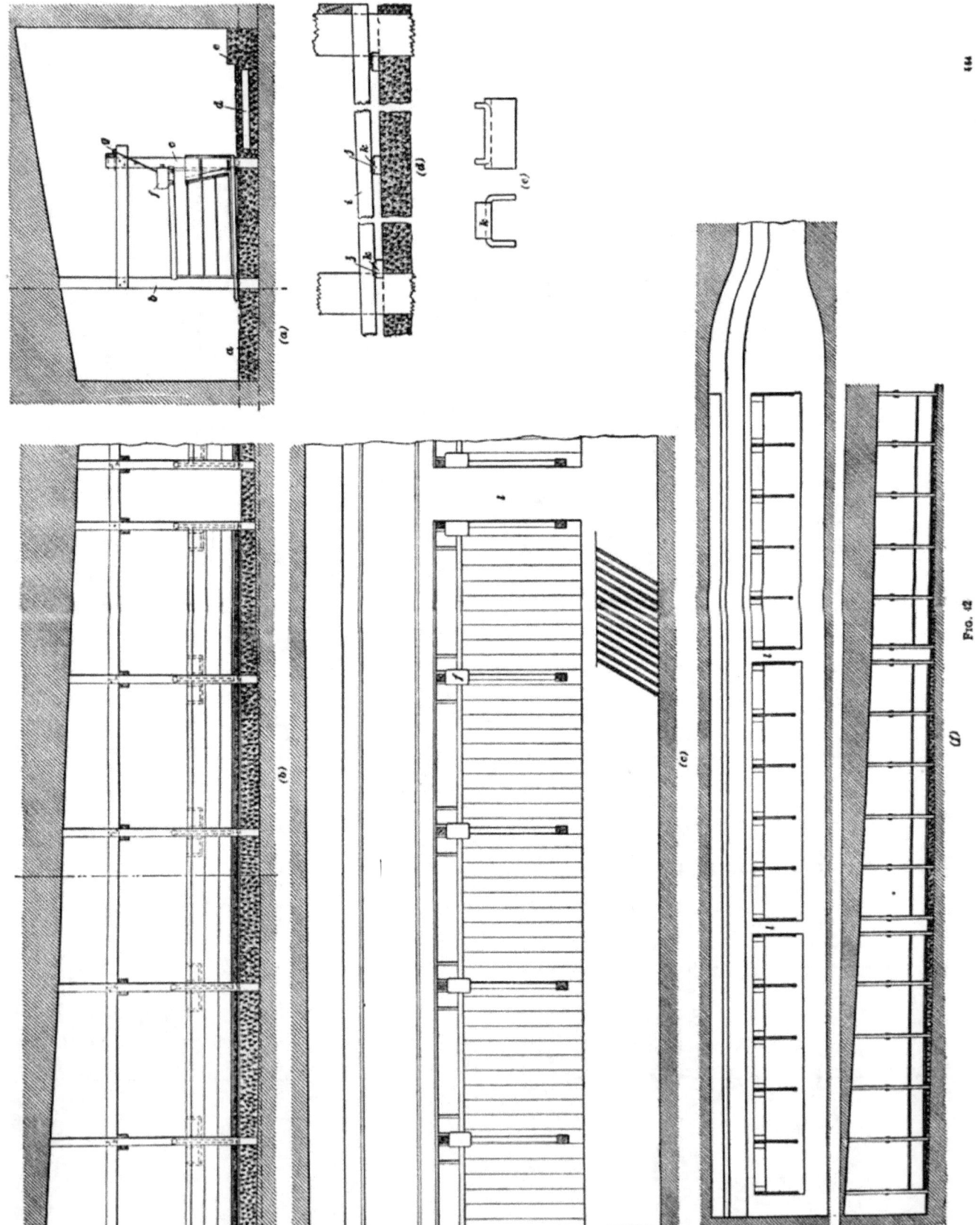

Fig. 12

the mule, for, at their best, mine roads are poor traveling-ways for animals, as they are usually uneven and covered with fine sharp coal. A mule's legs should, therefore, be thoroughly washed with a hose when he comes into the stable.

MINE STABLES

68. It is economy to provide a comfortable housing for the mine stock, either on the surface or below ground, depending on local conditions. These stables should be well ventilated and drained, and when on the surface should be well lighted.

69. Surface Stables.—It has been found that with a conveniently arranged stable, one man can care for forty head of stock. Fig. 41 shows the standard barn of the H. C. Frick Coke Company in Western Pennsylvania. It is 42 ft. × 100 ft. and 9 feet high in the clear on the stable floor and is designed for forty head of stock, thus giving each animal about 100 square feet of floor space. The feed-entries a, Fig. 41 (a), along each side are 3 feet 10 inches wide between timbers and a center passageway b is 12 feet wide. Each stall c is 9 feet 2 inches square between timbers and is intended for two mules, being divided by a pole. In front of each stall is a window. The construction of the floor for drainage purposes is shown in the end elevation (c) and the stall section (e). Ventilation is secured by air-shafts, which lead to ventilators d in the roof. These ventilating shafts are also used for hay chutes from the feed-floor to the manger. The detailed construction of the manger is shown in (d) and of the stall partition in (e).

The piping of a stable is sometimes arranged so that the water buckets in each stall can all be filled or emptied at once by the turning of a single cock.

70. Fig. 42 shows a barn below ground in a space excavated out of the solid coal, as built by the Lehigh and Wilkes-Barre Coal Company. The walls are of coal, and the roof the ordinary top rock of the mine, which will stand without a support. The floor a is of concrete in which the posts b and c are bedded, also the ties d on which the rails are supported, on which small cars run for bringing in feed for the animals. The back part of the floor is raised, forming a platform e on which the feed is stored so that it will not be injured when the barn is flushed out with the hose. Each stall provides space for two mules, and between each two stalls is a water box f, which can all be filled at one time from the 2-inch water pipe g, or

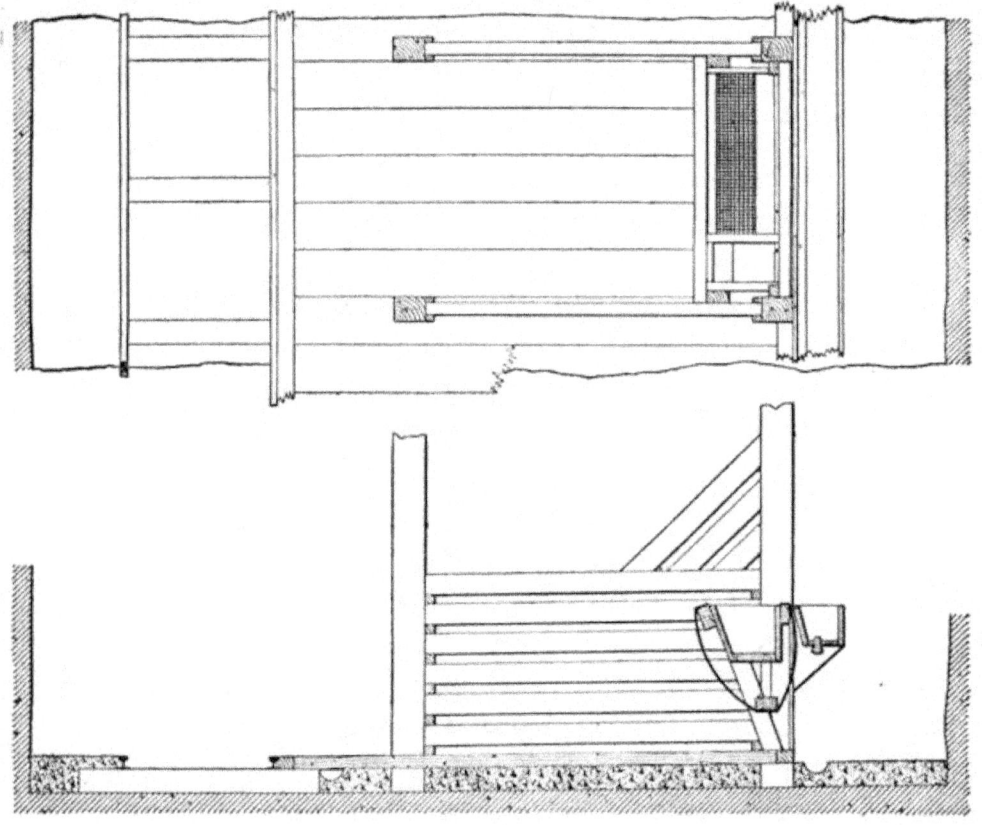

Fig. 43

each may be arranged with a separate cock. The water troughs can also be emptied singly, or at once. The partitions between the stalls are made of 3-inch oak planks set with a 1-inch space between planks as shown in Fig. 42 (a). The top of the partition is a $4'' \times 4''$ timber. The floor of the stalls is of novel construction and consists of 3-inch oak planks i, Fig. 42 (d), laid on steel stringers j. These steel stringers rest in cast supports k that are set in the concrete and are shown in detail in Fig. 42 (*e*). A space is thus provided underneath the floor, which can be flushed without removing the floor. The stable is constructed with five stalls in a section between cross-aisles *l*, as shown in Fig. 42 (*f*). In order to facilitate the removal of the droppings, another track might be placed back of the animals so that a car can be run the full length of the stable both in front and back of the stalls.

In locations where the walls or the roof will not stand, concrete is extended up the walls, and the roof is supported by steel I beams laid on top of the concrete, and filled in between with concrete.

71. Fig. 43 shows a slightly different arrangement used by the Lehigh Valley Coal Company, in which the floor of the stall consists of loose boards laid loosely on top of the concrete so that they can be easily raised for flushing out underneath the floor. The feed-bin is of metal, and the water trough is placed outside the stall.

HAULAGE
(PART 2)

LOCOMOTIVE HAULAGE

INTRODUCTION

1. The most common application of underground locomotive haulage is to coal mines, where the coal is usually gathered from the rooms or working places by animals and hauled to a parting or turnout. These partings or turnouts are usually on side or cross-entries, though they may be on main haulage ways. Here a number of cars are coupled into trips and are hauled by locomotives to the outside, or to the shaft bottom. In extensive mines, where cross-entries are long, the cars are gathered on the partings by mules and hauled by locomotives from gathering partings to the main parting, which is usually just off of the main haulage way. From this latter point, the trips are taken to their destination by the same or other locomotives. Small compressed-air and electric locomotives are now also used for gathering the cars from the rooms and collecting them at the partings, where the trips are made up and hauled out by the locomotives.

2. Advantages of Locomotive Haulage.—Under certain conditions, locomotive haulage possesses advantages over other systems of mine haulage. Its chief advantages over animal haulage are speed and ability to haul larger quantities of material at one time. One advantage that it possesses over rope haulage is that the motive power

Copyrighted by International Textbook Company. Entered at Stationers' Hall, London

accompanies the cars to be hauled and consequently there is a saving in time and greater safety in coupling and uncoupling cars. Another feature favorable to locomotive haulage is that locomotives can pull from the main haulage road directly to partings on cross-entries without stopping to couple and uncouple as is the case with rope haulage. Also, for gathering cars from rooms, small electric and compressed-air locomotives can be employed, while gathering from the rooms is impracticable with a rope haulage. Locomotive haulage is better adapted than other systems of mechanical haulage to mines that are being almost continuously extended, as it can readily be kept up with the development without extra expense.

3. Tracks for Mine Locomotives.—To secure all the advantages to be derived from locomotive haulage, it is necessary to arrange the mine tracks, partings, and switches so that cars may be handled in either direction without delay. For underground locomotive haulage, proper construction and alinement of tracks are more important than for surface locomotive haulage because a wreck or even a derailment in the mine usually stops the whole haulage system and the mine work as well, since there is not usually room to lift the locomotive off the track and set it to one side so that other locomotives, if available, may take care of the haulage, nor can temporary tracks be placed alongside the wreck. A soft and muddy roadbed, light rails, ties too small or placed too far apart, and fish-plates too small or carelessly put in will neutralize the economies of the most carefully designed and expensive equipment. Frogs and switches should be carefully constructed, and turnouts should be easy and gradual, in order that neither cars nor locomotives may be derailed. The roadbed should be firm and solid. When there is fireclay bottom, it must be kept dry. Ties should not be less than 6 inches wide and 6 inches deep and wider ties are preferable. They should be spaced so as to be about 1 foot apart and the rails should be firmly spiked to them. Splice bars or fish-plates should be firmly bolted to the rails. Toe-plates and guide rails should be used on curves. Rails should be spaced accurately to gauge, with the proper allowance on curves. The weight of the rail for locomotive haulage should never be less than 40 pounds to the yard on main haulage ways. Locomotive tires wear badly when the rails are narrow, and this wear is particularly noticeable on grades and where the tracks must be sanded, as, for instance, at the partings where the trips are started. In order to give additional surface contact for the wheels at partings and on grades, the loaded track is sometimes laid with 60-pound rails, while a 40-pound rail is used generally throughout the mine. If the track is dirty and greasy, the locomotive wheels slip and the pulling power of the locomotive is greatly decreased. It is therefore very desirable that the track be not only well laid, but that it be kept in good condition in order that good results may be obtained from the locomotives. As the hauling power of a locomotive is greatly reduced when it hauls up a grade, the grades of the mine tracks should be such that the locomotives may be operated most advantageously.

4. Power for Mine Locomotives.—Steam, compressed air, and electricity are the motive powers used in driving locomotives in mines. Gasoline has been tried but has never been used to any extent underground. The kind of power to be adopted at any mine will depend entirely on local conditions and no one power can have universal application.

The sizes and capacities of the various appliances used in connection with a haulage plant are ascertained from such data as the following: the height and width of the mine entry; gauge and weight of track; weight of empty and loaded cars; output to be handled per day; a profile of the road giving the amount and length of grades; the sharpest curve and the grade on which this curve occurs; the length of the haul; and, if compressed-air or electric locomotives are to be used, the distance between the compressor or generator and the terminal of the haulage line.

Naturally, where there are wide margins of difference in respect to the size of mine passages, special locomotives must be constructed; as a usual thing, however, one of the stock patterns will meet most requirements.

STEAM-LOCOMOTIVE HAULAGE

5. Advantages and Disadvantages.—Steam locomotives should be used only on return airways as the gases given off from the coal burned to produce the steam vitiate the air, and a number of deaths have occurred from the use

TABLE I

Mule Haulage		Steam-Locomotive Haulage	
Items	Cost	Items	Cost
Three mules' feed, harness, shoeing, care, etc., for 365 days, each per day, $.50 . . .	$547.50	Oil and repairs per year	$165
One driver's wages, 300 days at $1.75 per day	525.00	Fuel, 400 to 1,000 pounds coal, or one-third to three-fourths cord wood per day for 300 days, per day $2	600
Six per cent. interest, mules worth $150 each . . .	27.00	Engineer's wages, 300 days, per day, $2.50	750
		Car runner	450
		Interest, 6 per cent.	150
Total	$1,099.50	Total	$2,115

of such locomotives in places where there was not a sufficient air-current. They should not be used in gaseous mines, as steam locomotives should be used only on return airways where the greatest danger from explosive gas exists.

Steam locomotives should not be run on grades higher than $2\frac{1}{2}$ to 3 per cent., as they require forcing, and as water supply is likely to be scant there is danger, from the high temperatures, of burning the boiler tubes. In general, however, a steam-locomotive system is cheaper, both in first cost and in operation, than an electric or compressed-air haulage system. A steam locomotive requires but one man to operate it, while a compressed-air or an electric locomotive requires an engineer and fireman at the power generating plant in addition to the engineer or motorman on the locomotive.

6. Relative Cost of Steam Locomotive and Animal Haulage.—Table I shows an estimate of the difference between the cost per year of operating three mules in a string team and the cost of operating a mine locomotive, the prices for labor and supplies being assumed at an average rate.

The cost of fuel at a coal mine is very small. From Table I, it is evident that the cost of operating two mule teams of three mules each is about the same as that of one locomotive. The advantage of steam-locomotive haulage over animal haulage is due to the greater speed and capacity of the locomotives by which one locomotive can accomplish the same work as a number of mule string teams.

CONSTRUCTION OF STEAM MINE LOCOMOTIVES

7. Steam mine locomotives are similar in their general construction to an ordinary railroad locomotive; but in size and arrangement they are built to meet the requirements of the mine in which they operate. They are usually small, as they often have to run in low and narrow passages. When determining the height that a locomotive may have, 1 foot 2 inches should be deducted from the minimum height of the haulage way. This allows for the use of ties 6 inches thick; 40-pound rails, which are 4 inches high, and leaves 4 inches of clearance between the top of the locomotive and the roof. The makers of mine locomotives claim that a clearance of 2 inches is sufficient, but this is not enough if there is any creep or sag. Draw slate, when thin, will be brought down

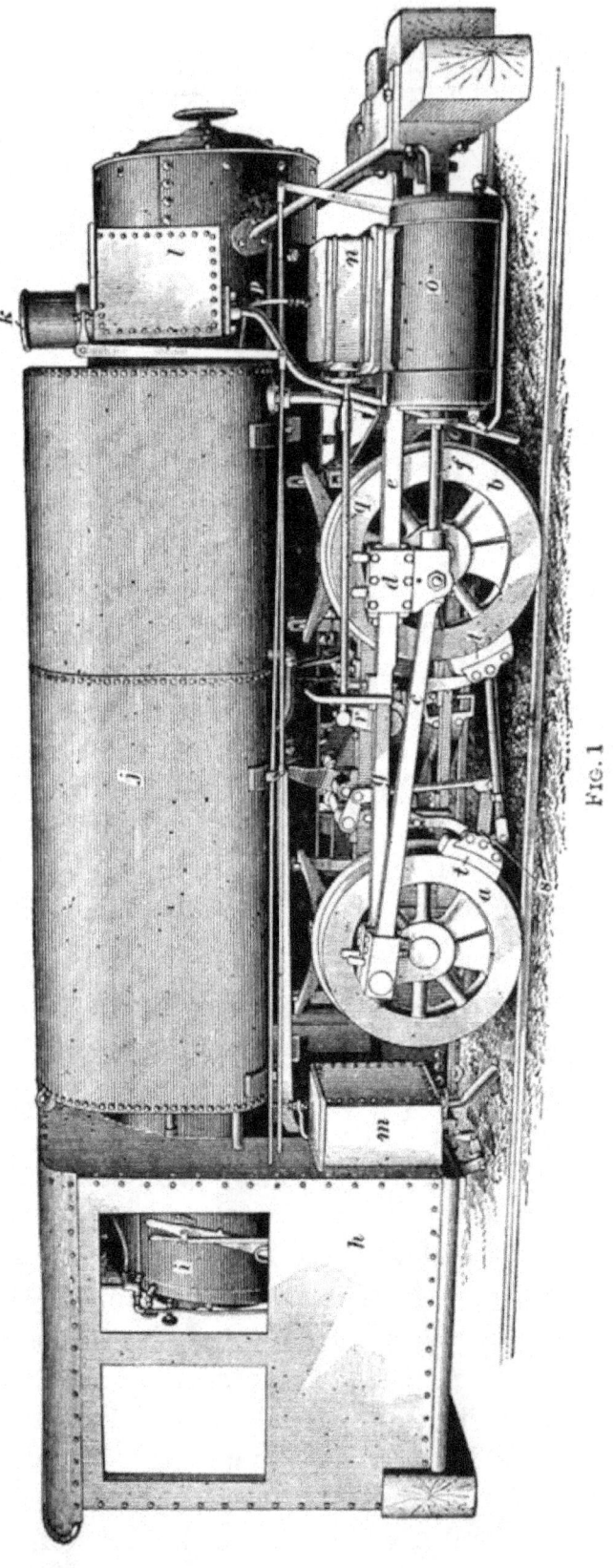

Fig. 1

by the air and the exhaust from the locomotive; for this reason, it is often advantageous to take down the draw slate and thus allow more space for the locomotive. Two types of steam locomotives are ordinarily used about mines the *four-wheel connected*, and the *six-wheel connected*.

8. Four-Wheel, Connected, Steam, Mine Locomotives.—Fig. 1 shows a modern **four-wheel, steam, mine locomotive** having two pair of driving wheels a and b. The wheels vary from 20 to 36 inches in diameter according to the height of the locomotive. The *wheel base* on mine locomotives or the distance between the centers of the axles of the outside driving wheels is a matter of considerable importance and it should be as short as possible, owing to the sharp curves mine locomotives must travel. The wheels are therefore made as small as practicable and the axles are placed as near together as the arrangement of the parts of the locomotive will permit. The wheel base varies from 3 feet on small locomotives weighing 8,000 pounds to 6 feet 3 inches on locomotives weighing 62,000 pounds. All the driving wheels of locomotives of the four-wheel type are flanged on the inside to make them follow the rails. The length of the crank is equal to the distance of the center of the crank-pin from the center of the driving wheel and depends on the length of the piston stroke. Since the crank acts like the power arm of a lever, the greater its length the greater the power it will exert. The driving rod or main rod c is connected with the driving wheel a and the crosshead d, which moves over a slide bar e as the piston rod f moves backwards and forwards. The parallel rod or side rod g transmits the power to the drivers b.

The *cab* h, Fig. 1, is covered with sheet iron to protect the engineer from sparks and falling rock. To increase the weight of the locomotive and to do away with a tender to hold fuel and water, a steam mine locomotive has its boiler i covered with an iron water tank j known as a *saddle tank*. In calculating the pulling power of a mine locomotive, the weight is taken on the assumption that the tank is half full of water and a *tender* is not reckoned in. A tender pulled after a mine locomotive would not only lessen the number of cars that could be pulled but it would be in the way when coupling or uncoupling them. With the saddle tank over the boiler, as shown in Fig. 1, the engineer can back to a train of cars and watch the coupling, which results in greater safety for the helper, less hard bumping of the cars, and a saving of time in coupling. The stack k is flush with the top of the water tank. Since locomotives run either way and the rails are at times wet, it is necessary to have sand boxes at each end, as at l and m. Each box is supplied with a pipe that directs the sand to the rail and the flow of sand is controlled from the cab by means of levers and reach rods.

The *steam chest* n is above the steam cylinder o. The pipe p above the steam chest carries lubricating oil to the valves. The valve rod q is attached to a rocker crank r on a shaft rocked by the eccentrics not shown but placed usually on the axle of the rear driving wheels. The movement of the rocker gives the reciprocating motion to the valve. The *togglejoint* s for applying the brakes t is worked from the cab by means of a lever or brake wheel. The fuel supply is sometimes carried in a box in the cab, but in most coal mines the supply is generally taken from the car nearest to the locomotive. The *brake shoes* t for locomotives are of cast iron, cast and wrought iron, or cast iron and steel. The brake shoes cover the entire width of the tread and sometimes the flange of the wheel as well. Wide shoes offer a large frictional surface and exert dressing action on that part of the tread of the wheels that ordinarily, with narrow rails, is left unworn. Steel brake shoes are set into cast-iron bodies and so arranged that they may be readily removed by loosening a single setscrew. A complete set can be exchanged in 10 or 15 minutes. One set of these shoes is said to outwear twelve sets of cast-iron shoes.

9. Six-wheel, connected, steam, mine locomotives have three pair of connected driving wheels. The first and last pairs are flanged, but the center ones are not flanged in

TABLE II

PRINCIPAL DIMENSIONS AND IMPORTANT DATA RELATING TO COMMON SIZES OF MINE LOCOMOTIVES

Cylinders: Diameter, in inches	5	6	7	8	9	10	11	12	6	8	10	12	14
Cylinders: Stroke, in inches	8	10	12	14	14	14	14	16	10	14	14	16	18
Diameter of driving wheels, in inches	20	23	24	26	28	30	30	33	20	24	28	33	36
Wheel base, in feet and inches	3′ 0″	4′ 0″	4′ 0″	4′ 0″	4′ 6″	4′ 6″	4′ 6″	5′ 9″	4′ 8″	5′ 0″	5′ 6″	7′ 0″	8′ 6″
Length over bumpers, in feet and inches	10′ 0″	11′ 7″	14′ 4″	14′ 9″	16′ 6″	4′ 9″	17′ 9″	19′ 6″	12′ 6″	14′ 7″	18′ 0″	21′ 0″	24′ 0″
Excess of width at cylinders over gauge of track, in feet and inches	1′ 11$\frac{1}{4}$″	2′ 1$\frac{1}{2}$″	2′ 3$\frac{3}{8}$″	2′ 5$\frac{1}{4}$″	2′ 7$\frac{3}{8}$″	2′ 9$\frac{7}{8}$″	3′ 0″	3′ 1$\frac{7}{8}$″	2′ 1$\frac{1}{2}$″	2′ 5$\frac{1}{4}$″	2′ 9$\frac{7}{8}$″	3′ 1$\frac{7}{8}$″	3′ 8$\frac{1}{8}$″
Height above rail, least desirable, in feet and inches	4′ 10″	5′ 2″	5′ 4″	5′ 7″	5′ 9″	5′ 11″	6′ 3″	6′ 6″	5′ 0″	5′ 9″	6′ 3″	6′ 8″	8′ 0″
Height above rail, least practicable, in feet and inches	4′ 4″	4′ 7″	4′ 9″	5′ 0″	5′ 3″	5′ 5″	5′ 8$\frac{1}{2}$″	6′ 0″	4′ 6″	5′ 4″	5′ 9″	6′ 0″	6′ 10″
Weight in working order, in pounds	8,000	14,000	16,500	21,000	25,000	28,000	34,000	42,000	14,500	23,000	32,000	45,000	64,000
Capacity of tank, in gallons	80	125	200	250	300	400	500	600	175	275	400	600	900
Weight per yard lightest rail advised, in pounds	12	16	20	20	25	30	35	45	12	16	25	30	45
Radius of sharpest curve advised, in feet	20	25	25	25	30	30	30	45	30	45	50	65	100
Radius of sharpest curve practicable, in feet	12	15	15	15	16	16	16	25	20	25	35	50	75
Boiler pressure per square inch, in pounds	140	140	140	140	140	140	140	140	140	140	140	150	150
Tractive power, in pounds	1,190	1,860	2,915	4,100	4,820	5,560	6,720	8,320	2,140	4,440	5,950	8,915	12,500
Hauling capacity, in tons of 2,000 pounds (exclusive of locomotive): On absolute level	175	280	440	620	725	840	1,015	1,255	320	670	900	1,345	1,890
On $\frac{1}{2}$-per-cent. grade = 26$\frac{4}{10}$ feet per mile	65	105	165	230	275	315	390	480	120	255	345	518	730
On 1-per-cent. grade = 52$\frac{8}{10}$ feet per mile	35	60	100	140	165	190	235	290	70	155	210	310	440
On 2-per-cent. grade = 105$\frac{6}{10}$ feet per mile	18	30	50	70	85	100	125	155	40	80	120	165	240
On 3-per-cent. grade = 158$\frac{4}{10}$ feet per mile	9	20	30	45	55	65	80	100	25	55	75	110	160

order that the locomotive may take the curves more easily. The distance between the centers of the axles of the first and last pairs of drivers determines the wheel base. Theoretically, the smallest size of locomotive, which has a wheel base of 4 feet 8 inches, would take a curve of 20 feet radius, but the makers give 30 feet as the radius of the sharpest curve advisable. The six-wheel locomotives have one main rod and two side rods on each side. The main rods are connected with the rear drivers and the crossheads, while the side rods connect the three pair of driving wheels. In case the driving wheels are of small diameter, the cylinders are inclined; but when the drivers are of sufficient diameter, the cylinders are placed in a horizontal position. The brakes are applied to the rear and front wheels, but not to the center wheels. Six-wheel mine locomotives have a greater pulling capacity than the four-wheel type, but the latter are probably more generally used.

The general dimensions and hauling capacities of some of the common sizes of steam mine locomotives are given in Table II, based on a frictional resistance of $6\frac{1}{2}$ pounds per ton. While these are the dimensions ordinarily used, locomotive builders can supply a locomotive to serve under nearly any conditions.

POWER OF A LOCOMOTIVE

10. Tractive Power.—The **tractive power** of a locomotive, often called **tractive force,** is the power developed by the machinery of the locomotive that is available to move the locomotive and the load attached to it. The item of speed is disregarded. It is customary to rate the tractive power of a locomotive at from one-eighth to one-fifth of the weight on the drivers, the exact coefficient varying with the weight of the locomotive and the different manufacturers.

11. The **drawbar pull** of the locomotive is the pull on the locomotive drawbar and is the portion of the tractive power available for hauling a load; that is, it is the total tractive power developed by the locomotive less the amount of power required to propel the locomotive itself. To determine the drawbar pull, the drawbar may be attached to a dynamometer and the pull directly measured; or, if a rope attached to the drawbar of a locomotive is passed over a pulley and a weight attached to it, the drawbar pull is equal to the weight the locomotive can lift.

The terms tractive power and drawbar pull are often used to mean the same thing, and from the practical standpoint only a small error results from so considering them, particularly with light locomotives such as are commonly used in the mines, since the amount of power consumed in running the locomotive itself on a level is only a small part of the total power generated by the locomotive. On an up grade, however, while the tractive power of the locomotive remains the same, the drawbar pull decreases rapidly with increasing grade on account of the increased amount of power required to run the locomotive.

12. The power of a locomotive to pull a load depends on the adhesion between the driving wheels and the rails. The amount of this adhesion, often called the *maximum starting effort*, *slipping point*, or *adhesive power* of a locomotive, depends largely on the weight on the drivers, and varies with the conditions under which the locomotive operates. Under ordinary conditions, and also on a wet, sanded rail, it is about one-fifth the weight of the locomotive; with favorable conditions and a dry rail without sand, it is about one-fourth; and on a well-sanded dry rail, about one-third the weight. The adhesive power is greatest on a level track and decreases as the inclination of the track increases.

If the tractive power of a locomotive exceeds the adhesive power, the wheels will slip and part of the tractive power will be wasted. The power driving the locomotive and the weight on the driving wheels must, therefore, be properly proportioned to obtain satisfactory results, for if the cylinders of a steam or an air locomotive are too large the driving wheels will slip, showing that there is too much power for the weight. A locomotive in this condition is said to be *overcylindered*. On the other hand, if there is not enough power for the weight, so that the wheels will not move, the locomotive is said to be *undercylindered*.

13. The steam or air pressure, the diameter of the cylinders, and the diameter of the driving wheels are factors that influence the tractive power of a locomotive, and this may be increased by decreasing the diameter of the driving wheels or by increasing the steam or air pressure. On the other hand, the tractive power may be decreased by increasing the size of the driving wheels or decreasing the steam or air pressure in the cylinders.

A large cylinder presents more piston area than a small one, and therefore, with the same steam pressure per square inch, more force will be exerted on the crank by using the larger cylinder. The cylinders and cranks of two locomotives can be so proportioned that with large driving wheels for the large cylinders and small driving wheels for the small cylinders the tractive power will be equal for the two locomotives.

A similar condition could be brought about by increasing the steam or air pressure in the small cylinder and decreasing it in the large cylinder without changing the diameter of the wheels.

14. The following is a general formula given by writers and manufacturers for determining the tractive power of a locomotive:

$$T = \frac{D^2 \times L \times .85p}{d}$$

in which
- T = tractive power, in pounds;
- D = diameter of cylinder, in inches;
- L = length of piston stroke, in inches;
- p = boiler pressure, in pounds per square inch;
- $.85p$ = 85 per cent. of boiler pressure, in pounds per square inch, which has been found by practical tests to be about the effective pressure in the cylinders;
- d = diameter of driving wheel, in inches.

EXAMPLE.—What will be the tractive power of a four-wheel, connected, mine locomotive whose steam cylinders are 7 inches in diameter and whose piston stroke is 12 inches, with steam pressure at 140 pounds per square inch and driving wheels 24 inches in diameter?

SOLUTION.—Substituting in the formula,
$$T = \frac{7^2 \times 12 \times .85 \times 140}{24} = 2,915.5 \text{ lb.} \quad \text{Ans.}$$

15. By a similar solution, it may be shown that if the steam pressure is increased to 150 pounds, the tractive power will be increased to 3,123.75 pounds; or, if the steam pressure is reduced to 130 pounds, the tractive force will be reduced to 2,707.25 pounds. It is evident also, from the same formula, that if the diameter of the driving wheels is increased, the steam pressure remaining the same, the tractive force will be diminished; for instance, if the diameters of the driving wheels are increased from 24 to 30 inches the tractive force will be reduced from 2,915.5 pounds to 2,332.4 pounds. Locomotives are so designed that those of equal weight should have approximately equal loads under the same conditions.

Although the formula in Art. **14** is based on practice and not on theory, that is, it is empirical, it has given satisfaction for many years to locomotive manufacturers. The .85 per cent. of the boiler pressure allows for the friction of the locomotive machinery; hence, for a locomotive on the level, the tractive power calculated by this formula will be very nearly the same as the drawbar pull and may be used instead of the drawbar pull in locomotive haulage problems.

In most of the tables for tractive power and drawbar pull published by the builders of mine locomotives, the values given are calculated by the formula in Art. **14** and no distinction is made between tractive power and drawbar pull. The frictional resistance of a locomotive is usually considered by the makers as $6\frac{1}{2}$ to $7\frac{1}{2}$ pounds per ton of weight on the drivers, while the tractive power varies from 300 to 450 pounds per ton of weight on the drivers. On account of the variable condition of the mine tracks, this resistance may vary within wide limits.

16. Drawbar Pull on a Grade.—Although the tractive power and drawbar pull may be taken as equal on the level with practically accurate results, this is not the case on a grade; for while the theoretical power generated by the locomotive is always the same under like conditions, the greater the up grade the greater is the proportion of this power required to move the locomotive and consequently the less is the drawbar pull, or the amount of power available for pulling a load. The drawbar pull, or the net tractive power, of a locomotive on an up grade is determined by deducting from the rated tractive power on a level track the product obtained by multiplying the weight of the locomotive by the per cent. of grade, or, which is the same thing, the amount to be deducted may be taken as 20 pounds per ton of weight for each per cent. of grade.

EXAMPLE.—On a 2-per-cent. grade, what is the drawbar pull of a four-wheel mine locomotive weighing 8,000 pounds and having $5'' \times 8''$ cylinders and 20-inch drivers and a boiler pressure of 140 pounds per square inch?

SOLUTION.—Substituting in the formula in Art. **14**,

$$T = \frac{D^2 \times L \times .85\, p}{d}; \quad T = \frac{5^2 \times 8 \times .85 \times 140}{20} = 1{,}190 \text{ lb.}$$

The resistance due to the grade, taken as 20 lb. per T. of weight for each per cent. of grade, is $20 \times 2 \times \frac{8000}{2000} = 160$ lb. Hence, the drawbar pull is $1{,}190 - 160 = 1{,}030$ lb. Ans.

17. Table III gives the drawbar pull for a number of the locomotives given in Table II. The drawbar pull on a level is assumed to be the same as the tractive power given in Table II and the drawbar pulls on the several grades are calculated from the drawbar pull on the level.

If the haul is of great length, the average drawbar pull required should be well within the rated drawbar pull of the locomotive; while, if the haul is short and the service intermittent, the locomotive may be operated at its rated drawbar pull. The steepest grade against the load and not the average grade should be considered in determining the size of a locomotive for a given haulage system. On a

§ 55　　　　　　　　HAULAGE　　　　　　　　15

short grade, the locomotive may be worked very close to the slipping point of the wheels or maximum starting effort.

TABLE III

Locomotive Weight	Maximum Starting Effort Under Ordinary Conditions 20 Per Cent. of Weight of the Locomotive	Drawbar Pull, in Pounds				
		Level*	½-Per-Cent. Grade	1-Per-Cent. Grade	2-Per-Cent. Grade	2½-Per-Cent. Grade
8,000	1,600	1,190	1,150	1,110	1,030	990
10,000	2,000	1,290	1,240	1,190	1,090	1,040
14,000	2,800	1,860	1,790	1,720	1,580	1,510
16,500	3,300	2,915	2,833	2,750	2,585	2,503
21,000	4,200	4,100	3,995	3,890	3,680	3,575
25,000	5,000	4,820	4,695	4,570	4,320	4,195
28,000	5,600	5,560	5,420	5,280	5,000	4,860
32,000	6,400	5,950	5,790	5,630	5,310	5,150
34,000	6,800	6,720	6,550	6,380	6,040	5,870
36,000	7,200	6,980	6,800	6,620	6,260	6,080
42,000	8,400	8,320	8,110	7,900	7,480	7,270
52,000	10,400	10,450	10,190	9,930	9,410	9,150
62,000	12,400	12,500	12,190	11,880	11,260	10,950

*Ratings as given by manufacturers in Table II.

HAULING CAPACITY OF LOCOMOTIVES

18. The **hauling capacity** of a locomotive is the weight of cars and load that the locomotive can haul exclusive of the weight of the locomotive.

Rule.—*To calculate the hauling capacity of a locomotive, divide the tractive power of the locomotive, in pounds, by the sum of the resistances due to gravity and friction, expressed in pounds of resistance per ton of weight, and from the quotient deduct the weight of the locomotive, in tons. The result will be the weight, in tons of 2,000 pounds, of the train that the locomotive can haul.*

EXAMPLE 1.—What would be the hauling capacity of a 10,000-pound or 5-ton locomotive on a level track, assuming that the total frictional resistances are 10 pounds per ton, and the tractive power of the locomotive is 1,290 pounds?

SOLUTION.—Applying the rule,
$$\frac{1{,}290}{5 \times 10} = 25.8 \text{ T.}; \quad 25.8 - 5 = 20.8 \text{ T. Ans.}$$

EXAMPLE 2.—What will be the hauling capacity of a 16,500-pound locomotive when the grade is 1½ per cent. and the frictional resistance due to cars and roadbed is 10 pounds per ton of 2,000 pounds, assuming that the cylinder is 7 inches in diameter, the piston stroke 12 inches long, the steam pressure 150 pounds per square inch, and the driving wheels 24 inches in diameter?

SOLUTION.—First, find the tractive force from the formula given in Art. **14**,
$$T = \frac{D^2 \times L \times .85\, p}{d}; \quad T = \frac{49 \times 12 \times (.85 \times 150)}{24} = 3{,}123.75 \text{ lb.}$$
Since the frictional resistance is 10 lb. per T., and the resistance due to gravity is $2{,}000 \times .015 = 30$ lb. per T., the total resistance is 40 lb. per T., and $3{,}123.75 \div 40 = 78.09$ T. The weight of the locomotive is 16,500 lb. or 8.25 T.; hence, the locomotive's hauling capacity is $78.09 - 8.25 = 69.84$ T. Ans.

19. Resistance Due to Curves.—It is not possible to formulate a rule for the resistance on curves that will apply under all conditions, owing to the numerous factors involved, such, for instance, as the elevation of the outer rail, the speed and length of the train, the degree and length of the curve, and the condition of the track and rolling stock. It may be stated generally that a curve with a long radius offers less resistance than one with a short radius.

Theoretically, every axle in the train rounding a curve should point to the center of the curve, and the wheels on the outside of the curve should be larger than on the inside; this, however, is not practicable, but by reducing the wheel base and coning the tread of the wheels the same result is accomplished in a measure, so that short curves are rounded without derailing the locomotive or cars. In passing around curves, the centrifugal force tends to tip the cars and crowd the wheels over against the outer rail. This crowding increases with the speed and is greater on a curve having a short radius than on one having a long radius. To counteract this crowding due to change in direction, it is customary to elevate the outer rail just before the curve is reached and along the curve. On sharp curves, the gauge of tracks is sometimes widened $\frac{1}{16}$ inch for each $2\frac{1}{2}°$ of curvature, to prevent the wheels from binding against the rails.

20. Grade Reduction on Curves.—Steep grades in connection with sharp curves are to be avoided, as they greatly decrease the load that a locomotive can pull. It is better to increase the distance of the haul, if by so doing such a combination can be avoided, for, although the cost of construction will be greater, there will not be a continuous loss through high operating expenses due to the smaller loads that can be hauled and the increased repairs required for cars and rolling stock. Where a curve occurs on a grade, it is customary to reduce the grade on the curved portion of the track, so that the resistance of the flattened grade and curve will not exceed the resistance of the steeper grades on the straight part of the track. A reduction of $\frac{2}{100}$ foot per 100 feet of track, or .02 per cent. for each degree of curvature is sufficient for standard-gauge tracks, but for mine tracks with 42-inch gauge and less the rate of compensation should be increased to $\frac{3}{100}$ foot per 100 feet, or .03 per cent., for each degree of curvature, particularly with sharp curves.

EXAMPLE.—What should be the compensated grade on a 40° curve in a mine track so that the resistance on the curve will be the same as that on a straight track having a grade of 4 per cent.?

SOLUTION.—Since a reduction of .03 per cent. should be made in the grade on the curve for each degree of curve, the reduction for a 40° curve will be $40 \times .03 = 1.2$ per cent. Since the straight track has a grade of 4 per cent., the compensated grade on the curve will be $4.0 - 1.2 = 2.8$ per cent., or a rise of 2.8 ft. per 100 feet of track. Ans.

21. For surface railroad practice, where the locomotives have fixed wheels and a wheel base of about 6 feet, it is customary to allow .5 pound per ton of weight per degree of curvature for the curve resistance, that is,

$$R = .5\,x$$

in which R = resistance due to curve, in pounds per ton;

x = degree of curvature.

This formula is used for mine locomotives by many locomotive manufacturers, but it does not give accurate results owing to the fact that mine locomotives have a much smaller wheel base and are of narrower gauge than locomotives for use on the surface. Owing to the wide variation in the conditions of mine track, car wheels, etc., at different mines and to the lack of experimental data relating to friction of mine cars, it is impossible to give a formula for the resistance of mine cars that will be of general application; hence, the resistance must be determined experimentally in each case.

22. Resistance on Curves.—From experiments with mine cars, the average frictional coefficient on curves has been found to be approximately

$$\frac{\text{wheel base}}{5 \times \text{radius of curve}}$$

The resistance to a car or locomotive on a curve is then expressed by the formula,

$$R = \frac{BW}{5r}$$

in which R = resistance, in pounds per ton per degree of curvature;

B = wheel base, in feet;

r = radius of curve, in feet;

W = weight of locomotive or car, in pounds.

EXAMPLE.—A locomotive weighing 8,000 pounds has a wheel base of 5 feet and is to run on a curve having a radius of 50 feet; what will be the resistance due to curvature?

SOLUTION.— $R = \dfrac{5 \times 8,000}{5 \times 50} = 160$ lb. Ans.

23. Speed of a Locomotive.—It is impossible for any locomotive to haul the heaviest load of which it is capable, and at the same time travel at its fastest speed. The speed at which a locomotive can haul a given trip is dependent on a number of factors. Some of these, such as the condition of track and cars, grades, curves, etc., are variable; others, such as the dimensions of the cylinders and boiler, the diameter of driving wheels, and general design, are constant. In any case, the speed increases as the load is diminished; the ratio between the load and the speed in any given case is more a matter of estimate than of precise calculation, and it is impossible to give any rule or formula that will cover correctly all sizes and designs of locomotives and all rates of speed. At a low rate of speed, it is practicable to exert nearly or completely the full boiler pressure in the cylinder during almost the entire stroke of the piston. As the speed increases, this becomes more difficult and the mean effective pressure in the cylinders decreases, which amounts to a decrease of the tractive force and of the load that can be hauled. The reason for this is twofold: the mechanical properties of steam are such that even if the supply were unlimited the cylinders could not be filled with steam approximating the boiler pressure as often as excessive speeds would require, nor could the exhaust steam be got rid of quickly enough; and, also, if these mechanical difficulties could be overcome, the locomotive could not carry a boiler large enough or evaporate water fast enough to supply the quantity of steam needed. It is also necessary, as a matter of economy, to use steam expansively. In mines, locomotives are run at such slow speeds that the resistance of the air-current need not be taken into account.

24. It requires greater power to start a train than to keep it moving after it has been started, and although it is impossible to give any exact rule it is possible to give the following *approximate rules* as to ratios between speeds and loads.

Rule I.—*Under usual conditions, steam locomotives can be depended on to haul the heaviest trains they can start at a speed, in miles per hour, equivalent to about one-fifth of the diameter of the driving wheels, in inches.*

For example, a locomotive having $9'' \times 14''$ cylinders, with driving wheels 28 inches in diameter, can be expected to haul trains very closely up to its full power at about 5.6 miles per hour.

Rule II.—*If the weight of the trip is reduced to about two-thirds or three-fourths of the full capacity, a speed, in miles per hour, equivalent to about one-half of the diameter of the driving wheels, in inches, is practicable.*

For example, a locomotive having $7'' \times 24''$ cylinders, with 24-inch driving wheels, can be expected to handle up to about two-thirds or three-fourths of its full capacity at about 12 miles per hour.

Rule III.—*With very light trips, say about one-eighth of its full capacity, a locomotive is capable of a speed in miles per hour approximating or exceeding the diameter of the driving wheels, in inches.*

For example, a locomotive with $12'' \times 16''$ cylinders, and 33-inch driving wheels, can handle trains of about one-eighth its full capacity at a speed of 33 or more miles per hour.

25. Work of a Locomotive.—The work performed by a locomotive is expended in overcoming the frictional resistance, gravity resistance, and the resistance due to curves in the track. This work, in foot-pounds, is equal to the resistance, in pounds, overcome by the locomotive multiplied by the distance through which it moves. The work may be calculated by the following formula:

$$F = DW(C + R + G)$$

in which F = foot-pounds of work performed;
D = distance, in feet, passed over;
W = weight of train and locomotive, in tons;
C = rolling friction, in pounds per ton;
R = resistance due to curves, in pounds per ton;
G = grade resistance, in pounds per ton.

When the train is going up grade, G is positive; when the train is going down grade, G is negative. The train will roll down grade by gravity and without locomotive assistance when G is minus and is greater than the rolling friction and the resistance due to curves. The grade resistance for each per cent. of grade is equal to 1 per cent. of the total weight, that is, 20 pounds per ton of 2,000 pounds.

EXAMPLE.—A mine track has a length on the grade of 6,200 feet, 3,000 feet of which is on a grade of 1.5 per cent. and includes a curve 500 feet long having a radius of 4,000 feet; the remaining 3,200 feet of track is straight and is on a grade of 2 per cent. How many foot-pounds of work will be required for an 8-ton. locomotive having a

5-foot wheel base to haul a trip of twelve loaded mine cars weighing 4 tons each over this track against the grade, the wheel base of each car being 3 feet? Assume the rolling friction of the cars to be 1 per cent. and that of the locomotive to be .5 per cent.

SOLUTION.—As this track is on two grades, the work on each grade and the curve must be found separately for the cars and for the locomotive by applying the formula $F = DW(C + R + G)$. Work required to haul the cars over the 1.5-per-cent. grade:

(a) Over the straight track, $D = 3,000 - 500 = 2,500$ ft.; $W = 48$ T.; $C = .01 \times 2,000 = 20$ lb. per T.; $R = 0$ lb. per T.; $G = 20 \times 1.5 = 30$ lb. per T. Substituting in the formula,
$$F = 2,500 \times 48 \times (20 + 0 + 30) = 6,000,000 \text{ ft.-lb.}$$

(b) Over the curve, $D = 500$ ft.; $W = 48$ T.; $C = 20$ lb. per T.; $R = \dfrac{3 \times 8,000 \times 12}{5 \times 4,000} = 14.4$ lb.; $G = 30$. Substituting in the formula,
$$F = 500 \times 48 \times (20 + 14.4 + 30) = 1,545,600 \text{ ft.-lb.}$$

Work required to haul the locomotive over the 1.5-per-cent. grade:

(c) Over the straight track, $D = 2,500$ ft.; $W = 8$ T.; $C = .005 \times 2,000 = 10$ lb. per T.; $R = 0$ lb. per T.; $G = 30$ lb. per T. Substituting in the formula,
$$F = 2,500 \times 8 \times (10 + 0 + 30) = 800,000 \text{ ft.-lb.}$$

(d) Over the curve, $D = 500$ ft.; $W = 8$ T.; $C = 10$ lb. per T.; $R = \dfrac{5 \times 16,000}{5 \times 4,000} = 4$ lb. per T.; $G = 30$ lb. per T. Substituting in the formula,
$$F = 500 \times 8 \times (10 + 4 + 30) = 176,000 \text{ ft.-lb.}$$

(e) Work required to haul the cars over the 2-per-cent. grade, $D = 3,200$ ft.; $W = 48$ T.; $C = 20$ lb. per T.; $R = 0$ lb. per T.; $G = 20 \times 2 = 40$ lb. per T. (See Art. **16**.) Substituting in the formula,
$$F = 3,200 \times 48 \times (20 + 0 + 40) = 9,216,000 \text{ ft.-lb.}$$

(f) Work required to haul the locomotive over the 2-per-cent. grade. $D = 3,200$ ft.; $W = 8$ T.; $C = 10$ lb. per T.; $R = 0$ lb. per T.; $G = 40$ lb. per T. Substituting in the formula,
$$F = 3,200 \times 8 \times (20 + 0 + 40) = 1,536,000 \text{ ft.-lb.}$$

The total amount of the work is equal to the sum of these separate items,

```
(a) =  6 000 000
(b) =  1 545 600
(c) =    800 000
(d) =    176 000
(e) =  9 216 000
(f) =  1 536 000

Total = 19 273 600 ft.-lb.  Ans.
```

26. Surplus Power for Locomotives.—It is a good plan to provide a reasonable amount of surplus power and not to work a locomotive to its full capacity. A reserve of power will be found economical because it cuts down the cost of repairs, fuel, and oil to the lowest point, besides lengthening the life of the machine. In case of emergency, it also provides an opportunity for an increased output.

COMPRESSED-AIR LOCOMOTIVE HAULAGE

27. Advantages and Disadvantages.—Compressed air does not vitiate the mine atmosphere but rather improves it by the exhaust from the locomotives. If the haulage way is a passageway through which the miners travel to and from their working places, compressed-air locomotives are of positive advantage, as they are continually adding fresh air from the exhaust. There is also, with them, no possibility of setting fire to the timbers, nor of setting off explosive gases by sparks.

There are no boilers or fireboxes in compressed-air locomotives, and hence fewer repairs are needed and there is no danger from boiler explosions, which sometimes occur when steam locomotives are used. Compressed-air locomotives are probably more easily and cheaply kept in repair than steam or electric locomotives, and their operation requires no greater, if as great, skill than an electric or steam locomotive.

Compressed-air locomotives, in common with other locomotives, are not economical on steep grades for any distance owing to the fact that so much of their power must be used in taking the locomotive up the grade. They may be run, however, for short distances on grades as steep as 8 per cent.

A compressed-air haulage system costs more to install than a steam-locomotive haulage system. The relative cost of a compressed-air and an electric haulage installation should be determined in each case by securing bids on the same specifications, as there is a wide difference in the cost of either of these equipments under different conditions, and owing to the rapid developments in the practice and constant changes in the machinery it is impossible to give definite figures or comparisons that will apply for any length of time, or for all conditions.

An advantage of a compressed-air system is the fact that the compressor can be shut off automatically when a certain pressure is reached in the storage tanks or transmission pipes, and in this way the power-generating plant is operated only when the power is needed by the locomotives.

28. The equipment of a compressed-air haulage plant consists of boilers, compressors, suitable piping for transmitting the air, receivers, valves, charging stations, and compressed-air locomotives.

29. Power for Air Compressors.—Usually, the power adopted for driving air compressors is steam, and that requires a boiler plant located near the compressor building. The horsepower of the compressor governs the size of the boiler plant, which should have a rated capacity greater than that of the compressor. If waterpower is available for running the compressor, this will be cheaper than steam power; or if water-power is available for running a dynamo, electricity may be transmitted to a motor attached to a compressor. Such an arrangement may be used in preference to steam power. Gas engines might be adopted to advantage as a source of power for compressors wherever there is a large supply of natural gas, coke-oven gas, or producer gas available.

30. Air Compressors.—Where compressed-air locomotives are used in mines, it is customary to place the air compressor on the surface near the mine mouth, and convey the air through pipes to stations in the mine where the locomotives are charged. The locomotives have pressures of from 600 to 1,000 pounds per square inch in their storage tanks; and for such high pressures, the air compressors are usually of the three- or four-stage type. The size of the air compressor will depend on the number and size of locomotives in use, together with the frequency with which they are charged.

31. To Calculate the Compressor Capacity.—The capacity of an air compressor for any pneumatic haulage plant is based on the number of cubic feet of free air required per minute. This may be found from the formula

$$C = \frac{nc}{t}$$

in which C = required capacity of compressor, in terms of cubic feet of free air per minute;

c = cubic feet of free air required to charge the locomotive;

n = number of charges required to do the work in a specified time;

t = specified time, in minutes.

EXAMPLE.—During a run of 20 minutes, the pressure in a locomotive tank of 100 cubic feet capacity is reduced from 800 to 100 pounds per square inch; what must be the capacity of a compressor to furnish a pressure of 800 pounds in 20 minutes?

SOLUTION.—To find the volume c, of free air required, subtract the pressure of the air at the end of the run from that at the beginning and multiply the difference by the volume of the tank, in cubic feet; divide this product by 14.7 lb., the atmospheric pressure. At the beginning of the run, this volume is $\frac{(800 - 100) \times 100}{14.7} = 4{,}762$ cu. ft.

Substituting in the formula $C = \frac{cn}{t}$,

$$C = \frac{1 \times 4{,}762}{20} = 238.1 \text{ cu. ft. of free air per minute.} \quad \text{Ans.}$$

PIPE LINES

32. The method of constructing **pipe lines** and of carrying them into the mine will depend on circumstances at the mine. The size of pipe necessary for a haulage plant must be determined from the requirements of each installation and the system of charging adopted. Usually, the pipe line is laid along the entry or haulage way parallel with the track on which the locomotive runs and is left uncovered in order that leaks may be detected. There may be several charging stations at intervals along the pipe line if the run is long, but for a short run there is usually only one.

The pipes that connect the compressor with the underground charging station are carried into the mine through the regular openings or through bore holes and then extended through the underground passages to the charging station. It may be necessary to place an expansion joint in some part of the pipe line where it is exposed on the surface or carried into a shaft, but this is not necessary in the mine, since the temperature remains nearly uniform in most mines of moderate depth.

The pipe used in compressed-air haulage plants is of wrought iron, and is made extra heavy and tested up to 800 or 1,800 pounds per square inch by hydrostatic pressure. As a rule, the pipes are used as storage tanks and are given a diameter to correspond with the demands for air made by the locomotive. If the pipe is to be used as a storage reservoir, it should have heavy, threaded, screw couplings provided at each end with an annular calking groove, into which a strip of soft metal can be driven, to prevent leakage. For extending a pipe line beyond a central charging station, 2- or 3-inch pipes may be used; and if storage tanks are placed at charging stations, the entire pipe line may be of pipe 2 or 3 inches in diameter.

33. Flanged couplings, or union couplings, should be used at all charging stations and at intervals along the pipe line to provide means for breaking the pipe line or repairing the same. These couplings consist of heavy cast-iron flanges 12 inches in diameter and $2\frac{1}{2}$ inches thick, with finished faces; the flanges are rough bored to take the line pipe, which is expanded into them by a special tube expander. The ends of the pipe are riveted into recesses in the flanges and are hammer-faced flush with the center bore. The flanges are counterbored $\frac{1}{32}$ inch deep and 7 inches in diameter to retain soft-metal or vulcanized-fiber gaskets. By using pipe lines as reservoirs instead of the large receivers, an economy in charging is secured, as the charging stations can be located at any point along the pipe lines, and a number can be installed; also extensions can be made with smaller pipes, which draw from the larger pipes as reservoirs.

34. Blowing Off the Pipes.—When, in long pipe lines, charging stations are arranged at intervals and gate valves are provided at the charging stations, any section between two gate valves may be blown off in case of accident without the necessity of blowing off the entire line and stopping all operations. This arrangement will also save the expense of refilling the pipe line. Where the pipe line is run through a shaft, a heavy cast T with several feet of pipe below it is located at the bottom of the shaft to collect water, a waste valve being inserted at the bottom of the pipe to allow the water to be blown off.

35. Friction in Pipes.—There is very little difference in pressure at the two ends of a pipe line of sufficient size, because while the locomotive is at work the pipe line slowly is receiving its full supply of air from the compressors and is of a capacity sufficient to always give the locomotive its full pressure of air when necessary.

CHARGING COMPRESSED-AIR LOCOMOTIVES

36. The Direct System of Charging.—The **direct system** of charging compressed-air locomotives consists in connecting the locomotive tank with the compressor and pumping in air until the desired pressure is reached. This system requires a compressor of large capacity and keeps the locomotive out of commission until its tank is filled and also keeps the compressor out of commission when no tank is being filled. For light work, when trips are made only at long intervals of time, direct charging might be practiced with economy. The cost of installation of the direct system is less than that of the stationary storage system, but it has been discarded almost entirely, as the latter is more economical and satisfactory in the majority of cases.

37. Stationary Storage System.—The **stationary storage system** may consist of a pipe line or one or more storage tanks especially constructed and tested to withstand a high pressure. With this system, the compressor is kept in very nearly continuous operation, and the time required to charge the locomotive is very short, in fact almost instantaneous.

In the earliest compressed-air locomotive installations, even when the runs were long, it was the general practice to use large reservoir tanks connected with the compressor by small pipes. The tanks were of a somewhat larger capacity than the locomotive storage tanks and in order that the pressures in the two tanks might equalize to the desired pressure without loss of time, the pressure was carried considerably higher in the reservoir tank than was needed in the locomotive tank.

In modern haulage plants, the stationary storage usually consists of a suitably proportioned pipe line, although for short hauls the reservoir tank is probably more convenient.

38. Stationary Storage Capacity.—When the locomotive storage tank is connected to the stationary storage to be filled with compressed air and the valves are turned, it is evident that as the pressure rises in the locomotive tank it must fall in the stationary storage until the pressures in each are equal. The operation is nearly instantaneous. The stationary storage should be proportioned to the locomotive storage, and the capacity of the compressor to the frequency with which the locomotive storage is charged.

In order to calculate the stationary storage,

Let V = volume of stationary storage, in cubic feet;
v = volume of locomotive storage, in cubic feet;
P = pressure in stationary storage, in pounds per square inch;
p = pressure desired in locomotive storage, in pounds per square inch;
p' = pressure in locomotive storage, in pounds per square inch previous to charging;

then, according to Mariotte's law,

$$V(P - p) = v(p - p') \qquad (1)$$

and since the stationary storage capacity V is sought, by transposing,

$$V = \frac{v(p - p')}{P - p} \qquad (2)$$

EXAMPLE.—What must the volume of a stationary storage be in order to instantly charge a locomotive tank of 150 cubic feet capacity to a pressure of 800 pounds per square inch, assuming there was 140 pounds per square inch residual pressure left in the tank, and 900 pounds per square inch in the stationary storage?

SOLUTION.— $v = 150$, $P = 900$, $p = 800$, $p' = 140$; substituting these values in formula **2**,

$$V = \frac{150 \times (800 - 140)}{900 - 800} = 990 \text{ cu. ft. Ans.}$$

39. Storage Capacity of Pipes.—The proper capacity for storage of compressed air may be had by means of properly proportioning the diameter of the transmission pipe to the length, provided that the line is of considerable length. To illustrate this point, suppose that the locomotive tank contains 150 cubic feet of compressed air at a pressure of 140 pounds per square inch. The pressure desired is 800 pounds per square inch and the pressure in the stationary storage is 900 pounds per square inch. By previous calculation, the capacity of the stationary storage was found to be 990 cubic feet. Suppose the compressor is to be located 5,280 feet from the charging station, then 6-inch pipe will give the required capacity, as 5.09 lineal feet of this pipe contain 1 cubic foot and 5,280 feet will give a capacity of 1,037 cubic feet. The operation of charging causes the air nearest the station to flow into the locomotive tank and when the pressures are equalized the flow stops. To increase the pressure in the locomotive tank from 140 to 800 pounds per square inch will require $\dfrac{(800 - 140) \times 150}{14.7} = 6{,}734.7$ cubic feet of free air.

40. Charging Valves.—Fig. 2 shows the method of charging compressed-air locomotives when the tanks are to be filled from the pipe lines. Attached to the stationary storage transmission pipe by a flange a, is a specially made

T to which is fitted a 1½-inch gate valve *b* having a short nipple into which is screwed a *Moran flexible joint c*. This flexible joint, shown in section in Fig. 3, has a ball-and-socket joint that permits a vertical and horizontal movement and allows

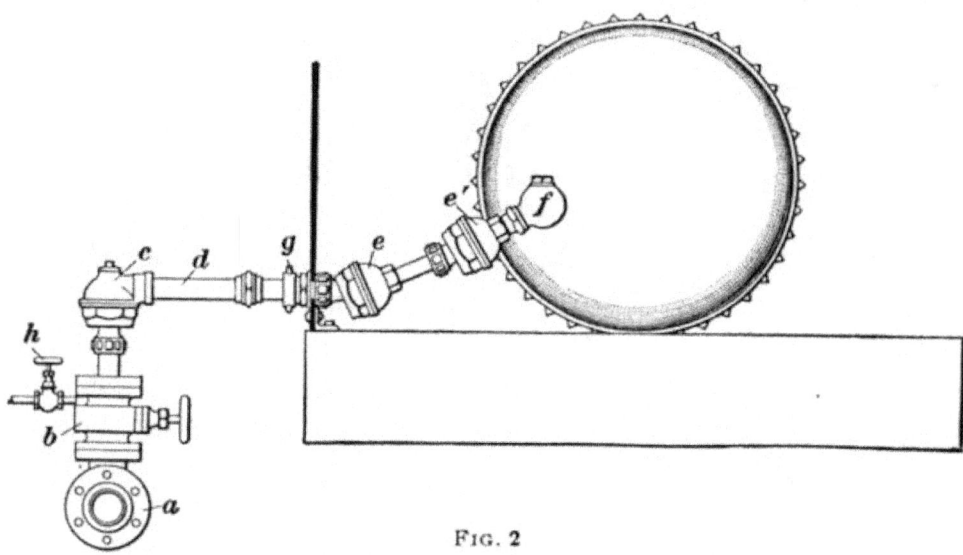

Fig. 2

the pipe *d*, Fig. 2, to be turned so as to be parallel with the storage pipe when not in use.

Attached to the locomotive are two similar ball-and-socket joints *e*, *e'*, also a check-valve *f*, opening into the tank but closing as soon as the gate valve is closed to cut off the pressure from the pipe-line storage. A special coupling *g* is needed to connect the pipe *d* with the tanks. After the coupling is made, the gate valve *b* is opened and the air pressure forces together the parts of the flexible joints, thus making perfectly tight connections. As soon as equilibrium is established between the pressure in the stationary storage and that in the locomotive storage tanks, the gate valve is closed. To break the coupling *g*, it is necessary to

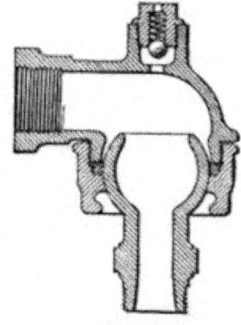

Fig. 3

first release the compressed air remaining in the connection between the gate valve and check-valve *f*. This is done by bleeding the pipe *d* by means of the globe valve *h* located above the gate valve *b*. As soon as the pressure is released the joints become loosened and the couplings are easily manipulated.

41. Pipe Line and Compressor Service.—When the locomotive is connected with the pipe line and the charging valve is opened, the pressure in the locomotive tank and in the pipes on equalizing should not fall much, if any, below the stated pressure that the locomotive is intended to carry. With this end in view, it is desirable that the volume in the stationary storage should be at least double the tank capacity of the locomotive. When several locomotives are served by the same pipe line, it is rarely necessary, for mine work, to design the system for charging more than one at a time. The relatively slight drop in gauge pressure after charging is soon recovered by the compressor. In case additional locomotives are required after the installation of the system, the same pipe line may still serve, provided that the compressor is of sufficient size to charge it to full pressure at shorter intervals of time.

COMPRESSED-AIR LOCOMOTIVES

42. General Design.—To meet the varying conditions of height and width of the mine passages, there is considerable diversity in the design and size of **compressed-air locomotives.** Some have six driving wheels and some four; some have the cylinders outside the frame and others have them inside the frame. For easy grades, short hauls and light loads, the single air-tank locomotive of modern dimensions and designed for a pressure of 600 pounds per square inch may be sufficient, and in some cases even lower pressures may be used, but there is more economy in the use of reasonably high pressures. For longer hauls and greater tonnage, larger air capacity is usually needed and two or more storage tanks, designed for a pressure of 800 or 900 pounds per square inch, are required. Very limited height and width of mine entries may compel the use of tanks of small capacity and the use of high-pressure air.

43. Fig. 4 shows a compressed-air locomotive intended for service where the haul is not long and is therefore fitted with a single storage tank. These locomotives weigh from

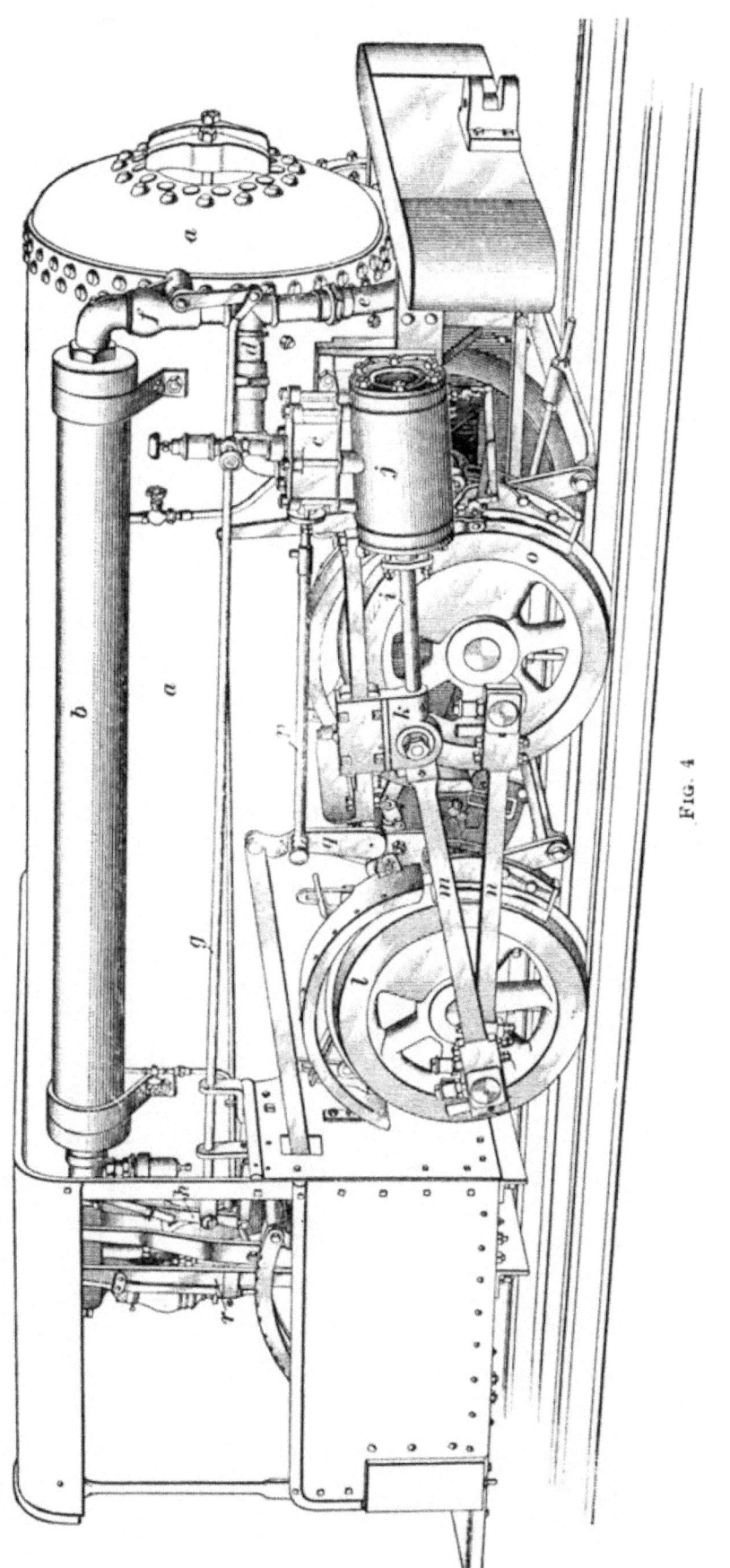

Fig. 4

$2\frac{1}{2}$ to 10 tons, the lighter weight locomotive having $4'' \times 8''$ cylinders and the heavier weight having $8'' \times 14''$ cylinders. The smaller sizes are adapted to the narrow gauges and sharp curves common in gold, silver, lead, and copper mines, while the larger sizes are mainly used in main and cross-entry haulage of coal mines.

In the illustration, a is the main storage tank and b the auxiliary tank. Air is admitted to the air chests c through a branch pipe d, leading from the pipe e, connected with the auxiliary tank b by means of the throttle valve f worked by the reach rod g and lever h in the cab. The piston rod i reciprocates in cylinder j, and is attached to a crosshead k that slides on the guide bar as shown. The back driving wheels l are connected by the main rods m, with the crossheads k and by the side rods n with the front drivers o. The valve rods p are connected by the rocker q with the eccentric shaft. The lever r shown in the cab is for reversing the engine and for shortening the throw of the valves, thus making the machine use air expansively and economically. The locomotive is provided with brakes, oil cups, pressure gauges, and tools necessary for keeping the machine in working order.

44. Compound compressed-air locomotives, although very efficient when tested in the shop, are not so well adapted to mine service as the single-expansion type, owing to the fact that there are twice as many cylinders, pistons, stuffingboxes, etc., on a compound as on a single-expansion locomotive, thus giving twice the chance for leakage. Furthermore, it is well known that mine machinery is used under adverse conditions and does not generally receive the same care and attention as that used in daylight, hence the simpler the machine the better.

45. The compressed-air locomotive shown in Fig. 5 has two storage tanks a with a total storage capacity of 133 cubic feet and the following general dimensions: diameter of drivers, 24 inches; cylinders, 7 in. $\times$ 12 in.; wheel base, 4 feet 6 inches; gauge, 3 feet; weight, 19,700 pounds.

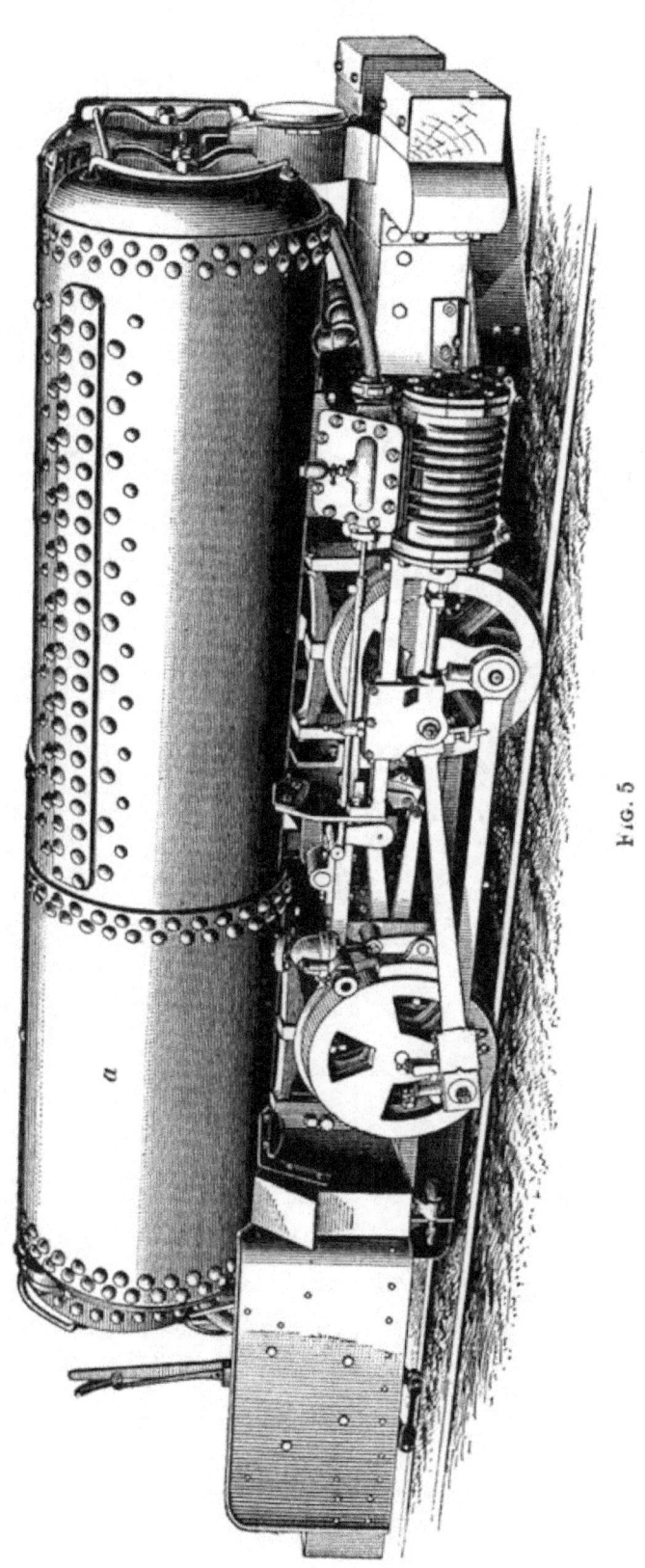

Fig. 5

46. Locomotives With Air-Tank Tenders.—Where unusually long hauls or heavy grades make it advisable to provide a greater reserve of energy than can be stored in the one or two tanks of an engine, additional storage tanks can be mounted on a truck and serve as a tender for the locomotive. The tender can be uncoupled and the locomotive operated separately when it is needed on short hauls; but where the haul is long and the air storage pipes are not con-

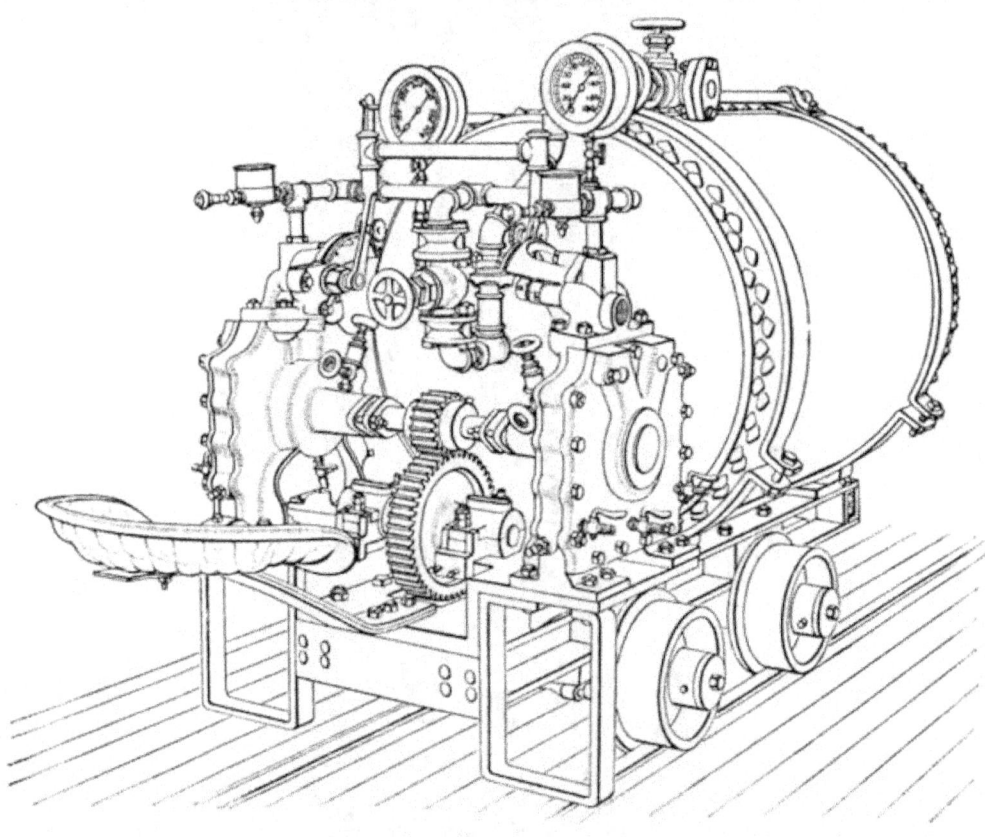

Fig. 6

tinued to the end of the run, the locomotive may, by carrying a tender, make the round trip with one charge.

47. Special Geared Locomotive.—Fig. 6 shows an unusual form of locomotive used at Grass Valley, California, for hauling a trip of five cars, each carrying 1 ton, over a round trip of 5,000 feet. The air tank is 36 inches in diameter and 48 inches long, and carries an air pressure of 500 pounds per square inch. The locomotive is 5 feet long over all, 30 inches wide, 4 feet 4 inches high, and runs on a track only

18 inches wide. It is operated by a pair of geared engines enclosed as shown, which is of particular advantage in wet and muddy mines.

48. Compressed-Air Gathering Locomotives.—Owing to the sharp curves made from entries into the rooms in coal mines, special small compressed-air locomotives are used in some cases to deliver empty cars to the miners and remove loaded cars to the partings. Fig. 7 shows a gathering locomotive built for the Consolidation Coal Company of Georges Creek coal field in Maryland. The machinery on this locomotive is placed inside the frame

Fig. 7

so that it will not extend beyond the ties and graze the ribs. This is made necessary because the rooms are only from 12 to 15 feet wide and are timbered with a collar and post, the latter standing in the center of the room, or just far enough to one side to permit a car to pass. The tracks in the rooms for these locomotives are not much better laid than for the cars where mules do the gathering. A loaded car weighs 7,000 pounds and a locomotive 8,000 pounds, so that track that would be safe for the car would probably be safe for the locomotive.

In the actual placing of cars in the workings, a gathering locomotive has no advantage over a mule, but a broken

Fig. 8

locomotive can usually be repaired over night, while an injured mule can only be replaced by a new one, so by the use of locomotives the work is less likely to be interrupted. The speed with which the cars can be hauled along the entries to the parting is in favor of the locomotive.

49. Fig. 8 shows a gathering locomotive used in the anthracite mines of the Delaware & Hudson Company near Scranton, Pennsylvania. This locomotive is intended to gather the cars from rooms worked in very thin coal and into which it is impossible to take a mule without blowing down the top or lifting the bottom. It is therefore made

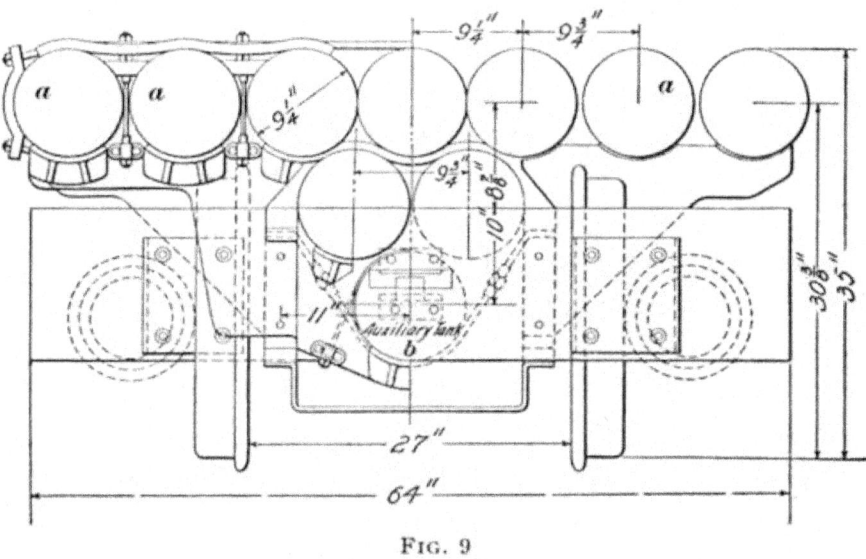

Fig. 9

very low. Its general dimensions are as follows: diameter of cylinder, 7 inches; stroke of piston, 14 inches; diameter of driving wheels, 24 inches; wheel base, 3 feet 4 inches; total length, 12 feet 3 inches; height, 2 feet 11 inches; width, 5 feet 4 inches; storage capacity for air, 40 cubic feet; storage pressure, 700 pounds per square inch; working pressure, 125 pounds per square inch; weight, 12,600 pounds.

The main storage tanks consist of nine steel tubes a laid and bound together as shown in Fig. 8 and in section in Fig. 9. These tubes are connected by copper pipes and extra heavy brass fittings to a pipe that has two connections with a pressure auxiliary tank b, Fig. 9. The storage tanks are arranged in two series, each with a suitable shut-off valve and each having independent filling connections and check-valves. By this arrangement, if the connection of either series fails, it can be cut off and the other used. The location of the storage tubes between the side frames prohibits the use of a link valve motion and eccentrics, and instead the valve gear is placed outside the frame. This gear, Fig. 8, is operated by a lever through the crank b and rod c. The operator's cab is necessarily small and is placed at the cylinder end of the locomotive and between the side frames. In case of derailment, it is protected by the bumpers.

50. Locomotive Storage Tanks.—The air-storage tanks are made of thick, flanged steel plates. The longitudinal seams are sextuple- or octuple-riveted with butt joints and welt strips inside and out; the girth seams are double-riveted. The welt strips add stiffness to the tank plates, at the same time, as arranged, they strengthen the longitudinal joints. They are more useful in long tanks than in short ones, also in tanks of large diameter than in those of small diameter, as they counteract the tendency to bulge. The heads of the tanks are made convex and in one piece; the front head, however, is sometimes cut for a manhole and the tank plate about the hole is reenforced by a steel casting. The manhole cover is of cast steel and is held in place by steel yokes, bolts, and nuts, and by this means the tank may be entered either for inspection or for repairs. Each tank is made absolutely tight and is tested by hydraulic means, to a considerably higher pressure than it is to carry.

The thickness of the tank plates is proportioned to the pressure to be carried and to the design of the tanks, and is such as to secure a large factor of safety. When pressures above 1,000 pounds per square inch are required, tanks of large diameter cannot be safely employed and a series of heavy seamless steel tubes are used as storage tanks. Tubes of this kind, 9 inches in diameter and $\frac{11}{32}$ inch thick, when made of high carbon steel will safely carry a pressure of from 2,000 to 2,500 pounds per square inch; such high pressures are not employed for mine haulage. It is possible to charge the storage tanks to a higher pressure than is absolutely necessary, thus accumulating a reserve power. This may be advisable, since on up grades locomotives use more power than on the level.

51. Size of Locomotive Storage Tanks.—In order that compressed-air locomotives may be able to make their trips from a charging station to the end of the haulage system and back, or from one charging station to another, their storage tanks must be of such proportion as to hold the proper quantity of compressed air.

In order that locomotive storage tanks may be designed correctly with a minimum storage space for the maximum work to be done, it is necessary to have a complete profile of the road over which the locomotives are to travel, and from this a tabulated statement of the air consumption on the various grades is calculated. To the total amount of air thus found 20 per cent. is to be added for contingencies. In order to provide storage for this quantity of air, it is usually the practice to compress it until the tank pressure is from four to six times that which will be used in the cylinders.

All tank fittings are screwed into flanges of cast steel, which are riveted to the body of the tanks, and in this way a good thread hold is obtained and the tank stiffened at that point.

The tanks rest on a steel locomotive frame from which the weight is transmitted by means of springs and equalizers to the journals of the driving wheels.

52. Auxiliary Tanks.—From the high-pressure tanks of the locomotive, the air is taken into an **auxiliary tank** before it is allowed to enter the locomotive cylinders. This tank is a section of wrought-iron pipe with closed ends about two-thirds of the length of the high-pressure tanks and 8 or 9 inches in diameter. It is placed in a convenient position parallel with the storage tanks. Its object is to hold air at a reduced pressure to be used directly in the cylinders. By means of an automatic reducing valve and stop-valve, the pressure in the auxiliary reservoir is adjusted to the requirements of the engines.

53. The Working Pressure.—From the auxiliary tank, the air passes through a throttle valve to the cylinders. The pressure in the auxiliary tank, which is called the **working pressure,** varies from 125 to 150 pounds per square inch.

54. Reasons for Reducing Cylinder Pressure.—The reduction of the storage-tank pressure permits of the maintenance of a constant working pressure in the cylinders; prevents the waste of air that would be likely to ensue if air at full storage-tank pressure were admitted to the cylinders, and makes the locomotive more manageable. The cylinders, moreover, need not be made so heavy as would be required for the high pressure direct from the storage tanks. This reduction in pressure is obtained by the use of a reducing valve, which may be quickly adjusted to regulate the pressure in the auxiliary tank.

55. Automatic Reducing Valve.—In order to maintain a uniform pressure at the throttle valve, a uniform pressure must be maintained in the auxiliary storage tank of the locomotive. This is accomplished by a special reducing valve that works automatically and feeds just sufficient air to the auxiliary tank to maintain a steady pressure. This valve is shown in Fig. 10. It consists of a double-seated balanced valve and an actuating piston controlled by the air pressure in the auxiliary tank and a spring so adjusted that the two regulate the air supply. The air pressure in the auxiliary tank acts on one side of the piston, tending to close the valve. This action is opposed by the spring adjusted to hold the valve off the seat until the maximum allowable pressure in the auxiliary tank is reached, when the pressure of the air overcomes the spring and the valve closes.

56. The Stop-Valve.—It is important to have valves, called **stop-valves,** between the auxiliary tank and the cylinders to prevent any leakage of air by the throttle to the cylinder when the former is closed. This valve is necessary, as any leakage of air would make it impossible to keep the locomotive at rest. For additional precaution, a single-seated stop-valve is sometimes placed between the storage reservoir and the reducing valve; this valve is controlled by the lever that controls the throttle valve. When the latter is open the stop-valve is open, thus admitting air to the auxiliary tank only as air is drawn from that for the cylinders. On some locomotives, stop-valves are only placed between the auxiliary tank and the cylinders.

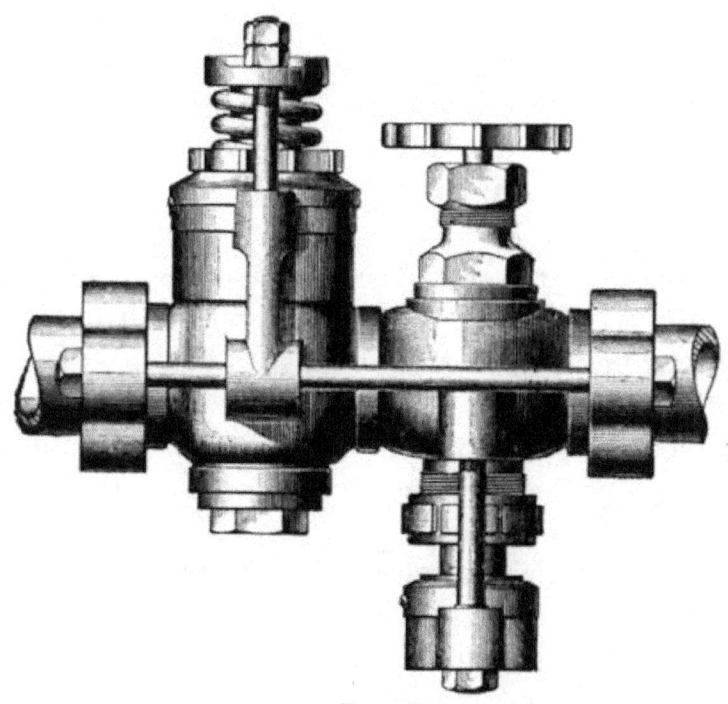

Fig. 10

57. Reheating Compressed Air.—The efficiency of compressed air can be greatly increased by reheating before admitting it into the cylinders, but the added complication of the locomotive machinery needed to reheat the air is seldom justified in haulage machinery by the saving in fuel or increased efficiency of the locomotive, except in special cases where a very long mileage must be made or where the price of fuel is very high.

Compressed air used in locomotive cylinders may be reheated by being passed over hot water in the auxiliary tank. The water, which partly fills the tank, is kept hot by injecting steam into the tank while the locomotive storage tank is being charged. The air, as it passes on its way to the cylinders, is compelled to pass over this water and is thus heated. Indirectly this is another cause for increased efficiency, as the moisture taken up by the air in this way affords better lubrication in the valves and cylinders. Compressed-air cylinders should not be lagged with non-conducting covering, as is so necessary for steam cylinders to minimize condensation. By exposing the surface of the cold cylinders to the warm air of the mine, some heat is absorbed and added energy is imparted to the compressed air. Sometimes, the exterior surface of the cylinders is cast with corrugations, in order to present the largest possible surface to the warm surrounding air of the mine.

58. Use of Compressed-Air Expansively.—The air is sometimes admitted to the cylinders throughout nearly full stroke, and consequently, as the exhaust is at high pressure, the efficiency is lower than it should be. This practice is doubtless due to the tendency to use as small a motor as possible for the service required, on account of the limited headroom and narrow crooked gangways so common in mines. Better economic results are obtainable, however, by using the air expansively and increasing the size of the locomotive and the weight on the drivers; this is almost always done with large locomotives. Ample reserve power is available when necessary, since full tank pressure can be admitted to the cylinders in starting a heavy load, or in pulling on steep grades and sharp curves.

In using the air expansively, as can be done with properly proportioned cylinders, there should be no trouble from freezing of the moisture. Although the cold developed will produce a low cylinder temperature, yet, as the initial working pressure is so much higher than that employed for pumps and other compressed-air machinery, the expanded air becomes relatively dry, and the force of the exhaust will still be sufficient to keep the ports clear of accumulated ice. To this end, the exhaust ports should be large, straight, and short. If the high-pressure air from the main tank were used in the engines, both pistons and cylinders would have to be made excessively heavy, and any reasonable degree of expansion would produce a degree of cold difficult to deal with.

TRACTIVE POWER OF COMPRESSED-AIR LOCOMOTIVES

59. In steam locomotives, but 85 per cent. of the boiler pressure enters into the calculation of the tractive power; but with compressed air, the efficiency is increased by the pipes and cylinders absorbing heat from the atmosphere, which is at a higher temperature than the compressed air contained in them. At seven-eighths cut-off, which is practically full stroke of the piston, the air pressure in the cylinder is taken as 98 per cent. of the auxiliary-tank pressure. For earlier cut-offs, a further decrease in efficiency can be expected; but in any case, for an equal cut-off, the efficiency of the air is greater than that of steam.

60. The formula for the tractive power of an air locomotive is similar to that for a steam locomotive, except that the factor .85 is changed to .98; hence

$$T = \frac{D^2 \times L \times .98 \times p}{d}$$

in which T = tractive power, in pounds;
D = diameter of cylinder, in inches;
L = length of piston stroke, in inches;
p = working pressure of air, in pounds per square inch;
$.98\,p$ = 98 per cent. of pressure of compressed air in auxiliary tank;
d = diameter of driving wheel, in inches.

EXAMPLE.—What will be the tractive power of a compressed-air locomotive whose cylinders are 5 inches in diameter and whose piston stroke is 10 inches; the diameter of the drivers is 22 inches and the air pressure is 140 pounds per square inch?

SOLUTION.—Substituting in the formula,
$$T = \frac{25 \times 10 \times .98 \times 140}{22} = 1{,}559 \text{ lb.} \quad \textbf{Ans.}$$

HAULAGE
(PART 3)

ELECTRIC LOCOMOTIVE HAULAGE

INTRODUCTION

1. Advantages and Disadvantages.—One of the most important matters which the mine manager has to deal with is the transmission of power. It cannot be stated that one kind of energy is better than another for all conditions, nor for any particular service without an examination of all the conditions under which it is to be used. The chief advantage of electric power for mine haulage is the readiness with which it can be transmitted long distances. It cannot, however, be used in some mines or in parts of many mines, because of the presence of explosive gases or mixtures of gas and impalpable coal dust that may be ignited by electric sparks from switches, controllers, broken wires, or blowing fuses. Moreover, physical conditions or the method by which the mine is worked may also oppose the use of electrical energy as a motive power.

Electric locomotives of the ordinary type can be operated on short grades of 12 to 15 per cent., but they are not recommended when the grade exceeds 3 per cent. for long distances, or 5 per cent. for short distances; beyond this grade lies the field of rack-rail, or sprocket-type, locomotives, or of rope haulage. Electric locomotives of the rack-rail type can be used to advantage on grades as high as 18 per cent.

Copyrighted by International Textbook Company. Entered at Stationers' Hall, London

2. An electric locomotive can usually go only where a conductor has been put in place along the haulageway, excepting in the case of gathering locomotives, where, for short distances, a cable is unreeled as the locomotive goes from the main haulage road and is reeled up again as the locomotive returns. An electric locomotive is not, therefore, as independent in its action as a locomotive driven by compressed air or steam, which can go wherever there is a suitable track, although the distance run with one charge of an air locomotive is limited by the capacity of the motor tanks. A steam locomotive, however, gives off smoke and gases that are objectionable in the mine.

3. Steam or water-power is the prime source of both compressed-air and electric power and comparison between the two last-named systems of haulage should be made only by considering their relative cost of maintenance, depreciation, operation, repairs, etc. The advantages claimed for electric haulage over compressed air are lower cost and greater flexibility. Electric energy can be generated cheaper than compressed air; wire for transmitting the current is cheaper than pipe and can be put in place more cheaply and quickly than pipe. The disadvantages of electric power are the dangers of igniting firedamp in gaseous mines; the annoyance and danger of shocks to men and animals; and the limited distance that the locomotive can be moved from the main trolley wire.

4. Current and Voltage.—Direct current is generally used for electric haulage; the pressure most commonly used is about 250 volts, although 500 volts has been tried and is still used in some places. The objections to the higher pressure is the greater danger of injurious or fatal shocks, as well as the greater difficulty of insulating the wires from ground. The higher pressure can be profitably used only where all the passages through which the wires are strung are high or roomy enough to permit placing the wires where there will be little danger of contact with them, and dry enough to preserve the insulation.

5. Electric Generators.—If the power house is near the mouth of the mine, direct-current dynamos are generally used to generate the electric energy for haulage purposes, and at the pressure used in the mine, 250 or 500 volts as the case may be. It is frequently advantageous to locate the power house at some distance from the mine, so as to take advantage of a water fall to generate the power, or for other economic reasons; in such cases, in order to reduce the cost of line copper, it is customary to transmit a high-voltage current to the mouth of the mine, or sometimes to the interior near where it is to be used, there transform it in step-down transformers, and convert it by means of rotary converters to direct current at ordinary mine voltage, after which it is treated in the mine installation precisely as would be the case with a direct current generated at the mouth of the mine.

6. Ehrenfeld Plant.—One of the first and best known installations of this character is at Ehrenfeld, Pennsylvania, in the mines of the Pennsylvania Coal and Coke Company. This company had a rope system of haulage until the haulageway attained a length of nearly 2 miles. At this point, it became necessary to adopt a system of haulage that could be applied to the main-entry haulage, which would eventually extend 5 miles underground. Electric haulage was chosen and by the system adopted a three-phase, 25-cycle, 179-volt alternating current is generated at the pit mouth, and transformed in step-up transformers to 5,600 volts; at this pressure it is transmitted to a substation in the mine, 9,000 feet from the power house, where it is stepped down and converted into direct current at 275 volts. A direct-current system would have entailed a prohibitive investment for copper with the low voltage desired, namely 275 volts. This system was adopted with a view to the future requirements of the mine.

7. Fig. 1 shows a plan of the power house on the surface in which a is the boiler room; b, the dynamo room; and c, the transformer room. The dotted lines at e show the space provided for a larger direct-connected dynamo when needed.

The two dynamos d, each of 150 kilowatts and belt-driven, are double-current generators, separately excited and delivering direct current at 275 volts from one side, and three-phase alternating current at 179 volts from the other side.

The direct current is fed into the outer end of the haulage system, one terminal of the generator being connected to the

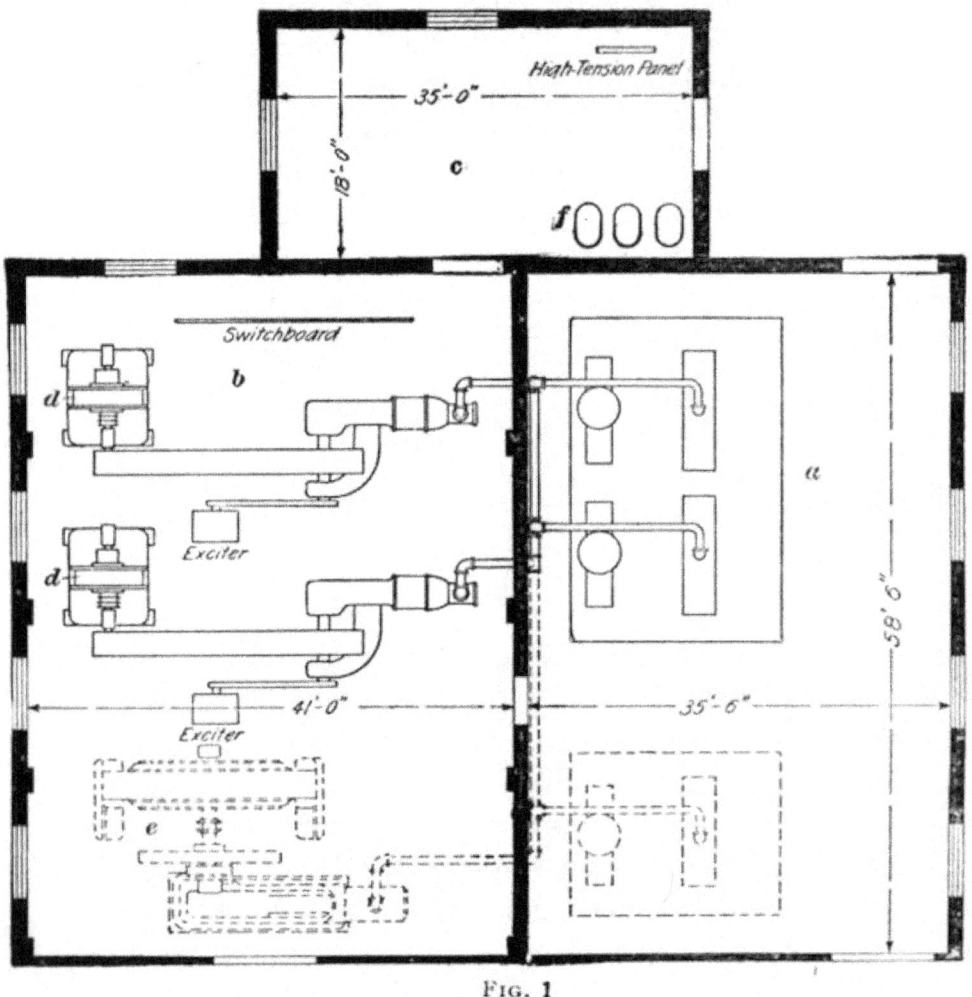

Fig. 1

trolley wire and the other terminal to the rail return. The step-up transformers for raising the alternating-current pressure from 179 volts to 5,600 volts are shown at f. At a substation in the mine, Fig. 2, step-down transformers a and 150-kilowatt rotary converter b provide a direct current at 275 volts for the haulage circuits. Both the step-up and the step-down transformers are oil-cooled and exact duplicates.

Fig. 2

The transmission line consists of a three-conductor, rubber-insulated, lead-covered cable laid in the main return airway parallel to the main entry. Each conductor of this cable is a No. 6, B. & S. gauge wire.

8. The double-current generators at first installed in the power house were simply rotary converters, exact duplicates of the converter in the substation, with their shafts extended for belt driving. It was proposed, as the system was extended underground, to remove these two generators from the power house and establish them in additional rotary converter substations in the mine, replacing them in the power house by large, engine-driven, revolving-field

TABLE I

Weights of Locomotives Pounds	Minimum Weights of Rail Pounds per Yard	Weights Advisable Pounds per Yard	Bonds, B. & S. Wire Gauge Number	
			250 Volts	500 Volts
6,000 to 8,000	20	20	0	2
10,000 to 12,000	20	25	0	1
14,000 to 16,000	25	30	0	0
18,000 to 20,000	30	35	00	0
22,000 to 24,000	35	40	00	0
26,000 to 28,000	40	45	00	00
30,000	50	60	000	00

alternators designed to deliver 5,600-volt three-phase current direct to the transmission line. Direct current for the outer end of the haulage circuit would then probably be supplied by a rotary converter located in the transformer room of the power house. By using the double-current generators, a very considerable saving in the original investment was effected. These machines have operated in the most satisfactory manner, and their regulation with a rapidly varying alternating- and direct-current load has been surprisingly good. In many mines, the conditions are such that double-current generators would constitute an ideal permanent installation.

9. Rails for Electric Mine Locomotives.—Electric locomotive builders recommend the use of the weights of rails in Table I for different weights of locomotives. The table also gives the size of bond, or electrical connection, from rail to rail past the joints.

10. Experience teaches that on **main** haulways, even with a good roadbed and ties, it is economy to use heavy rails. The first cost of heavy rails is higher than that of light ones, but the wear on the rolling stock is less and so is the danger of derailment caused by spreading rails. The weight it is advisable to use depends on the frequency with which the road is to be used, as well as on the weight of the locomotive. On a branch line over which the locomotive will pass but seldom it is permissible to use comparatively light rails; but where there is a great deal of traffic it is often advisable to use rails as heavy as 60 or 75 pounds per yard.

WIRING FOR ELECTRIC MINE HAULAGE

11. Carrying the Wires Into the Mine.—In a shaft mine or a steep slope, insulated feeder wires are run from the dynamo on the surface down the shaft or slope, or occasionally down a bore hole, into the mine, where they are connected to the trolley wire and rails in the gangways; or, the feeder wires may be continued along the haulage ways for a distance depending on the length of the haulage road and the amount of electric current that must be carried. Where the mine opening is a shallow slope or a drift, the power is sometimes carried into the mine by bare-wire conductors fastened at intervals to the caps or legs of the timbers. If the mine opening is wet, the power is transmitted through lead-covered cables. In shafts, the cables are held in position by wooden brackets placed on the sides of the shaft, and are sometimes suspended from the top by block and tackle, by means of which the cables may be moved up or down. In a wet shaft, the lead cable is carried far enough into the mine from the bottom of the shaft to be free from the shaft water and is then connected with the bare wire used in haulage. A main switch should be provided at the foot of the shaft or slope by which the power can be turned off or on instantly.

12. Shape of Trolley Wire.—Trolley wire is made with round, figure **8**, or grooved cross-section as shown, respectively, in Fig. 3 (*a*), (*b*), and (*c*). The round wire, shown in (*a*), is generally used for street railways in towns and cities. The ear by which it is suspended is tapered so as not to interfere with the trolley wheel. Wire with figure **8** section shown in (*b*) is used for high-speed roads where it is necessary to have a perfectly smooth running surface for the trolley. It is suspended by means of clamps *a*, *a*, leaving the lower part *b* free of obstruction. The preference for mine work now inclines to the grooved form shown in (*c*). This wire is also supported by clamp ears *a*, *a* that fit into grooves *b*, *b* in the sides of the wire just above the center. The figure **8** wire is apt to twist between supports and throw off the trolley; the round, or the grooved wire, which is practically circular in section, may be twisted without interfering in any way with the trolley. The figure **8** wire is also more apt to pull or twist out of shape or out of the clamps entirely, when rounding curves, than either of the other shapes.

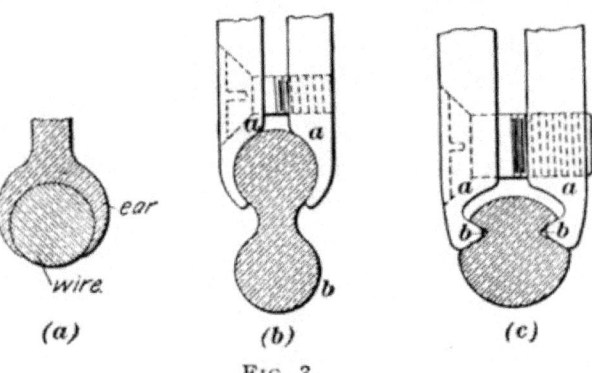

Fig. 3

13. Size of Trolley Wire.—Trolley wire varies in size from 0 to 0000. The heavier the wire, the harder it is to support and the greater is its cost. But if large carrying capacity is required, it is often better to use a heavy trolley wire rather than a light one with feeders, since a large wire possesses a very desirable rigidity and the trolley wheel is more likely to follow the wire and less likely to spark. In a 500-volt plant, the use of 0000 grooved trolley wire in connection with a well-bonded track, seldom requires the installation of separate feed-wires.

The importance of using conductors of large capacity is easily demonstrated: Suppose a locomotive to be operating on a grade 1,200 feet in length and 10,000 feet from the power house, and that the current consumed is 300 amperes. Assuming the generator voltage to be 500, the line drop of potential with insufficient copper might amount to 200 volts, leaving but 300 volts for the motors; the speed would be about 4 miles per hour, and 3 minutes would be required to ascend the grade. If the feeder and trolley wires were of proper size, the line drop would not exceed 100 volts, the voltage across the motors would be 400; and the speed would be about 6 miles per hour, or 2 minutes would be required to ascend the grade. In the former case, the demands on the generator and locomotive would be 50 per cent. more severe, since it takes 50 per cent. more time to ascend the grade.

14. Calculating Size of Wires.—The size of the feeder and trolley wire to be used is determined both by estimate and by calculation. Before calculations can be made, the quantity of current that the wire is to carry must be estimated. This is a widely varying quantity and to estimate it closely the local conditions must be very carefully considered. Some of these conditions are the individual capacity of the locomotives; their relations to one another if more than one is used, that is, the probability of a number of locomotives ever operating under heavy load at a distance from the power house; the profile of the road, etc.

Having estimated, as closely as possible, the average current required, it is usually considered safe to calculate the size of wire so that when this current is flowing the volts drop in the wire shall not exceed 10 per cent. of the total pressure at the locomotives. It is well also to estimate the maximum current ever likely to be needed at distant portions of the mine and to proportion the wires so that the drop with maximum current shall not exceed about 20 per cent.

15. In the case of electric locomotives equipped with series-wound motors, excessive pressure drop in the line is not such a serious matter, so far as its effect on the motors is concerned, as would be the case with shunt-wound machines, such as pump motors. The effect on the locomotive is merely to reduce the speed, the torque, or tendency of the motors to rotate about the armature axis, remaining practically constant through a wide range of voltage. For calculating the size of wire needed for feeder and trolley wires, the following approximate formula is used:

$$A = \frac{14 \times L \times I \times 100}{E \times D}$$

in which A = cross-section of conductor, in circular mils;
L = average length, in feet, of feeder over which the power is to be transmitted;
I = total current, in amperes;
E = electromotive force, in volts;
D = per cent. of drop of potential.

If the load is all bunched at the end of the feeder, L is the actual length of the feeder, in feet. If the load is uniformly distributed all along the line, as it would be if a number of locomotives were continually moving along the line, the distance L would be taken as one-half that used in the case where the load was bunched at the end; i. e., the whole current I may be considered to flow through an average of only one-half the length of the line.

16. The formula given in Art. **15** is the same as given in *Transmission, Lighting, and Signaling* for calculating the size of wires, except that the constant is here changed from 10.8 to 14 and the volts drop instead of being expressed direct is expressed as the total volts times the per cent. drop. In using the formula here, only the distance one way is considered; whereas, in calculating wires for metallic-return circuits, the distance both ways must be considered. Although the resistance of the track return is low, it is not safe to ignore it entirely, as would be the case if the constant 10.8 were used and the distance L considered but one way; hence, the constant is raised to 14 to allow for **track resistance**. This allowance is only approximately correct, as the track resistance depends on the weight of rails, the effectiveness of the rail bonding, and the quality of the steel of which the rails are made; some grades of steel contain impurities that greatly increase their resistance. The formula as given is, however, sufficiently correct for most purposes in calculating wires for mine transmission.

EXAMPLE.—In Fig. 4, *a b* represents a section of track 4,000 feet long. From the dynamo *c* to the beginning of the haulage section, the distance is 1,200 feet. The trolley wire is No. 00 B. & S. gauge, and is fed from the feeder at regular intervals. Two mining locomotives are operated,

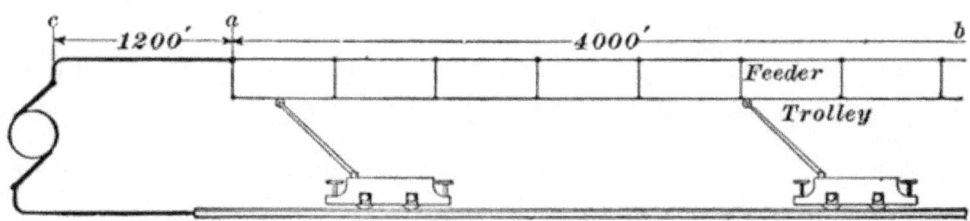

FIG. 4

each of which takes an average current of 75 amperes. The total allowable drop to the end of the line is to be 5 per cent. of the terminal voltage, which is 500 volts. Calculate the size of feeder required, assuming that the constant 14 in the formula takes account of the resistance of the return circuit.

SOLUTION.—Since the locomotives are moving from place to place, the center of distribution for the load may be taken at the center of the 4,000 ft. The distance L will then be $1,200 + 2,000 = 3,200$ ft. The total current will be 150 amperes; hence,

$$A = \frac{14 \times 3,200 \times 150 \times 100}{500 \times 5} = 268,800 \text{ cir. mils}$$

This would require either a stranded cable of sufficient sectional area or the use of two No. 00 wires in parallel from *c* to *a*. (See Table I, in *Transmission, Lighting, and Signaling*.) From *a* to *b*, the No. 00 trolley wire is in parallel with the feeder; hence, the section of feeder *a b* may be a single No. 00 wire. In most cases, the drop is allowed to run as high as 10 per cent., because the loads are usually heavier and the distances longer than in the example given. Ans.

17. Branch Lines.—Where electric mining machines, pumps, etc. are used along an electric haulage line, it is sometimes more convenient to lead the current to them by means of branch lines run from the trolley or haulage feeder lines, than by independent feeder lines. In such cases, each branch line should have a separate fused switch installed near its junction with the main line, so that, by opening the switch, the branch can be isolated when desired; and, also, so that by the blowing of the fuses the branch will be automatically cut off when it is short-circuited by a fall of rock, a careless or ignorant workman, or anything else. If no provision is made for automatically cutting off a branch line, a short circuit on it may open the circuit-breaker in the power house and thus shut down the whole system.

It is advisable in the majority of instances to have independent feeder lines for haulage and machines, with connecting lines at two or three points, which can ordinarily be used to run the lines in multiple, but which can be opened by switches should trouble on one line interfere with the operation of the other line. Individual circuit-breakers on locomotives, with the station breaker set slightly above their combined capacity, will tend to avoid interruption of service, which is important in both machine and pump work. In case of overload or a defect on a locomotive, its circuit-breaker will then open before that in the station.

18. Location of Wires.—The trolley wire is located above the track, preferably along one side and from 6 to 15 inches outside the rail, so as to be out of the way of men and animals passing along the road. Where the roof is good, the trolley wire may be supported directly from it. The trolley construction should be of the most substantial nature, and the work of installation should be in charge of an experienced man, as the care and thoroughness with which this is done determines largely the successful operation of the plant.

19. Methods of Supporting Wires.—Feeder wires, when necessary, are usually supported from the roof or carried on insulators fastened in the side of the entry. Where the roof is poor, posts or timber frames are set in the entry and the feeder wires fastened to these timbers by standard wooden pins or brackets with glass insulators. A convenient method, when the rib is firm and straight, is to bore a hole about 3 inches in diameter into the rib at the proper height, drive a large wooden plug firmly into the hole to a depth of about 6 inches, and nail standard wooden brackets to the plug. The plugs can be made of a piece of $3'' \times 4''$ wooden joist or mine rail, large enough so that both nails of a bracket will go into the same plug, one up and the other down, as in Fig. 5. Another method is to spike a piece of joist vertically to two plugs, one above the other, in the wall and then attach the brackets to the joist.

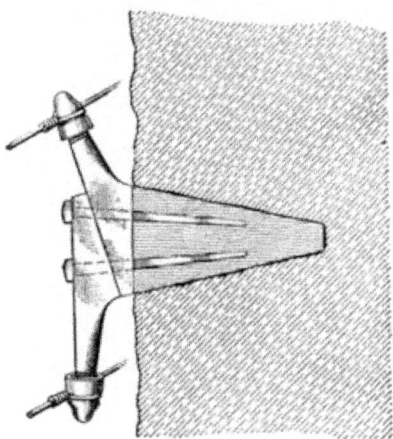

FIG. 5

20. There are various types of insulator pins suitable for fastening in the mine roof, where it is sufficiently firm and durable. One type consists of a tapering split iron pin, one half of which is provided with teeth to engage in the roof, when inserted in a hole of proper size; the second half, when forced or driven up, wedges the teeth of the first half into the coal or slate, and the insulator, when screwed on, holds the pin firmly and prevents the wedge from loosening. Other forms of ribbed pins are driven into the roof, depending on friction to hold them, and in some cases standard wooden pins are tightly driven into a hole in the roof and allowed to swell; or the end of the pin is split and a wooden wedge inserted so as to spread the ends apart and hold the pin fast when it is driven into the hole. These forms of hangers may be depended on when the roof is a firm, hard slate or rock.

21. Petticoat Hanger.—Fig. 6 shows a hanger designed to be attached directly to the roof of the mine. The upper portion, called the expansion bolt, consists of a corrugated split sleeve a in which is a threaded bolt b provided with a wedge-shaped plug c at its upper end and a nut and washer d

at its lower end. The bolt is prevented from turning by a lug that slips in the slot f. The expansion bolt is inserted

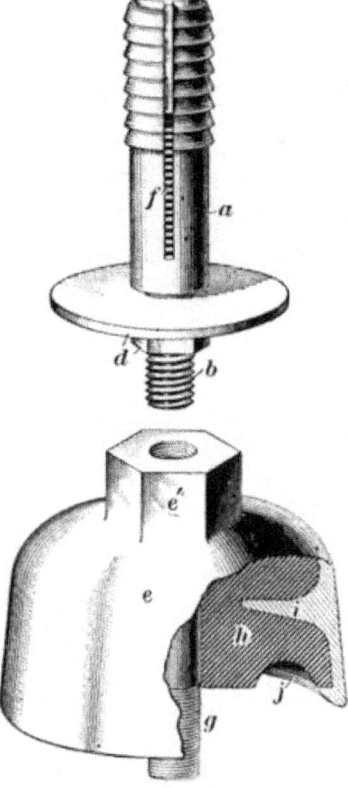

FIG. 6

in a hole drilled in the roof and the hexagonal nut d is screwed up, thus drawing down the wedge-shaped plug c and expanding the corrugated sleeve a so as to hold it firmly in the hole. The lower portion consists of an inverted cast-iron cup e provided with a boss e' in which is a tapped hole for screwing on to the end of bolt b. A stud bolt g with a flanged head is held firmly in place by the insulation h, which is molded into the casting e under heavy pressure, so that the bolt is entirely surrounded by the insulation, except the lower end, which is threaded to receive the trolley clamp, Fig. 8. The insulation is prevented from slipping out of the cup by the flange i. The height of the hanger from the top of the boss e' that fits against the nut b to the bottom of the insulation h is $2\frac{5}{16}$ inches. The circular groove j in the lower surface of the insulation is a drip groove to prevent water from reaching the center bolt g holding the trolley hanger; this form of hanger is often called a **petticoat hanger**.

22. Fig. 7 shows a slightly different form of this hanger designed to be attached to the timbers supporting the roof of the roadway. As before, the insulation a is covered and protected by a cast-iron hood, or cup, b. The insulation holds firmly in place and separates from each

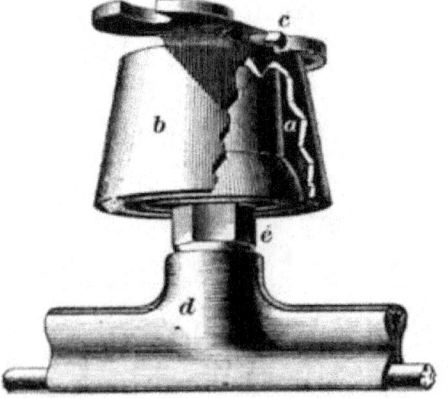

FIG. 7

other two steel studs, one projecting upwards and the other downwards. The upper stud passes through the hood and is

Fig. 8

provided with two arms c for attaching the hanger to the timbers. The lower stud extends somewhat below the cup b and carries the jaws, or clamps, d that grip the wire. Above the

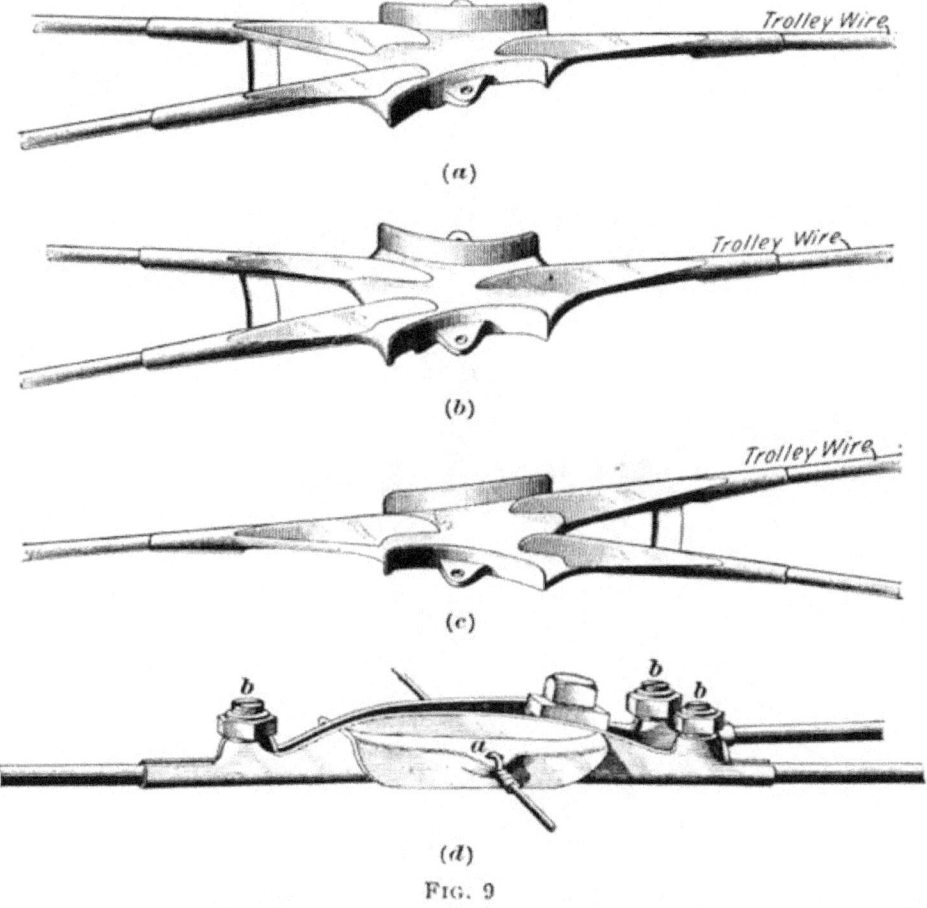

Fig. 9

clamps is a conical-shaped nut e threaded on the lower stud. The clamps are so hinged at the center that, by screwing the nut e down into the cavity in the top of the clamps, they are wedged apart at the top, forcing the jaws together so as to grip the wire firmly.

Fig. 8 shows a form of clamp often used with a grooved wire. One of the jaws of this clamp is provided with a screw socket by which it is attached to the lower end of the stud bolt g, Fig. 6, and three holes through which pass the screws that thread into the other jaw and clamp the two jaws firmly together.

23. Trolley Frogs.—Fig. 9 (a), (b), and (c) shows the under sides of three overhead switches, or **trolley frogs**, used to guide the trolley wheel from one wire to another; (a) is a simple **V** frog; (b), a right-hand frog; and (c), a left-hand frog. The **V** frog in its natural position is shown in Fig. 9 (d); the trolley wire is held by clamps b, b, b and the span or supporting wires are attached to the ears a. The frog must be placed with reference to the track so that the motion of the locomotive, as it takes the switch, will have given the trolley an inclination in the right direction before the wheel strikes the frog; it must also be hung level, or it will cause the wheel to leave the wire.

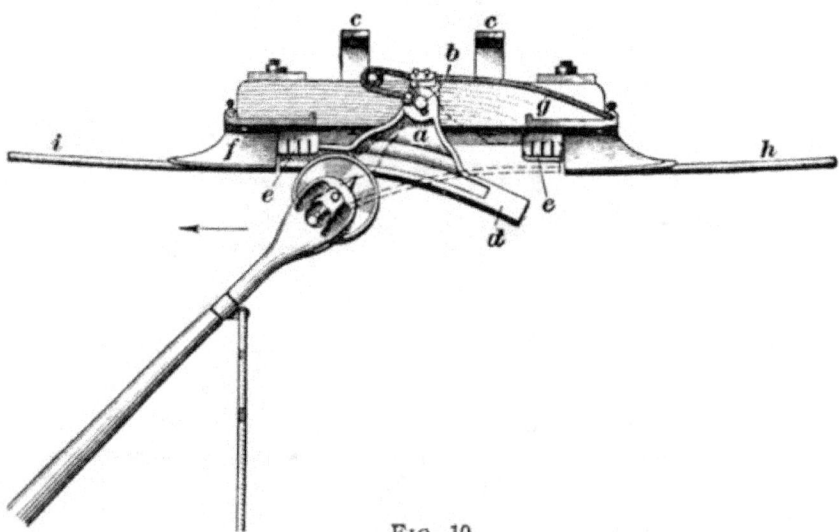

Fig. 10

24. Automatic Trolley Switch.—Fig. 10 shows an automatic trolley switch designed to reduce the amount of live wire in mines where but one locomotive is used. The switch consists of a rocker-arm a hung from the bolt b, which is supported by a timber frame attached to the roof or to the roof timbers by the hangers c. The rocker terminates at each end in a switch blade d, made to fit tightly in clips e that are attached to the metal piece f to which the trolley wires are also attached. A jumper, or short piece of flexible wire, g connects the rocker with the live part h of the trolley wire. When the trolley passes the switch in the direction indicated by the arrow the rocker is moved from the dotted- to the full-line position, thus closing the connection between parts h and i of the trolley wire. When the trolley passes again on the return trip, the rocker is moved back and left in the dotted-line position, thus leaving the part i of the trolley wire disconnected, or dead.

Such a switch may be placed at the mouth of a side entry; or, if desired, at any point along the main entry, so that the locomotive, on entering a mine or a branch, automatically closes connections to sections that it is approaching; and when leaving, it opens the connections so that there is no danger of shock from live wires to men and animals working in the sections a locomotive has left. The switch cannot be used in sections where more than one locomotive is employed.

25. Rail Bonds.—All electric mine haulage roads depend on the rails or the ground to carry the current from the locomotives back to the power house or substation. In order to keep the transmission losses as low as possible, the track resistance should be made low. The larger part of the track resistance occurs at the joints between rails. In many cases, it has been assumed that the fish-plates, between which the rail ends are clamped, afford sufficient electrical contact from rail to rail, but such is not the case. Competent engineers now insist on effective bonding of rail joints, that is, connecting the ends of the rails at the joints by a good electrical conductor.

There are a great many types of rail bonds, but for convenience they may be divided into two classes: *protected bonds*, or those placed between the fish-plate and the rail; and *unprotected bonds*, which either span the fish-plate or are placed under the rail. Rail bonds are usually made of copper and are attached to the rails in such a way that the contacts between the two metals, copper and iron, are clean and bright when made, and the two are forced so closely together that there is little likelihood of the connection becoming poor.

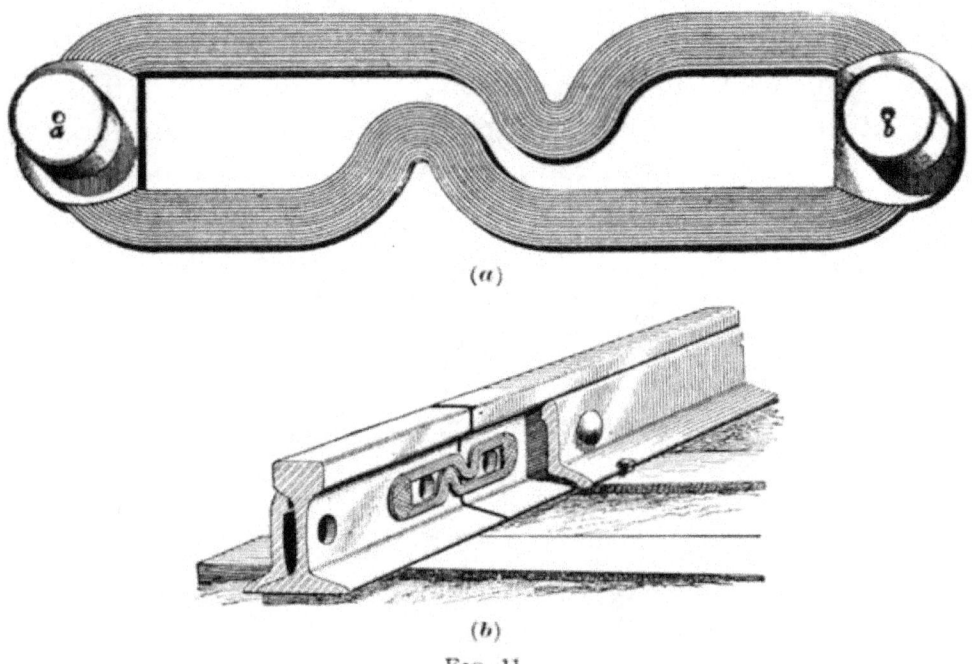

Fig. 11

26. Fig. 11 (a) shows a protected bond of the *double-loop type* shaped so as to give flexibility and at the same time allow openings for the track bolts. It is made of thin copper strips on which copper terminals a, b are cast. After the terminals have been passed through the holes in the rail, they are compressed by a special screw compressor that forces the metal out sidewise firmly against the sides of the holes.

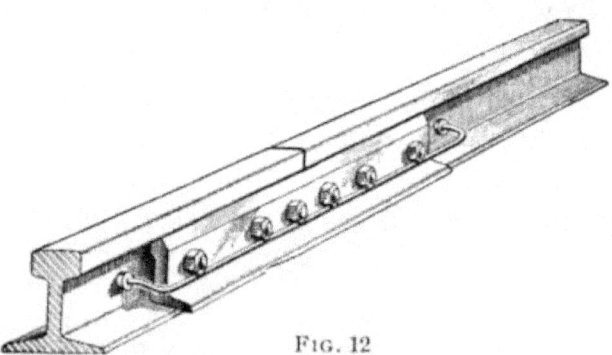

Fig. 12

Fig. 11 (b) shows such a bond in position, part of the fishplate being cut away. Holes through the rail for track bolts show through the loops of the bond.

27. Fig. 12 shows a type of unprotected bond which may be used where there is no danger of damage or theft. This bond is riveted in the rail by a hydraulic or screw press or, in some forms, by a tapered steel pin that expands the copper terminals in and around the holes.

It is advisable also to cross-bond the two rails of a track at about every fifth joint. All bonds should be inspected frequently, as they may work loose.

ELECTRIC MINE LOCOMOTIVES

GENERAL DESCRIPTION

28. An electric mine locomotive consists, in general, of a heavy rectangular cast-iron frame built up of side and end frames and carried on two or three pairs of driving wheels that are driven by series-railway-type motors through single reduction gears. An iron bracket, cast or bolted to each motor frame, is provided with bearings in which an axle runs, thus supporting one side of the motor; the other side is suspended from the frame of the locomotive by springs so as to reduce the jarring caused by traveling over uneven tracks. The motors are started, stopped, etc. by means of one or two controllers on each locomotive.

29. Baldwin-Westinghouse.—Fig. 13 shows a mine locomotive of the Baldwin-Westinghouse make. The massive construction of the side and end frames is plainly shown. Openings in the side frame show the ends of the main journals and their bearings, as well as the heavy coiled springs that support the frame above the journal bearings. The controller a operated by the hand wheel b and reverse lever b', the hand wheel c for operating the brake, the handles d, d of the sand-box levers, and handle e for ringing the gong are all within easy reach of the motorman. The headlights, mounted on each end of the locomotive, are usually electric lamps in iron cases, with reflectors. A locomotive is often referred to among miners as a "motor"; but, properly speaking, there are usually two motors on each

Fig. 13

locomotive, although on some there is but one, and on very heavy locomotives there may be more than two. When two are used, they may be arranged between the axles—that is, *central*; or one in the rear and one between the axles—that is, *tandem*. The following description is intended to be general and not to apply to locomotives of any particular make or type. There is considerable variation in details of the locomotives made by different manufacturers.

30. Frame.—The locomotive frame is made extra heavy and strong, for the double purpose of protecting the motors, controller, and motorman from the severe shocks and strains experienced in operation, and providing weight sufficient to utilize the full tractive power of the motors. Bumpers, to suit those on the mine cars, are either bolted to or cast with the end frames, which also carry a suitable coupling or drawbar to attach to cars. On some locomotives, where a rounded end frame answers for bumpers, projecting lugs or hooks are provided to prevent the cars from climbing on to the locomotive in case of collision or derailment.

The end frames are usually mortised into the side frames, to give greater stiffness, and the joint is firmly held by heavy bolts. Some makers prefer to bolt through the end frames, so that the bolts carry the entire tension of the trip, but get none of the shearing strain due to bumping or pushing the trip. Other makers bolt through side frames into end frames, and depend on the mortise joint for pushing, and the shearing strength of the bolts for pulling.

The side frames are slotted to receive journal-boxes, and are usually reenforced above with a block fitted in at the bottom of the slot to compensate for the weakening effect of the slot. Braces, which are also used to carry the heavy wooden covering that protects the machinery, connect the side frames at several points. Resistances, controller, brake stand, etc. are either suspended from these top braces or placed on iron plates bolted across the bottom of the frame.

31. Wheels and Axles.—Steel axles, on which chilled-iron wheels are firmly pressed and keyed, carry the frame and motors. The axles are of large diameter, because the power is applied to them through gears, instead of driving the wheels direct as in steam or air locomotives, and for the same reason the wheels must be keyed to the axles, besides being forced on by hydraulic pressure.

Journal-boxes fit loosely into the openings in the side frames, and support the frame, which rests on heavy spiral springs. Brass or bronze bearings are used, and the journal-box is filled with waste or wool saturated with oil. Journal bearings vary in size according to the weight of the locomotives. A 12-ton locomotive should have a journal not less than 4 inches in diameter, and 6 inches in length.

Axle bearings for motors are from $3\frac{1}{2}$ to $4\frac{1}{2}$ inches in diameter, and 4 inches to 8 inches long. Axle brackets are usually two in number and are either cast with or bolted to the field frame of the motor. When the motor frame is split, the axle brackets are usually cast to the upper half of the frame, allowing the lower half to swing down for inspection, as shown in Fig. 14 (*b*). Bearing sleeves are of bronze, or in some cases Babbitt lined, with slots for waste or wool lubrication carried in the bearing box. There is usually more or less space between the hubs of the wheels and the motor axle bearings, depending on the gauge of the track and the width of the motor. To hold the motor in its proper place, axle collars, with setscrews, are fitted between the motor and wheels, and the collars are adjusted to give desired clearance.

32. Gears and Gear-Cases.—Axle gears are of cast iron or, preferably, forged steel, made in halves and securely bolted together. Keys of ample proportions are fitted into the axle and gear. Both the gear and the armature pinion are machine cut; they are, in the best practice, completely enclosed in an oil-tight malleable-iron case, which protects them from dust and dirt and holds heavy oil or grease in sufficient quantity to thoroughly lubricate the gear at all times. Gear-cases should be made with the upper half readily removable, to allow for the frequent inspection of the gears. Serious and costly accidents sometimes result from carelessly allowing the bolts in the axle gear to become loose.

33. Brakes.—Some form of **screw brake** is now very generally used on mine locomotives, although for light locomotives or on moderate grades, the lever type, with notched quadrant, has an advantage in the quickness with which it may be applied and released. In the case of the screw brake, the pitch of the screw is such that a moderate turn on the brake wheel exerts a powerful leverage by which it is possible to slide all the wheels without excessive effort.

Brake shoes are ordinarily of cast iron, but cast-steel shoes have recently been used to great advantage. One set of the latter has outworn twelve sets of the ordinary iron shoes, with the additional advantage of leaving the wheels in better shape. If the brake shoe covers the entire width of the tread it wears down the outer part, which would otherwise with narrow rails be left unworn, thus causing grooved wheels.

In the latest locomotives, the steel brake shoes are set in cast-iron bodies, the steel shoe being readily removable by loosening a single setscrew. A complete set of these shoes can be exchanged in 10 or 15 minutes; but with the old types of shoe the operation sometimes consumed several hours.

34. Sand Boxes.—The side frames are usually cored out at each end to give a considerable capacity for sand. Positive-acting flat valves, made self-closing by strong springs, control the supply of sand to the track. The sand spouts are made of tubing and bent so as to apply the sand directly in front of the wheels, but the end of the spouts should be cut off parallel to the rail; otherwise, with a dirty or muddy track, there is liability of the wheels throwing mud into the spouts and clogging them. The sand should be clean, sharp, and dry, and the valves should permit an even, steady flow of a small amount, instead of requiring a wide-open valve. Sand on the rail increases the track resistance very materially, but it is important that it be used as sparingly as possible. In cases where the constant use of sand is necessary, an adjustable brush, or similar device, for cleaning the rail after the locomotive would be advantageous.

35. Headlights are mounted on the frame of the locomotive, and incandescent lamps of the proper voltage are commonly used. On 250-volt systems, two 125-volt 32-candle-power lamps are used, in series, one at each end. In some cases, where the voltage may run a little above 250, 270-volt lamps are connected direct, and other makers use two 125-volt lamps or three 90-volt lamps in each headlight. Lamps should be placed vertically, where possible, as the vibration and jolting of the locomotive is hard on the filaments, especially where they are long and of small section, as in the higher voltage lamps.

Oil lamps are sometimes used, and have some advantages, but as ordinarily made, they are liable to be broken, and in case of collision or accident might be dangerous to the motorman.

36. Motors.—The **motors** used on mining locomotives are of the series-enclosed type, such as are used on street railways. The armatures are preferably iron-clad, drum-wound, with formed coils; i. e., the coils are wound on suitably shaped forms and then thoroughly insulated before being placed in the slots on the armature cores. An armature on which the coils lie in slots is called iron-clad. Fig. 14 (*a*) and (*b*) shows two views of a motor suitable for use on a mine locomotive. The frame is split horizontally into two halves *A A*, hinged at one side and held together by bolts *a* at the other side, so that they can be opened, as shown in view (*b*), or closed, as in view (*a*), so as to completely enclose and protect all the working parts of the motor. The two arms, or brackets, *b, b*, Fig. 14 (*a*), receive the axle bearings; the axle bearing boxes *c, c* contain the oil or grease for lubrication; the bolts *d* hold a pole piece and field coil in place; the removable cap *e* covers an opening through which the commutator and brushes may be inspected and the brushes or brush holders renewed when necessary; the armature bearing boxes *f* also hold a quantity of grease. Openings with hinged covers are

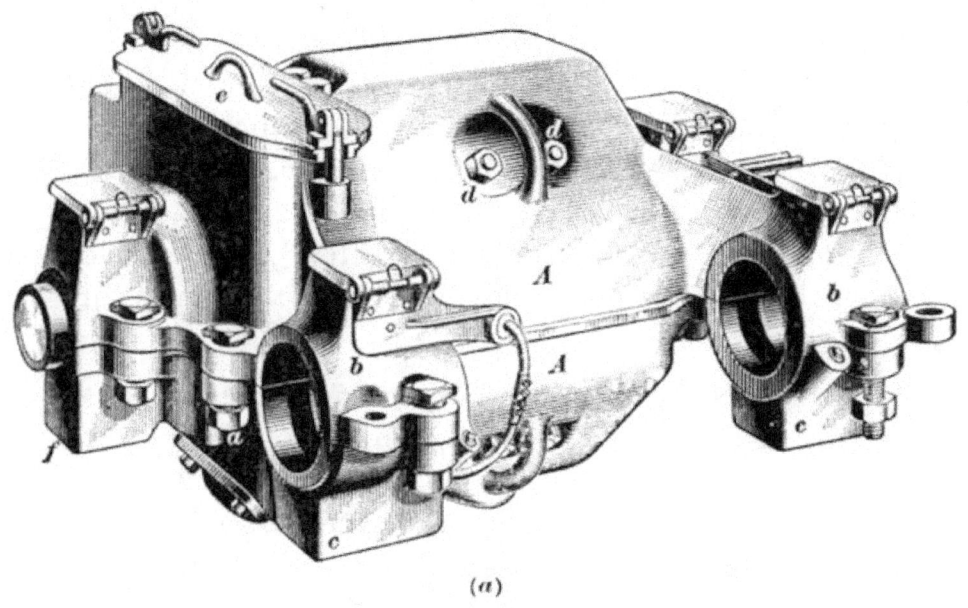

(a)

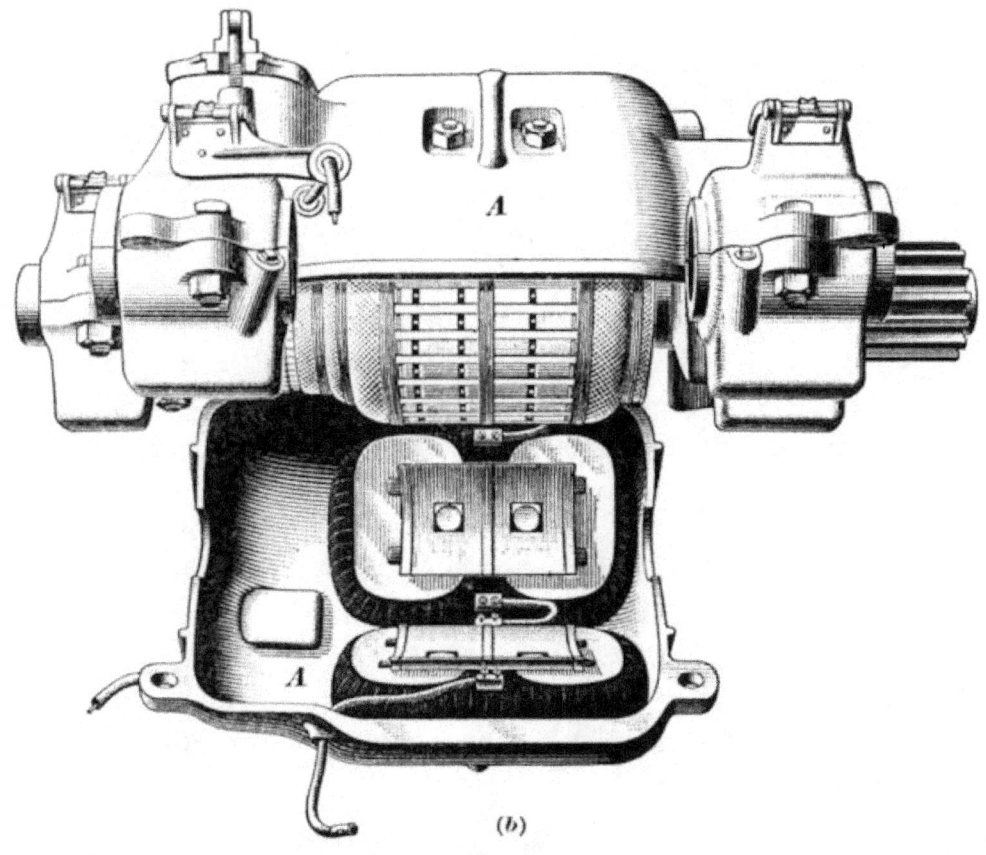

(b)

Fig. 14

provided over both the armature bearings and the axle bearings for the insertion of grease or oil. By leaving the armature bearing boxes f in position, the lower half of the field frame may be swung down, leaving the armature in position, as in view (a), or by removing these, the armature can be lowered with the field, thus giving access to the poles and field coils above the armature. A small handhole with a suitable cover is usually provided underneath the motor for removing dirt or making inspections.

37. The **field coils** are placed on the poles before these are bolted to the frame. The coils, wound on forms, are interchangeable, and the insulation is of asbestos and paper, or similar material not liable to damage except by exceptionally high temperatures. The field leads are brought out through the frame, as shown in the figure, and need not be disconnected to drop the lower frame.

38. The **armature** cores are built up of annealed laminated iron, assembled on spiders, or, as in the motor shown in Fig. 14, directly on the shaft. The **commutators** of motors used on locomotives usually have large diameters and safe wearing depths of about 1 inch; the mica segments are carefully selected to wear evenly with the copper, and the completed commutators are usually tested to ten times the working voltage between the segments and shell, and to full working voltage between the segments. Carbon brushes are used, and ordinarily two brush holders are held by a hardwood yoke supported from the upper frame. On some motors four brush holders are required, in which case the yoke is arranged to turn completely over, so as to bring the lower brushes up for inspection through the opening above the commutator.

The armature pinion is made with a taper bore, to facilitate its removal; it is keyed to the axle, and held on by a clamping nut.

39. Controller.—The **controller** used on the best types of mining locomotives is of the magnetic blow-out type, in which contacts are broken in a magnetic field, the strength of which varies as the strength or amount of the current to be broken. The arc, composed of copper vapor, is analogous to a wire carrying current in a magnetic field, and is forced across the field away from the contacts until it is broken or blown out. The resistance to this action is so slight that it cannot be measured, so that the blow-out is extremely rapid; there can be no arc of any consequence in a controller of this kind, except when the controller cylinder is moved very slightly, barely separating the contact fingers. In this case, the arc is so short that it will not blow out, and the contact tips may be burnt.

Series-parallel controllers, such as are used universally in street-railway practice, are not adapted to electric mine haulage, except where the work at starting is light; the reverse condition usually prevails, the starting effort often being the maximum load. In such cases, with the motors in series, when one pair of wheels slips, the counter electromotive force generated by this motor at its increased speed, limits the current passing, reduces the torque of the remaining motor, and thereby leaves but slight tendency to accelerate. With motors in parallel, however, the slipping of one pair of wheels has no effect on the others. It is usually of advantage when handling empty cars, switching, and other light work, where speed is not desirable, to run with the motors in series, thus avoiding constant or excessive use of the resistance points.

40. The R37 controller, used in General Electric and Westinghouse mine locomotives, is a typical one. It is shown in Fig. 15 (*a*) and (*b*). The cover A, view (*a*), is removed and the hinged pole piece C, normally held in place by the bolt e' in hole e, is swung back so as to show the interior of the controller. The changes in the connections are made by means of a *control cylinder*, or *drum*, D, often called the *power drum*, and a *reverse cylinder*, or *switch*, E. On both drums are contact segments k that make or break contact with stationary fingers f when the cylinders are turned by means of handle F and wheel G. B is a blow-out coil through which all the current must pass on its way from the trolley to the controller. The hinged pole piece C is provided with arc guards d; and when the controller is in

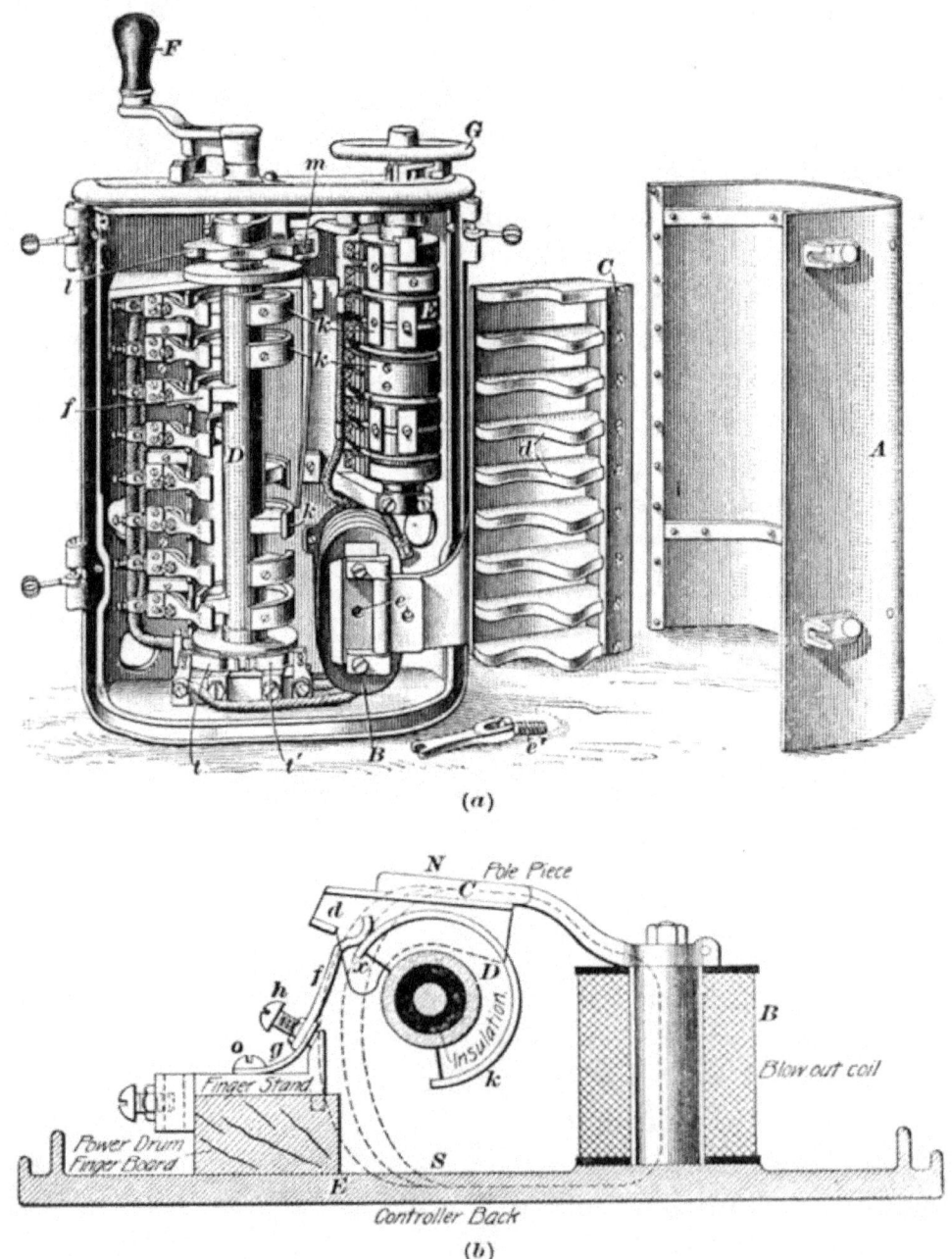

Fig. 15

use, the pole-piece is swung over the power drum as shown in the end view (*b*) of the controller, bringing the arc guards between the fingers so as to prevent arcing from one finger to another. In view (*b*), *C* is the pole piece, and *D* the end of the power drum. The arc guards *d* are made of vulcanized asbestos, called *vulcabeston*. The fingers *f* are stamped out of thick copper and are each attached to a phosphor-bronze spring *g*, which is fastened to a finger stand by a screw *o*. The springs hold the fingers firmly against the segments *k*, their tension being adjustable by means of screws *h*. The tension should be so adjusted that when the drum is turned until the ends *x* of the segment leave the fingers, these will drop about $\frac{1}{16}$ inch. The segments *k* are attached to the drum and insulated from it and from each other, except where connections are purposely made from one to another.

41. When the controller is in use, the flux caused by the current through blow-out coil *B* passes through the magnet core, through the pole piece *C*, the end of which thus becomes the north pole *N* of the magnet, through between the fingers and the retreating segments, thus blowing out any arc that may form, and into the controller back *E*, which thus becomes the south pole *S*, and so back to the magnet core.

42. Just above the power drum is a notched wheel *l*, Fig. 15 (*a*), called the **star wheel,** or **index wheel,** which turns with the drum and engages with a spring-actuated roller *m* that causes the drum to take definite positions corresponding to the various connections of the motors and resistances, thus, to some extent, preventing the motorman from running with the power drum between notches and thereby burning some of the contact tips and fingers. At the bottom of the power drum are two contact fingers *t*, *t'*, which both make contact with a segment on the drum when it is at the off-position. The use of these contacts will be explained in connection with Fig. 16.

43. The **reverse cylinder** *E* is smaller than the power drum and is usually mounted in the upper right-hand corner of the controller. It is provided with contacts and fingers for changing connections to the two motors, so that the

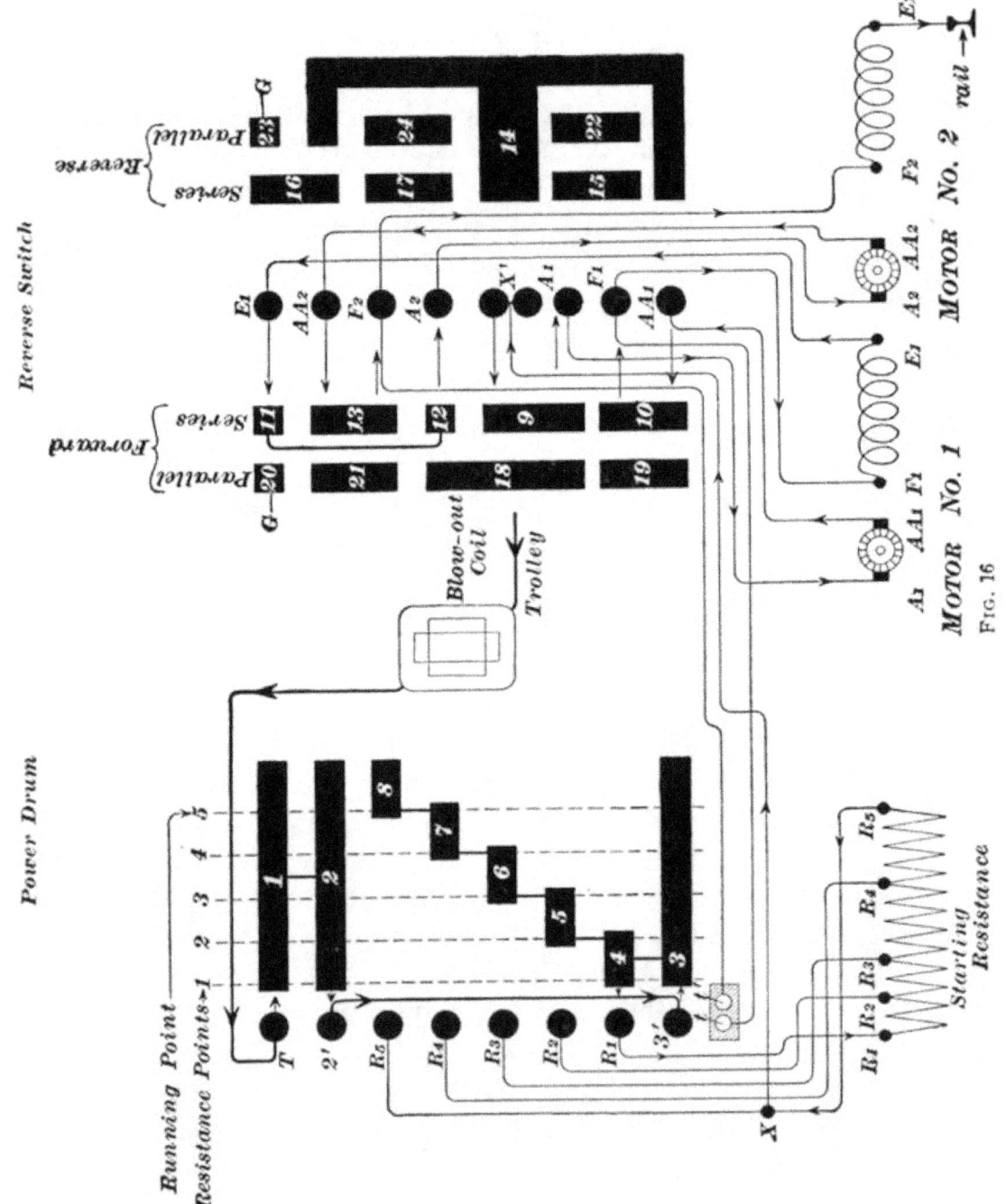

Fig. 16

locomotive will run forwards or backwards with the motors in series or in parallel in either direction. The reverse cylinder is not provided with a magnetic blow-out, arc guards, or other devices for taking care of the arcing; hence, it should never be turned while the current is on. All modern controllers are provided with an interlocking device, so that the reverse cylinder cannot be moved except when the power drum is at the off-position.

44. Connections.—Fig. 16 is a diagram of the connections of a mine locomotive operated by the controller just described. Mine locomotives are usually operated from one end only and are provided with but one controller; but if it is desired to have a controller on each end, the two are connected in parallel, the connections of both being exactly the same as those of the one shown. The diagram represents the cylinder segments by black bands; that is, as if the cylinder and its segments were developed, or laid out flat. The fingers are represented by black circles, and the different positions of the fingers by dotted vertical lines *1, 2, 3, 4, 5*. In studying the diagram, it may be considered that the developed cylinder slides under the fingers. In this controller, the segments of the power drum are made in two castings; contact segments *1* and *2* are cast together as one casting, and segments *3, 4, 5, 6, 7*, and *8* constitute the other casting. Each casting is insulated from the drum and the two are insulated from each other. The connection of the segments in each casting to each other is indicated by heavy vertical lines between them.

45. Assuming that the reverse switch is in the forward series-position and that the power handle is moved to the first notch, the path of the current through the motors from the trolley to the rails may be traced as follows: Through the blow-out coil–finger T–segments *1* and *2*–fingers $2'$ and $3'$ which are connected–segments *3* and *4*–finger R_1–all the starting resistance and back to point X–fingers X' on the reverse cylinder–segment *9*–finger A_1–armature of motor No. *1*–finger AA_1–segment *10*–finger F_1–field coils of motor

No. *1*–finger E_1–segments *11* and *12* which are connected –finger A_2–armature of motor No. *2*–finger AA_2–segment *13*–finger F_2–field coils of motor No. *2*–framework of the locomotive–wheels–rails; the framework and wheels are not represented in the diagram. The direction of the current flow through this path is indicated by arrowheads. Fingers R_1, R_2, etc. are connected, respectively, to points R_1, R_2, etc. of the starting resistance, and moving the power drum to the second notch so that segment *5* touches finger R_2 permits the current to pass directly from segment *3* through the casting, segment *5*, and finger R_2 to point R_2 of the resistance, thus cutting a section R_1R_2 of the resistance out of circuit. On the second notch, section R_1R_3 is cut out, etc. until on the fifth notch all the resistance is cut out, the current passing from segment *3* through the casting, segment *8*, and finger R_5 to X and thence as before. The finger positions represented by the vertical dotted lines *1, 2, 3, 4* are therefore called resistance points and that represented by line *5*, the running point.

46. When the reverse cylinder is on the forwards parallel position, the current passes through the power drum and starting resistance the same as previously described, but after reaching fingers X' on the reverse cylinder it passes to segment *18* and thence divides through fingers A_1 and A_2, a part going through each armature and field coil. The current through motor No. *1* returns through finger E_1 to segment *20*, which is connected to the frame of the locomotive; that through motor No. *2* passes, as before, from the field coil direct to the frame. By tracing the path of the current with the reverse cylinder in either of the reverse positions, it will be found to flow through the power drum, the starting resistance, and the field coils, in the same direction under all conditions; but it is reversed through the armatures, owing to the differently shaped segments on the reverse cylinder.

47. If the reverse switch is moved from forwards to reverse position while the locomotive is running at high speed and the controller handle is at off-position, the motors will act as generators; and if they are connected in parallel, will act exactly the same as is described in *Operation of Dynamo-Electric Machinery*, Part 2, for series-wound generators in parallel; i. e., unless provided with an equalizer, one will overcome the other and drive it in the reverse direction as a motor and the current flow between the two motors of a locomotive, when running with a heavy trip at high speed, may be so great as to damage the machines. This action of the motors is called *bucking*, and to prevent lazy motormen from bucking the motors to check the speed instead of using the hand-brake, an equalizer connection is provided, which connects the positive terminals of the two field coils when the power drum is in the off-position, as shown in Fig. 15 (*a*), by making contact between the two fingers t, t' at the bottom of the power drum.

48. The controller, like all other electrical apparatus, should be kept free from dirt. Dust, if allowed to accumulate, is liable to start arcs. A small amount of vaseline or vaseline mixed with a little graphite, applied occasionally to the contacts will tend to prevent cutting; and if applied sparingly will not interfere perceptibly with the quality of the electrical contact. The bearings of the two drums should be oiled occasionally; and if a heavy black deposit accumulates on the arc guards it should be scraped off. Sometimes, holes are burned through the arc guards, though this is not common unless the controller has unusually bad short circuits to handle. If this happens, the guard should be replaced by a new one as soon as possible, but a temporary repair can be made by plugging up the opening with a thick paste made of a mixture of shellac and asbestos.

49. Starting resistances for use with mine locomotives are much the same as those used with street-railway cars. There are numerous forms, the aim in all cases being to assemble the resistance in the least possible space that will permit the necessary radiation of heat. Fig. 17 shows a Westinghouse starting coil, or *diverter*, as it is sometimes called. It is made in sections, each consisting of a closely wound spiral of band iron with mica insulation between the layers. On the older types, the sections were assembled

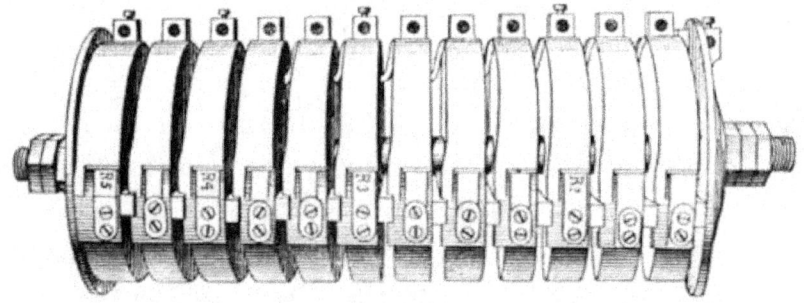

Fig. 17

close together, as in Fig. 18, but a ventilating space is now left between them and this is found to be a decided improvement, because the heat is dissipated so much more readily.

50. Other common forms of resistances used with mine locomotives are made with German-silver or iron wire wound as compactly as possible between asbestos paper insulation.

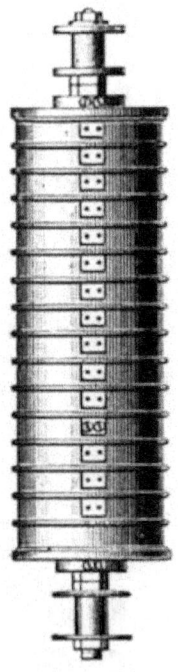

Fig. 18

The resistances are always made up in small units and several of these, assembled with suitable ventilating spaces between the units, constitute a complete resistance. For street-railway work, the General Electric Company uses a resistance made of cast-iron grids, but these are little used on mine locomotives, because the service is so rough that the grids break.

51. Locomotive Wiring.—The cables or wires from the controller to the motors and resistances are ordinarily carried along the side frame near the top. They should be bunched together and enclosed in a heavy canvas covering, to protect them from wear or injury, and should not be located over or against the resistances, nor so as to interfere with the inspection or removal of resistances or motors. To facilitate the location of trouble in wiring or motors, it is an advantage to have each cable lettered or numbered, both in the controller and at the motor, or resistance, end.

§ 56   HAULAGE   35

HAULAGE LOCOMOTIVES

52. General Electric.—Fig. 19 shows a standard 13-ton General Electric locomotive. The motors are central, that is, both are placed between the axles, and are spring-suspended from the frame so as to reduce the wear and shock on the bearings due to rough track. With the motors central, the wheel base is 56 inches; but to give a shorter wheel base of 44 inches with 30-inch wheels for use on sharp curves the

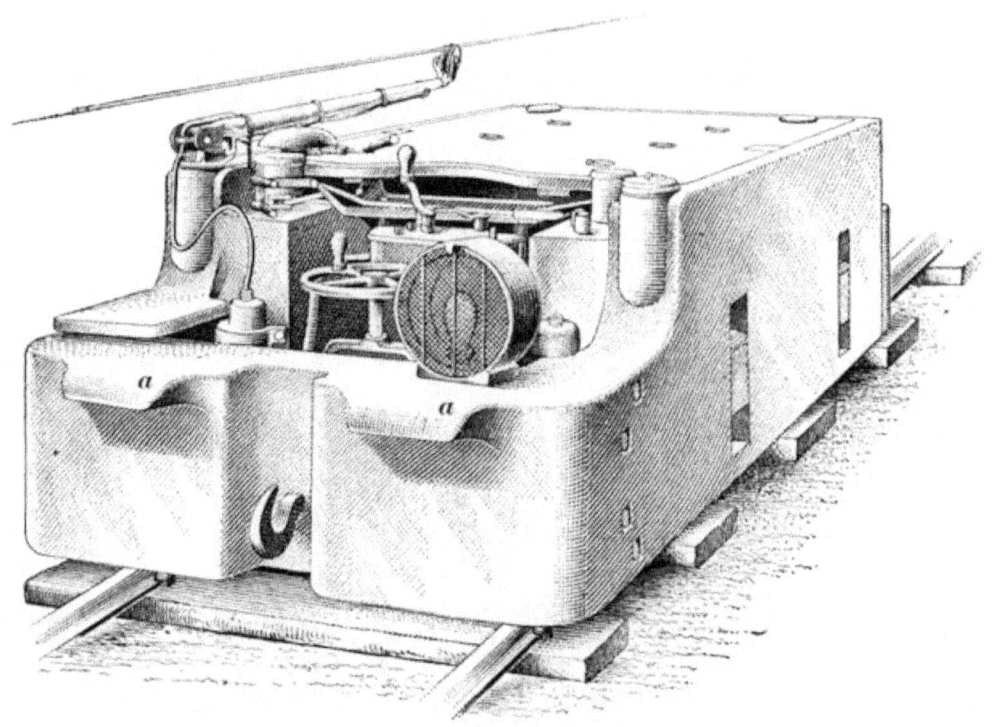

Fig. 19

motors are sometimes placed tandem. Connections are provided for placing the trolley on either side, to conform to the conditions in the mine. The trolley pole is from 3 feet to 7 feet long and swivels in a socket bored in the frame. The harp carrying the trolley wheel also swivels on the pole head; the wheel has a vertical range of about 4 feet and is held firmly against the trolley wire by a spring at the base of the pole. Electrical connection with the controller is made through an insulated flexible cable and contact plug; the plug on the side not in use is covered by an insulated cap, and there are no

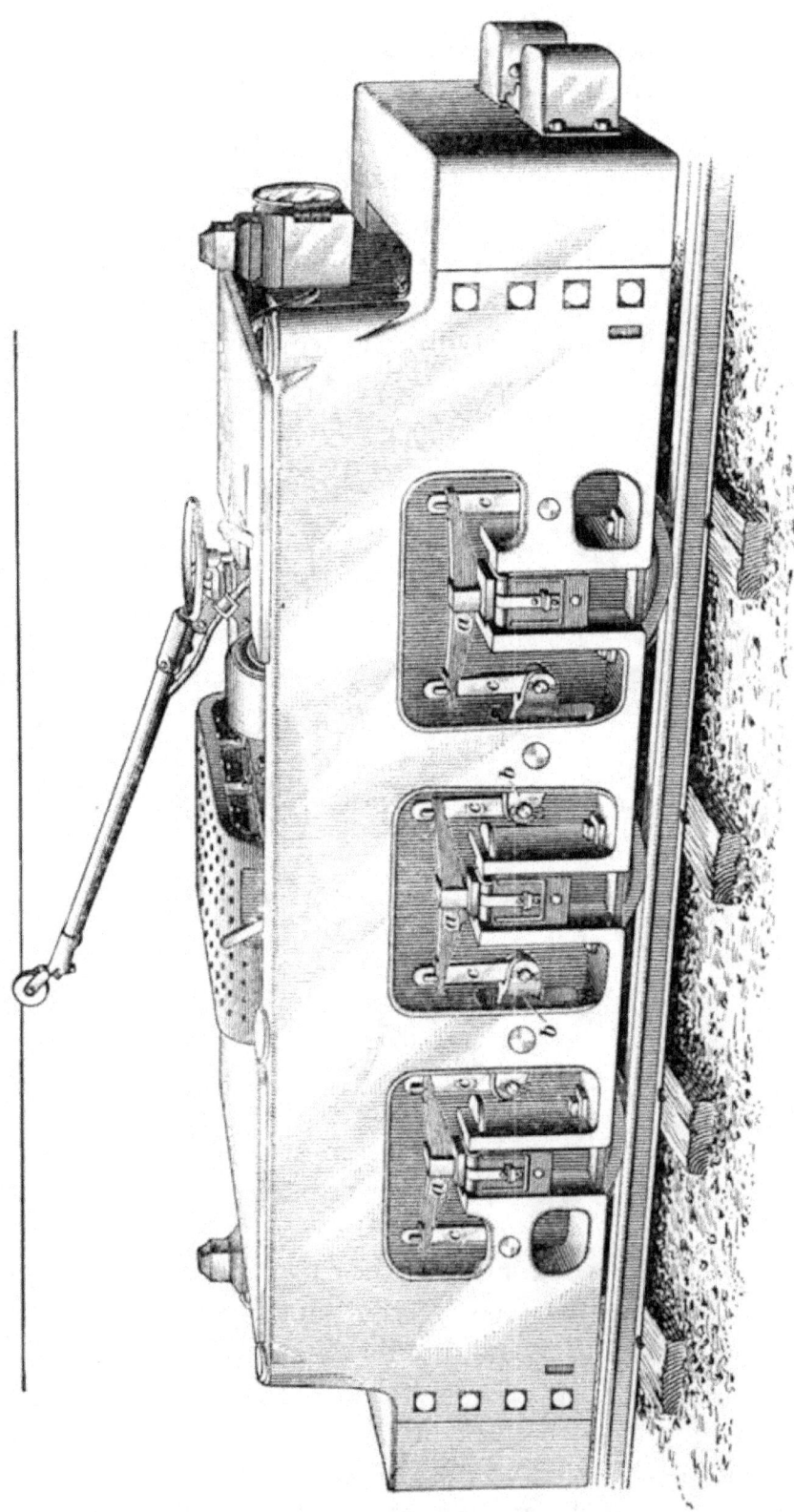

Fig. 20

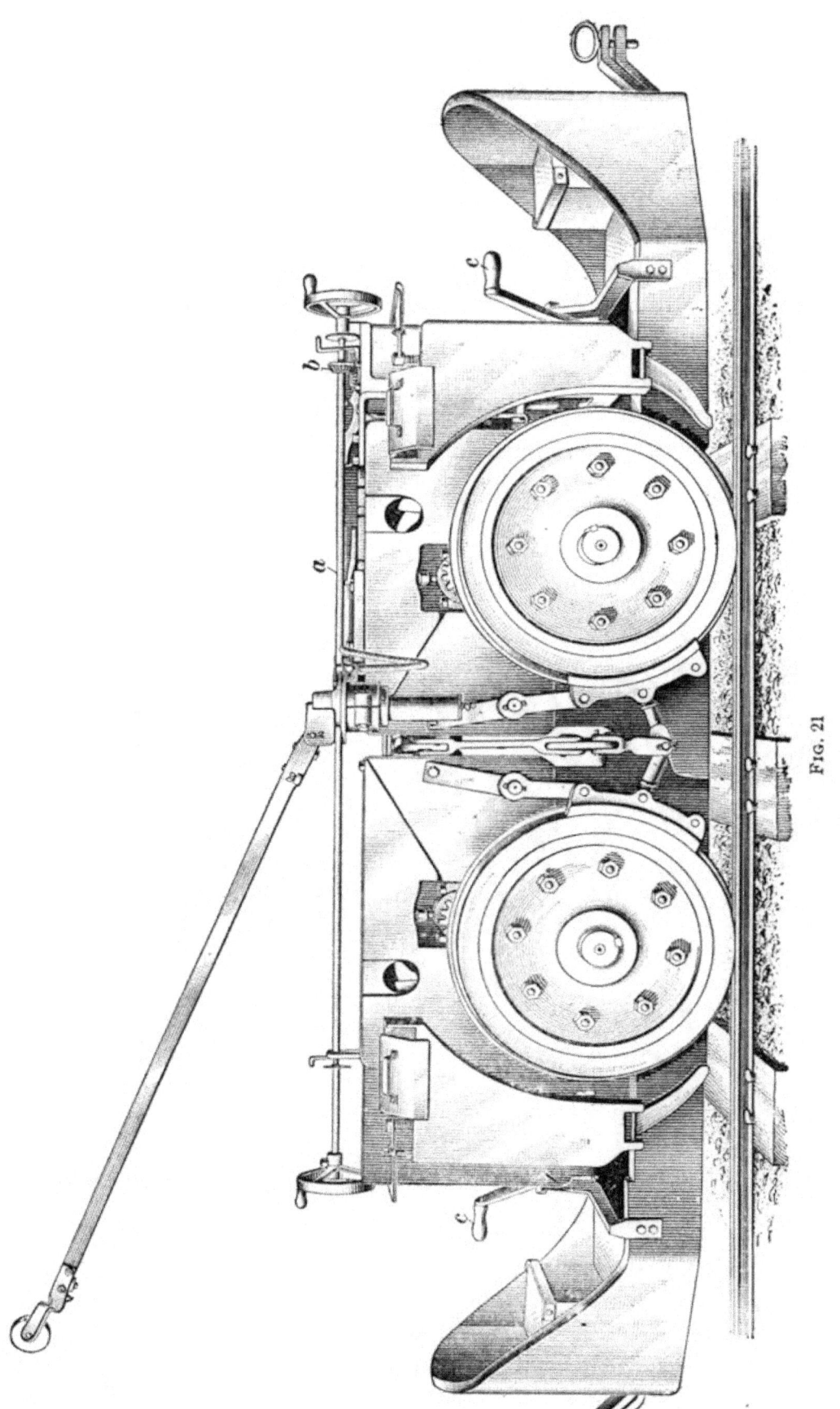

Fig. 21

live contacts exposed within reach of the operator. In this type, with motors central, the resistances are placed in the rear, and are protected by a removable ventilating grating. The general arrangement of controller, brake wheel, sand levers, gong, and headlight are shown. Projections a, a, cast on the frame, hook over the car bumpers and prevent the cars from climbing over the locomotive in case of a collision or a very sudden stop.

53. Jeffrey Six-Wheel.—Fig. 20 illustrates a three-motor, six-wheel **Jeffrey locomotive**. The operator's seat and the controlling mechanism are in the center. This places

FIG. 22

the driver in a position where he is in the least danger, because he is protected on all sides by the heavy framework. The middle drivers are without flanges and the wheel base with 28-inch drivers is 62 inches. The arrangement of journal springs a, equalizing bars b, and hangers c is well shown. This locomotive, weighing 15 tons, is equipped with three 35-horsepower motors.

54. Morgan-Gardner.—The **Morgan-Gardner locomotive**, Fig. 21, is of the double-motor type with four driving wheels and a flexible base, this latter feature making it possible for the wheels to follow the rails on narrow or uneven tracks. The main frame is cast in one piece and is closed across the bottom to prevent dirt and mud from getting into the working parts. The locomotive has outside wheels, and the motorman can ride on either end, both ends being provided with the necessary levers for operating the locomotive. There is but one controller operated by means of shaft a and miter wheel b. The brake is operated by means of brake levers c.

In this type of locomotive, the motorman has very little protection at the rear, although he can always take the end of the locomotive opposite the trip and thus avoid the danger of being crushed between the trip and the locomotive in case of a very sudden stop.

55. Goodman Single-Motor.—Fig. 22 shows the **Goodman single-motor locomotive** intended especially for use in very narrow entries and on sharp curves and uneven track. The wheels are outside and no part of the frame overhangs the rails. The axles are not rigidly connected but are free to take such position that all four wheels will always be on the track.

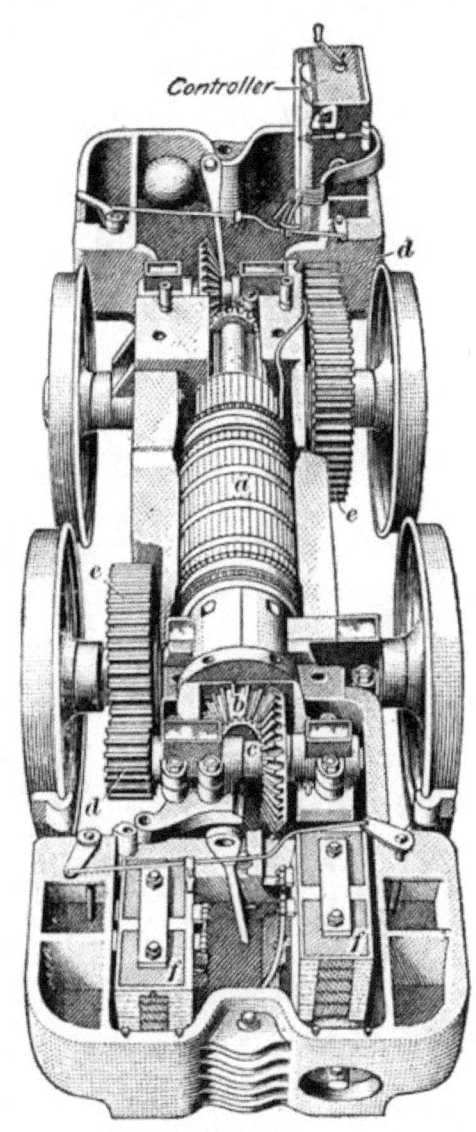

Fig. 23

56. Fig. 23 shows the interior of the locomotive shown in Fig. 22, with the cover and the upper half of the motor frame removed. The one motor is built into the frame, the armature a being placed lengthwise of the locomotive and geared at each end through miter wheels b to a counter-shaft c

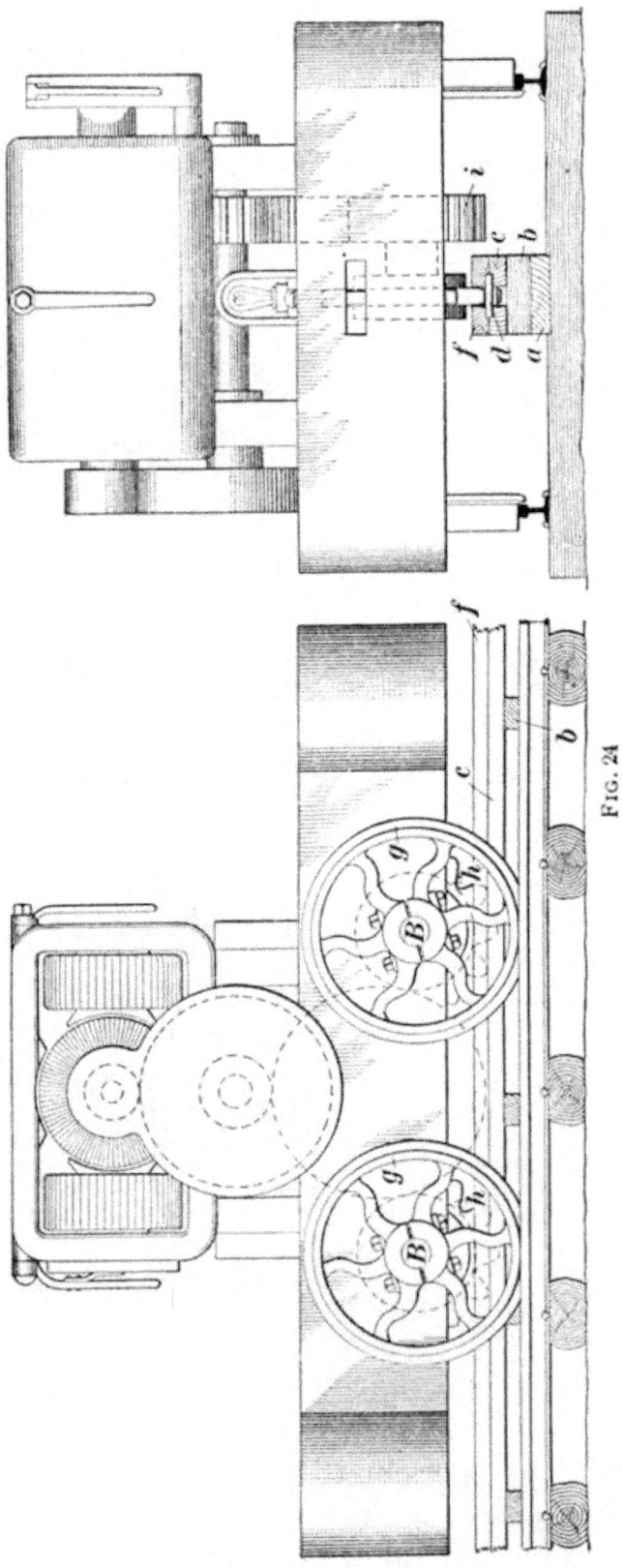

Fig. 24

that carries a pinion d meshing with a gear e on the axle. The locomotive is ordinarily operated from one end, the resistance f being located in the other end.

57. Morgan-Gardner Combined Third and Traction Rail.—For grades of from 5 to 18 per cent. locomotives of the rack-rail type can be used to advantage. Fig. 24 shows the arrangement of the track, gears, sprockets, etc. of a Morgan-Gardner combined third- and rack-rail type locomotive. The third rail is laid centrally between the two regular track rails. There are three sizes of third rail: standard, heavy, and special, and the component parts of each are made in 16-foot lengths. The standard size only will be described.

58. A $6\frac{1}{2}'' \times 1\frac{1}{2}''$ white-pine, bottom stringer a, Fig. 24, is securely spiked to the track ties, which are first trimmed, if necessary, to receive it. On this stringer are spiked, at intervals of about 18 inches, pine blocks b of sufficient thickness to bring the height of the third rail 4 inches above the height of the steel rails. Two longitudinal pine strips c each $2\frac{1}{2}$ in. $\times$ $1\frac{1}{2}$ in., are spiked to the blocks, leaving a $1\frac{1}{2}$-inch space between them. They are trimmed on top in such a way as to allow the iron track d, constituting the third rail, to be partly countersunk. This track is 4 inches wide and $\frac{5}{8}$ inch thick, and consists of a flat bar of iron with $1\frac{5}{8}$-inch square holes punched through it at intervals of $1\frac{5}{8}$ inches. It is made continuous by means of perforated fish-plates securely bolted to the ends of the two bars at a joint. Two pine strips f similar to the strips c, except that they are trimmed on the lower side instead of the upper side to cover the iron rail, are fastened to the strips c with 4-inch spikes. An insulated copper cable is connected with the perforated rack, or third rail d and conducts the positive electric current to it. The wooden strips on which the third rail is mounted act both as a carrier and as a medium for insulating it. The negative current is carried by the bonded steel track rails.

59. Locomotives of the sprocket type are made in two standard sizes; one, known as class B, shown in Fig. 24, is

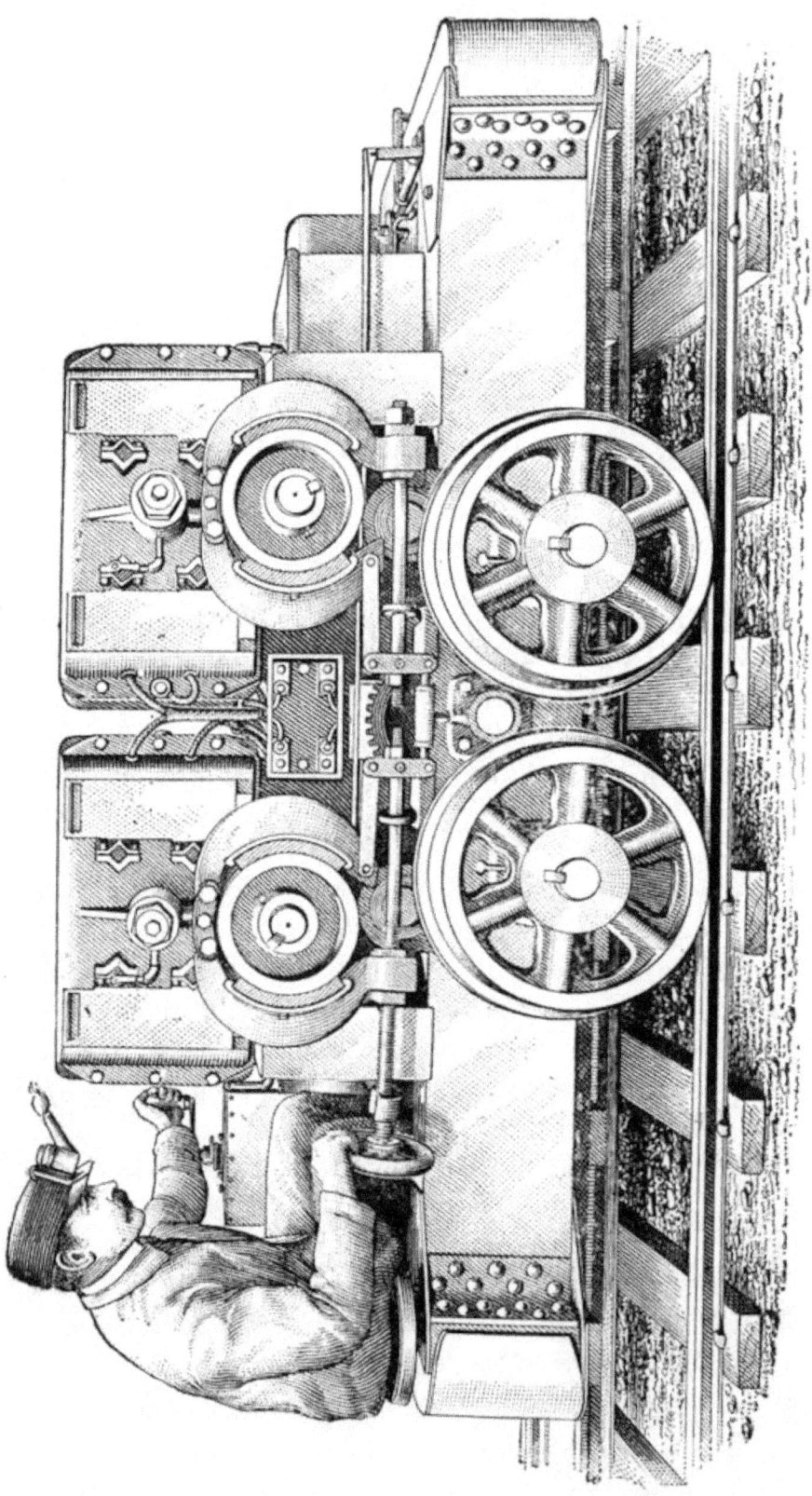

Fig. 25

§ 56　　　　　　　　HAULAGE　　　　　　　　43

equipped with a single 75-horsepower motor, and the other, known as class BB, shown complete in Fig. 25, has two motors of the same size. In principle and in all important points of construction, the two locomotives are essentially the same, and a detailed description of one will suffice for both. The smaller locomotive is 7 feet long, $3\frac{1}{2}$ feet high, and weighs 3 tons, and the larger one is 10 feet long, 4 feet high, and weighs 5 tons. The motors are wound for 250 or 500 volts, and the gauge is made of any desired width from

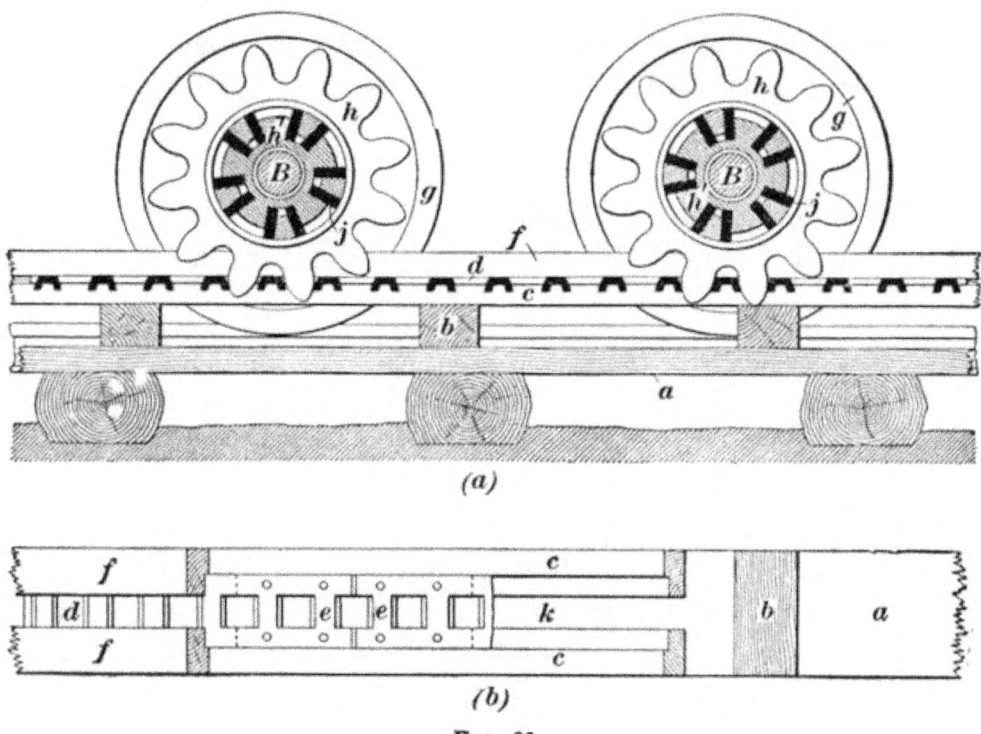

Fig. 26

18 inches up. A small, low locomotive is also made to be used in very low entries.

On each of the two axles B, Figs. 24 and 26 (a) are two track wheels g, a steel sprocket h, and a gear i (not shown in Fig. 26), which is attached to the sprocket hub h'. The track wheels and the sprocket hub and gear are attached direct to the axle, but the sprocket is insulated from its hub by means of hard maple blocks j, Fig. 26 (a), shown in solid black. The blocks separate and insulate the inward projections cast on the sprocket rim from the tooth-like projections on the spider, but at the same time the sprocket and its hub must turn together. The teeth of the two sprocket wheels, which are geared to run in unison, run in the slot k, Fig. 26 (b), between the wooden strips cc and ff holding the third rail d. In coming in contact with the third rail, the sprockets take up the current, and through copper contact springs, which rub against the sprocket rims, impart the current to the motor or motors. The insulation of the sprockets is necessary, because the gear and shaft are in contact through the track wheels with the rails or the negative side of the circuit. Insulation by means of wooden blocks in the hub has given no trouble on voltages as high as 550. The motors drive the sprockets, which mesh with the perforations in the rigid third rail and thus impart motion to the locomotive. As all the transmission of mechanical power is through cut-steel gear-wheels, there is little loss. The location of one of the fish-plates is indicated at e, e, Fig. 26 (b), bolted under a joint in the rail.

60. The loss by leakage from the third rail in wet and muddy mines has caused little trouble unless the third rail is entirely submerged. Both third and track rails are of low resistance, and it would doubtless be possible to operate on low voltages, even in such an extreme case. In one case, with 500-volt current and water dripping on the track for a distance of, possibly, 100 feet, with a total length of 1,500 feet, a recent test showed 20 amperes leakage, when the current was turned on, after 12 hours of idleness. This leakage reduced to 5 amperes within 5 minutes, and within 15 minutes was not perceptible on the ammeter, showing that the dampness on third-rail insulation quickly dried out. The locomotive used on this track weighed about 5,000 pounds, and ran on rails weighing 20 pounds per yard. It had a capacity of hauling eight cars, weighing 3 tons each, against a grade of 10.5 per cent.

GATHERING LOCOMOTIVES

61. General-Electric.—It is not practicable to gather cars from the face of a room with an ordinary electric locomotive, for to do so would require each room to be wired, and the increased cost would be prohibitive. It would also be necessary to constantly keep increasing the length of the trolley wires as the rooms increased in length. Then the locomotive could not work close to the face, as the trolley wire could not be kept in place there, owing to the constant blasting. In overcoming these objections, several forms of **gathering locomotives** have been put into use.

Fig. 27 shows a General Electric gathering locomotive. A flexible insulated cable a is hooked on to the main-haulage-way trolley wire and as the locomotive goes into the chamber the cable unwinds from the horizontal reel b, which is of large diameter and placed on top of the locomotive frame and which, as the cable unwinds, is retarded from turning too fast by friction. The electric current passes through a suitable contact at the center of the reel to the controller of the locomotive. If iron rails are used in the rooms, one or both will be used as a return; if not, the cable may be made double, one conductor being for the return current. A double cable is necessary when the locomotive is run on wooden rails, for example, close to the working face.

62. When a double cable is used, the conductors are usually connected to the rails and the trolley, as shown in Fig. 28. One of the cross-bonding wires a near the point where the locomotive is to enter the chamber is made longer than usual, so as to form a loop b into which the negative conductor is hooked, the positive conductor c being hooked over the trolley wire as usual. This method removes all strain from the trolley.

63. The mechanism for driving the reel, Fig. 27, is actuated by a chain and gear-connection to one of the main axles. A sprocket chain from the axle drives a horizontal intermediate shaft that carries two miter pinions, each of

Fig. 27

which is provided with a clutch that may be thrown in or out at the will of the operator. Each clutch will drive its pinion in but one direction; that is, when the clutch is in, the shaft will turn freely in the pinion in one direction, but it cannot turn in the other without turning the pinion also. When the clutches are both out, the shaft can turn freely in either

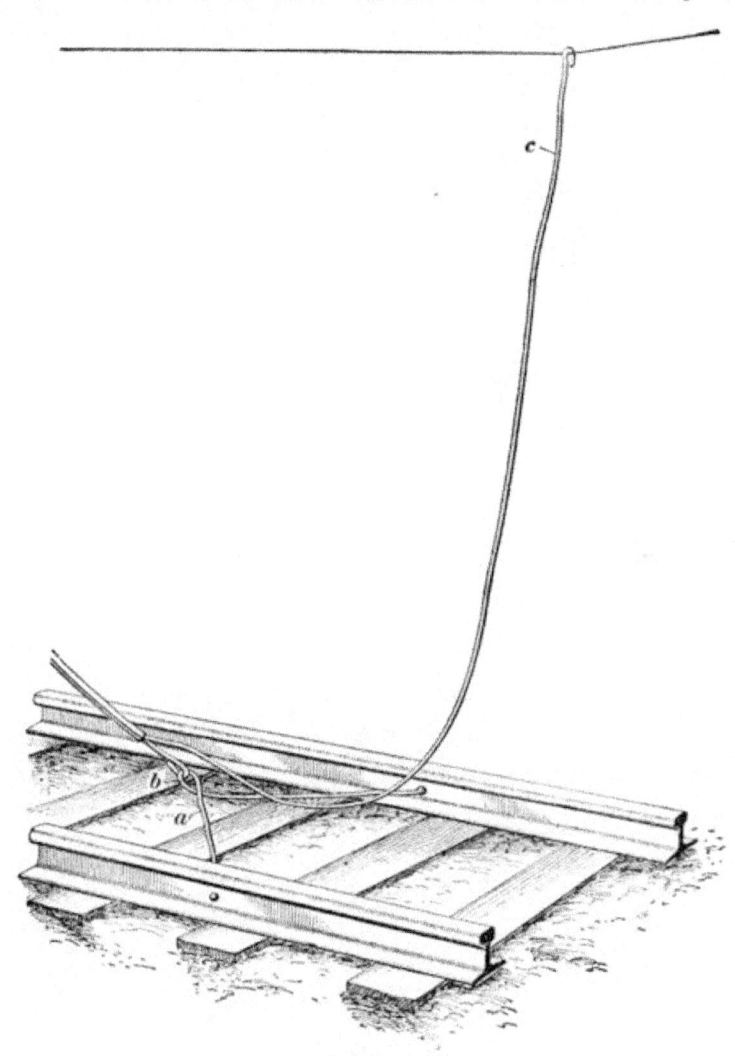

Fig. 28

direction without moving either pinion. Between the two miter pinions is another small pinion on the end of a vertical intermediate shaft, the other end of which carries a large wheel that rubs against the reel resting on it and thus drives it by friction. When it is desired to enter a chamber where there is no trolley wire, the end of the cable is hooked or clamped to the trolley wire in the main gangway, the trolley pole is fastened down, and the reel clutch is moved from the center, or off-position to the side that will produce the desired rotation of the reel in rewinding; as the locomotive proceeds the reel unwinds, the friction of the various parts being sufficient to maintain the proper tension, about 30 pounds.

The clutch does not act so long as the locomotive continues to move away from the main haulage way, but the instant it starts to move in the reverse direction the pinion is positively driven so as to wind the cable back on the reel and the speed of the reel is high enough to keep the cable under tension as before. The weight of the reel resting on the friction wheel prevents slipping, unless the tension on the cable becomes too great. If it is desired to enter a room with the locomotive headed in the opposite direction, the reel clutch is moved to the other side so as to use the other miter pinion; that is, the cable can be unwound or wound up automatically whichever way the locomotive is moving.

64. These locomotives have been used successfully on grades as high as 14 per cent., but on high grades they must be used with considerable caution, especially in returning toward the main heading; because, in order to keep the speed down, the brakes must be set very hard and if the wheels are skidded, or slid, the reel mechanism will not work and the cable will be run over.

65. Jeffrey.—Fig. 29 shows a Jeffrey gathering locomotive, a type in which the reel for carrying the flexible insulated double-conductor cable is placed on the end of the locomotive. When it is desired to run the locomotive into places where there is no trolley wire, one conductor of the cable is connected to the rail directly under the trolley wire and the other to the trolley. The cable is automatically paid out as the locomotive runs from the parting and is automatically rewound as the locomotive returns. The reel is operated by a friction clutch driven by a chain and sprocket from the motor, the clutch being controlled by the motorman at the operating end of the locomotive. The

Fig. 29

49

speed of the reel when the cable is being wound up is slightly faster than that of the locomotive on the track, and the cable is thus kept taut, but when the tension becomes too great the friction clutch allows the reel to slip, thus preventing injury to the cable. The cable runs between small metal guide rollers mounted on an iron box that travels back and forth on a threaded rod the full length of the cable reel at such a speed as to wind the cable closely on the reel. By means of the trolley, the locomotive is also used on portions of the haulage system where the trolley wire is in place; and by means of a double-throw switch, the motors are connected with either the trolley or the cable circuits. Locomotives of this kind are made that are only 29 inches in height.

66. To facilitate making the connections of the cable to the main-trolley circuits, a board is sometimes placed at the entrance to the room and provided with one or two terminals, depending on whether a single or a double conductor is to be used. These terminals are mounted on a slate base and are so protected that it is impossible to come in contact with them accidentally. The free end of the cable has a plug or connector fastened to it, and this may be quickly attached to the terminal on the board.

67. The Goodman Manufacturing Company also makes a very compact gathering locomotive that may be run anywhere a mine car can go. It is no larger than a car, the wheels are outside the base, and the cable reel is on one end and is operated entirely independently of the locomotive gearing. The wheel base is very short and there is but one motor, which is built as a unit with the locomotive frame and geared to each axle similar to the locomotive shown in Fig. 23. The wheel base is flexible so that the locomotive will readily follow a rough track. The total height of the machine from the rails to the top of the frame is $32\frac{1}{2}$ inches and the total length 7 feet 9 inches.

TABLE II

COMPARATIVE SIZES OF 10-TON, FOUR-WHEELED, INSIDE-WHEELED, TWO-MOTOR LOCOMOTIVES

Maker	Weight Short Tons	Minimum Gauge Inches	Amount to Add to Gauge to Give Width Over All Inches	Height of Frame Above Rails Exclusive of Trolley Inches	Length Over All Inches	Wheel Base Inches	Diameter of Wheels Inches	Weight of Rail Pounds per Yard
General Electric Company	10	24	22	37	150	46*	33	30
Jeffrey Manufacturing Company	10	30	20½	36	136	44	30	30
Baldwin-Westinghouse	10	30	20	36	139	48	30	30
Goodman	10	26	21½	39	140	48	33	30
Morgan-Gardner	10	36	9	41	174	48	30	30

*This wheel base is for locomotives with gauges between 24 inches and 36 inches. For gauges over 36 inches, the wheel base is 2 inches less.

DIMENSIONS OF ELECTRIC MINE LOCOMOTIVES

68. There is considerable range in the sizes and general dimensions of locomotives, and these dimensions are constantly being changed as new designs replace the old ones, but the general dimensions and weights of medium-sized locomotives given in Table II will be useful for reference. The trolley stand is about 5 inches above the frame.

RATING AND CAPACITY

69. Weight of Locomotive and Capacity of Motors. The motors with which a mine locomotive is equipped should be of sufficient capacity to take advantage of the traction afforded by the full weight of the locomotive, and the weight should be great enough to take advantage of all the power of the motors. A guarantee to that effect from the manufacturer is of some value; but any statement regarding the power of the motors, unaccompanied by a statement of the tractive power or drawbar pull of the locomotive, is of little value unless it is for a comparison of motors. Street-railway motors are usually rated in horsepower at the uniform load that they will carry for 1 hour with a temperature rise in the windings not exceeding 75° C. (135° F.). These conditions do not approximate those under which the mine locomotive will operate. It is, therefore, advisable for the prospective purchaser to confine his dealings to a reliable manufacturer and to leave entirely in the latter's hands the electrical design of the locomotive, simply securing some such guarantee as the following: "With 250 volts at the terminals of the motors, the locomotive is guaranteed to be of sufficient capacity to exert a drawbar pull of 2,500 pounds at a speed of 6 miles per hour continuously for 1 hour, or 3,200 pounds at a speed of 6 miles per hour for 40 minutes with a temperature rise of not over 75° C. (135° F.) above the surrounding air, having a temperature of 25° C. (77° F.)."

70. Drawbar Pull.—The rated drawbar pull and speed of the locomotive should be used as a basis for the calculation

§ 56   HAULAGE   53

of mine haulage, but at the same time the general character of the service should be considered. If the haul is of great length, the average drawbar pull required should be well within the rated drawbar pull of the locomotive, while if, as is usually the case, the haul is short and the service intermittent, the locomotive may be operated at its rated drawbar pull.

The rated drawbar pull of an electric locomotive is commonly taken at one-sixth its weight, although it may vary from one-fifth to one-eighth the weight, depending on the kind and condition of the rails, size of wheels, the capacity of the motors, and the weight of the locomotive. The drawbar pull of a light locomotive is not so great, proportionally to its weight, as that of a heavy one. The tables of rated drawbar pull published by the manufacturers of locomotives are sometimes based on the same proportion of the weight for all sizes, while others vary the proportions for the different weights, and still others give the average results of a number of tests without regard to the weight.

71. The effective horsepower H of a locomotive is found by multiplying the drawbar pull P, in pounds, by the speed D, in feet, per minute and dividing by 33,000, the number of foot-pounds in 1 horsepower.

$$H = \frac{P \times D}{33,000} \qquad (1)$$

The speed, however, is nearly always expressed in miles per hour. As 1 mile per hour equals $5,280 \div 60 = 88$ feet per minute, if S is miles per hour, the formula may be expressed

$$H = \frac{P \times S \times 88}{33,000} = \frac{PS}{375} \qquad (2)$$

and the drawbar pull

$$P = \frac{375\,H}{S} \qquad (3)$$

in which
P = drawbar pull, in pounds;
H = effective horsepower;
S = speed, in miles per hour.

72. Drawbar Pull From Electrical Input.—The electrical input of a locomotive is the amount of power, in kilowatts, that is taken by the locomotive from the trolley line. It is assumed that 20 to 25 per cent. of this power is used to supply the copper and iron losses of the motors and in overcoming the friction of the motors, gearing, and other machinery of the locomotive, so that only 75 to 80 per cent. of the power furnished the locomotive, that is, 75 to 80 per cent. of its input, is available for tractive power.

73. It is commonly assumed that about 1 per cent. of the total work done by a locomotive is required to move itself; i. e., the work effective in moving the load is the total work multiplied by .99. In ordinary calculations, no account is taken of the power required to move the locomotive on a level and the effective work is taken to be the same as the total work; i. e., the total tractive power and the drawbar pull are considered to be the same.

74. If the input in kilowatts is expressed by KW, and the efficiency of the locomotive is 77.5 per cent., the power in kilowatts available for tractive power is $.775\,KW$.

Since 1 horsepower equals .746 kilowatt, the total horsepower developed by the locomotive equals $\frac{.775KW}{.746}$. If it takes 1 per cent. of the total horsepower to move the locomotive, the effective horsepower is

$$H = \frac{.99 \times .775KW}{.746} = 1.03\,KW \quad (1)$$

Substituting this value for the horsepower in formula **3** of Art. **71**,

$$P = \frac{375}{S} \times 1.03\,KW = \frac{386KW}{S} \quad (2)$$

$$KW = \frac{P \times S}{386} \quad (3)$$

75. Table III, given by the General Electric Company, shows the approximate input of a locomotive for different speeds and drawbar pulls.

TABLE III

Weight Pounds	P Drawbar Pull Pounds	Volts	S Speed, Miles per Hour	Approximate Kilowatts Input at Rated Drawbar Pull and Speed	Amperes
6,000	700	250	6.0	11	44
6,000	700	500	6.0	11	22
9,000	1,200	250	6.0	19	76
9,000	1,200	500	6.0	19	38
13,000	2,500	250	7.4	50	200
13,000	2,500	500	7.4	50	100
16,000	3,000	250	6.5	50	200
16,000	3,000	500	6.5	50	100
20,000	3,500	250	8.3	76	304
20,000	3,500	500	8.3	76	152
20,000	3,500	250	8.0	75	300
20,000	3,500	500	8.0	75	150
20,000	3,500	250	6.0	55	220
20,000	3,500	500	6.0	55	110
26,000	4,500	250	7.8	90	360
26,000	4,500	500	7.8	90	180
26,000	4,500	250	7.5	90	360
26,000	4,500	500	7.5	90	180
26,000	4,500	250	7.5	90	360
26,000	4,500	500	7.5	90	180
40,000	7,500	250	8.0	160	640
40,000	7,500	500	8.0	160	320

76. Drawbar Pull on a Grade.—The drawbar pull of an electric locomotive working on a grade is determined by deducting from the drawbar pull, expressed in pounds, on a level, the product obtained by multiplying the weight of the locomotive, in pounds, by the per cent. of grade, as was fully explained for compressed air and steam locomotives in *Haulage*, Part 2.

The steepest and not the average grade should be considered in determining the size of the locomotive for a given haulage system. The adverse grades of a haulage system usually determine the size of the locomotive, and it is apparent that a reasonable amount of money may be exceedingly well spent in reducing the grades. On a short grade, the locomotive may be worked very close to the slipping point of the wheels or maximum starting effort, as given in Table IV. The momentum that may be acquired on a straight section of level track preceding a grade will materially assist the locomotive in ascending the grade. The limiting grade is entirely a matter of operating expenses, as a traction locomotive will propel itself on as steep a grade as 12 per cent., or even 15 per cent., with good heavy rails. In practical service, however, the adverse grades should not exceed 3 per cent. for long runs, or 5 per cent. for short runs. The grades in favor of the loads, and against the empties, should not exceed 5 per cent. for long runs, or 8 per cent. for short runs.

77. Table IV gives the drawbar pulls on grades up to 5 per cent. based on a rated drawbar pull on the level and deducting for each grade the weight of the locomotive times the per cent. grade, as just explained.

78. There is a certain grade in favor of the load that requires the same tractive power to propel a loaded car as an empty car. This grade obviously depends on the weight of the entire trip including the locomotive when the cars are loaded and empty, and on the track resistance. To illustrate, assume the loaded trip to weigh 30 tons, the empty trip 10 tons, including the locomotive in each case, the track

TABLE IV
DRAWBAR PULL

Weight of Locomotive Pounds	Maximum Starting Effort on Good Track With Sand Pounds	Rated Drawbar Pull, Pounds					
		Level	1-Per-Cent. Grade	2-Per-Cent. Grade	3-Per-Cent. Grade	4-Per-Cent. Grade	5-Per-Cent. Grade
6,000	1,000	700	640	580	520	460	400
9,000	1,800	1,200	1,110	1,020	930	840	750
13,000	3,200	2,500	2,370	2,240	2,110	1,980	1,850
16,000	4,000	3,000	2,840	2,680	2,520	2,360	2,200
20,000	5,000	3,500	3,300	3,100	2,900	2,700	2,500
26,000	6,500	4,500	4,240	3,980	3,720	3,660	3,200
40,000	10,000	7,500	7,100	6,700	6,300	5,900	5,500

resistance to be 30 pounds per ton and the grade $\frac{3}{4}$ per cent. The resistance due to the grade is 15 pounds per ton, ($\frac{3}{4}$ per cent. of 2,000 pounds), and the net resistances ascending and descending the grade are, respectively, 45 and 15 pounds per ton and the total net resistance of the empty car will be $10 \times 45 = 450$ pounds and of the loaded car $30 \times 15 = 450$ pounds. Under these circumstances, however, the service does not possess the desired intermittency and the motors do not have as much time to cool as under ordinary conditions, where there is considerable difference between the drawbar pull of the empty and loaded trips. The locomotive must, therefore, be of greater capacity in order that the temperature of the motors may not become dangerously high. This condition will seldom be encountered, but the example shows the importance of a careful analysis of local conditions.

79. Overloads.—The heating effect of an electric current in a conductor is proportional to the square of the current; hence, when motors are subjected to overloads, they heat very rapidly. Knowing the maximum current the motors of a locomotive are capable of carrying without excessive heating and knowing the overloads that will be required in hauling up the grades of the road, it is possible to calculate how often the locomotive can be subjected to the overloads without injuring the motors. For example, if the safe rating of the motors for continuous duty is 40 amperes and they are subjected, at intervals, to a load of 100 amperes for periods of 3 minutes each, the frequency of the periods may be $3 \times (\frac{100}{40})^2 = 18\frac{3}{4}$ minutes; i. e., a load of 100 amperes for periods of 3 minutes each at intervals of $18\frac{3}{4}$ minutes will have the same heating effect as a continuous load of 40 amperes. In making this statement, it is assumed that during the $15\frac{3}{4}$ minutes remaining in each interval after the application of the overload, no current whatever is flowing through the motors. If the motors take a current of 100 amperes for periods of 3 minutes each at certain intervals and take, say, 20 amperes for the remainder of the time, the intervals must be longer. Suppose that x represents the length of an interval; then $x - 3$ is the time the 20 amperes is flowing and the heating effect may be represented by the expression $20^2 \times (x - 3)$ or $20^2 x - 20^2 \times 3$. The heating effect during the 3 minutes may be represented by $100^2 \times 3$; and for the whole interval, by the sum of the two expressions, or $20^2 \times (x - 3) + 100^2 \times 3$. But this heating effect must be the same as if 40 amperes were flowing during the whole interval x, or $40^2 x$; that is,

$$40^2 x = 20^2 x - 20^2 \times 3 + 100^2 \times 3;$$
$$40^2 x - 20^2 x = 100^2 \times 3 - 20^2 \times 3;$$
$$1{,}200 x = 28{,}800; \quad x = 24;$$

that is, the intervals must be 24 minutes.

80. Determination of Size Required for a Given Service.—In selecting a locomotive for a mine, there are many conditions to be considered, but the selection should always be made with a view to meeting the worst possible condition without crippling or injuring the locomotive. For example, suppose that it is desired to haul trips of thirty cars, the empties weighing 2,000 pounds, and loads 6,000 pounds, over a track all the grades of which favor the loads. The frictional resistance of the cars is assumed to be 2 per cent. and of the locomotive 1 per cent. The profile of the road is as follows:

Grade Per Cent.	Distance Feet	Grade Per Cent.	Distance Feet	Grade Per Cent.	Distance Feet
1.3	800	.3	700	2.4	400
2.0	600	1.77	1,025	3.5	425
1.3	800	.9	300	1.2	320

What must be the tractive power of the locomotive?

Since all the grades favor the loads, the maximum drawbar pull will be exerted in hauling the thirty empties up the steepest grade 3.5 per cent. The total resistance to be overcome by the locomotive is the sum of that due to frictional resistance and that due to grade, or, in case of the cars, $2 + 3.5 = 5.5$ per cent. of their weight and, in case of the locomotive, $1 + 3.5 = 4.5$ per cent. of its weight. The thirty empties, weighing 60,000 pounds, will then require $60,000 \times .055 = 3,300$ pounds drawbar pull. In general, a locomotive with cast-iron chilled wheels, such as electric locomotives usually have, should have a weight on the drivers of six times the drawbar pull or, in this case, $6 \times 3,300 = 19,800$, practically 20,000 pounds. The tractive power necessary to move the locomotive up the grade is then $20,000 \times .045 = 900$ pounds, and the total tractive power is $3,300 + 900 = 4,200$ pounds, which is below the slipping point, 5,000 pounds, given in Table IV for a 10-ton locomotive on dry rails with sand.

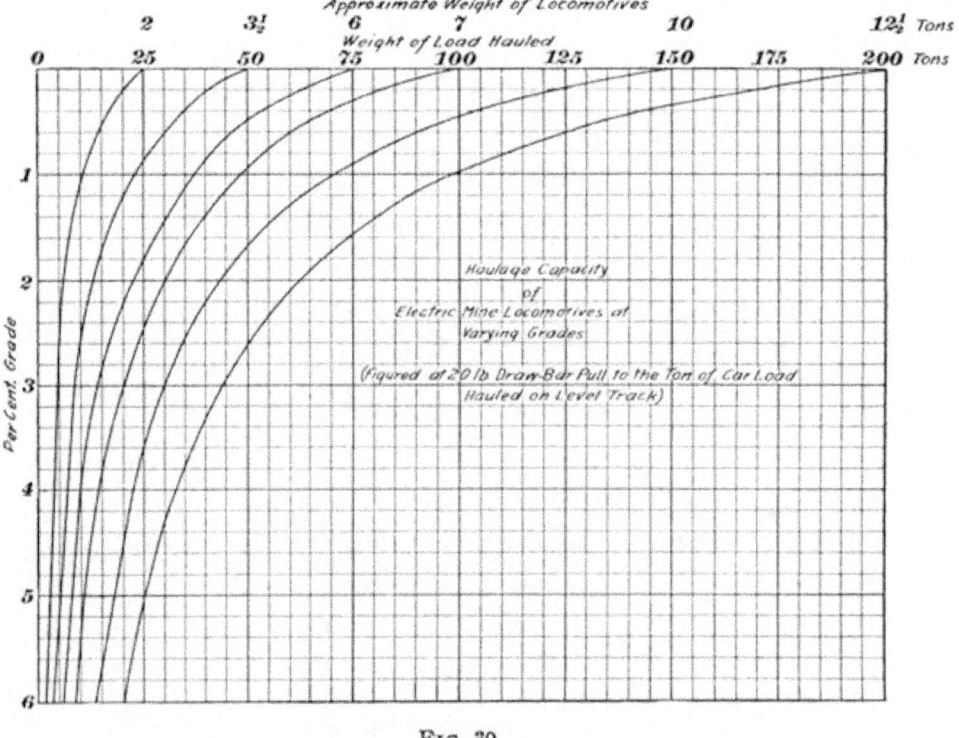

FIG. 30

81. Fig. 30 is a diagram for determining the hauling capacity of electric locomotives on different grades. The curves from left to right represent the capacities of 2-, $3\frac{1}{2}$-, 6-, 7-, 10-, and $12\frac{1}{2}$-ton locomotives, respectively, as indicated by the top row of figures along the upper margin. Distances from the top of the curve downwards represent the per cent. of grade, values being given along the left-hand margin. Distances from left to right represent weights in tons that different locomotives can haul, values being given along the upper margin. For example, if it is required to select a locomotive capable of hauling 50 tons up a grade of 2.5 per cent., find the figure 50 on the upper margin and follow directly downwards to a point midway between 2 and 3 per cent., which point rests very nearly on a curve that, being traced to its upper end, is found to represent a $12\frac{1}{2}$-ton locomotive. In like manner, to haul 50 tons up a grade of 1 per cent., it is found to require a 7-ton locomotive. From its curve, a 10-ton locomotive is seen to be capable of hauling 45 tons up a grade of 2 per cent., etc.

OPERATION OF ELECTRIC MINE LOCOMOTIVES

82. Motormen, or Motorneers.—It is well to emphasize the importance of employing experienced and reliable motormen. Each motorman should have a careful training. He should be cool and careful and should be so familiar with his machine and with the track over which he operates that he will almost instinctively know what to do in an emergency. He should keep his locomotive clean, especially the commutators, brushes, and windings of the motors; he should watch for unusual indications of burning or wear and should find and remove the cause or report to those who make repairs. The brushes should be kept clean, properly set, and at the proper tension. The locomotive wheels should not be permitted to slide.

An ammeter, which indicates the instantaneous consumption of power, should occasionally be placed on the locomotive, so that the motorman may learn to manipulate his controller in such a manner as to accomplish the desired results with minimum current. The careful and economical use of the current makes the service of the generators and motors much easier, thereby reducing the maintenance account and enabling the operation of more electrical machinery from a generator of given capacity, thus reducing the investment charges.

The fuse wire should be changed occasionally, but no larger sized wire should be used than that sent out by the manufacturers, or burned-out motors may result.

83. Shocks Caused by Grounds.—There is little danger of serious shock with the voltage ordinarily used in a mine; but if any portion of the electric circuit between the motors and the trolley should accidentally become connected with a part of the locomotive that is not in direct contact with the ground, a disagreeable shock may be had on handling that part. In such a case, the accidental connection, or ground, which may be caused by a nail or other piece of metal in a resistance box or controller, should be found and removed.

No powder, except it is in tight cans or boxes, should be carried in cars behind motors, as cases are on record where the locomotive has become insulated by dirt on the rails, and the current has passed along drawbars and couplings, and ignited powder in open cans on car bottoms, with fatal results.

84. Continuous Operation.—To use electric locomotives successfully in mine haulage, it may be necessary to have convenient partings and gathering stations. These should be arranged with easily operated switches so that running, or flying, switches can be made both inside and outside the mine. The secret of successful haulage by electric locomotive is in continuous operations. The motorman should learn to run his machine backwards for short distances without changing the trolley, or he will lose much time in switching, and he must always be on the alert to prevent loss of time by permitting the locomotive to stop. The haulage should be so arranged that there is no waiting for loads or empties, and it should be possible to start with a load the instant the last empty has passed into the parting.

85. Slipping of Wheels.—The folly of operating electric mine locomotives at loads great enough to cause their wheels to slip should be clearly understood, since the adhesion of slipping wheels is very small. For instance, a locomotive might be hauling a load at a speed of 6 miles per hour and exerting a drawbar pull of 5,000 pounds. If an attempt were made to accelerate the speed too rapidly to 9 miles per hour, the tractive power would increase until the wheels began to slip, when the drawbar pull would diminish and the train would be retarded instead of being accelerated. The slipping of the wheels relieves the motors and is a feature of proper locomotive design, but energy is worse than wasted in sliding friction, for in addition to the waste it wears out the wheels and rails. Considering the extreme hardness of the chilled-iron wheels, and the deep grooves often worn in them, it is evident that the amount of wasted energy is by no means insignificant. The rapid wearing of the wheels is also due to the too copious use of sand, which is a useful expedient, but one that should be adopted as seldom as possible. Sand is an insulator and it is possible to get so much on the track that it will insulate the motor entirely. However, the motorman should not allow wheels to skid and wear the rails when the use of sand will avoid such an occurrence.

86. Use of Controller.—If the locomotive is equipped with proper controlling apparatus, the remedy for too rapid wearing of wheels lies largely in the hands of the motorman. A heavy train of cars should always be started slack; that is, before starting ahead, the cars should be pushed together. Since the tractive power is sacrificed by the slippage of the wheels, the starting of a locomotive should be gradual, and too rapid acceleration should not be attempted. On an electric locomotive, the motors should be independent of each other, or in multiple when starting with a heavy load, as explained in Art. **39.** Resistances, or rheostats, of ample capacity should be provided and so arranged that the controller may be operated on the starting notches for considerable periods of time. A suitable controller in the hands of an intelligent motorman will enable him to start a heavy train with a minimum demand on the generator. On the other hand, if the controller is not properly designed, or if the motorman is inexperienced or careless, the results will be very unsatisfactory. If the controller is manipulated with sufficient rapidity, a locomotive, even with no load except its own weight, may be made to stand practically still and spin its wheels on the track, and the faster the wheels spin, the less their tractive effort, or tendency to propel the locomotive. This fact emphasizes the importance of approaching the other extreme, and of starting slowly, especially with a heavy train.

87. Speed.—The speed of an electric mine locomotive depends largely on the load, decreasing as the load or drawbar pull is increased. Conversely, a reduction of the load causes an increase in the speed, and a locomotive whose speed at its rated drawbar pulls is 8 miles per hour might run at 18 or 20 miles per hour with no load, unless resistances were inserted in the motor circuits, or unless the connections of the motors were changed from parallel to series.

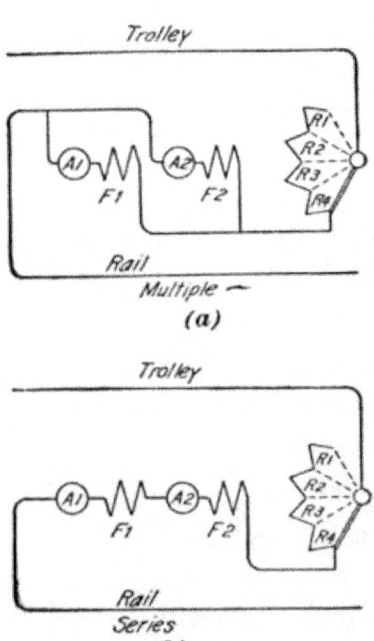

Fig. 31

88. Operating at Reduced Speed With Light Loads.—It is usually necessary to limit the speed when switching or running in the empties; otherwise the cars may be derailed. To keep the speed down, the motors could be run in multiple, with resistance in series; but this would not only cause the waste of a great deal of energy, which would be dissipated by heating the resistances, but the life of the resistances would be shortened. It is far better, when operating with light loads at slow speed, to run the motors in series; much less current will then be used and it will seldom be necessary to use the resistance.

The connections of the motors in parallel, or multiple, and in series were explained in Arts. **45** and **46,** but these may also be illustrated by referring to Fig. 31 (a) and (b). The armatures are designated as A_1 and A_2, the field coils as F_1 and F_2, and the sections of resistance as R_1, R_2, etc. Fig. 31 (a) shows the two motors connected in parallel for moving heavy loads; the rheostat is shown on the running notch, at the start it would be in the dotted position on the first contact R_1. Fig. 31 (b) shows the two motors in series and the rheostat on the running notch; this is the connection recommended for running with light loads at slow or moderate speeds.

89. Use of Circuit-Breakers.—Annoying delays are often experienced at mine haulage plants, where there are several locomotives, due to the frequent blowing of circuit-breakers. It is proper, of course, that the circuits should be automatically opened, when the generator is called on to deliver a current beyond its safe capacity, but it is annoying if the service is interrupted too frequently. The remedy for this trouble lies, to a very great extent, in the hands of the motormen, since the intelligent use of the controllers will minimize the demands on the generator. The trouble is aggravated by the simultaneous starting of all the locomotives when the circuit-breaker is reset by the engineer at the power house, since each locomotive requires a comparatively large current when starting, and the sum of their starting currents is often sufficient to immediately open the circuit-breaker again. In this way, it often happens that the circuit-breaker is thrown out several times in rapid succession, and the entire haulage system interrupted for some time. The engineer occasionally becomes indignant, allows the circuit-breaker to remain out a few moments before resetting it, and the relations between the employes of the power house and the motormen become strained, with no good results to the company. As a remedy, the following suggestions are offered:

1. Each locomotive should be equipped with an individual circuit-breaker, so set that it will open when the safe current

of the locomotive is exceeded, that is, at just about the slipping point of the wheels. The circuit-breaker should be installed in such a manner that it cannot be held in by the motorman and should be of such design that it cannot be tampered with, or *plugged*.

2. Each motorman should be given a serial number, and whenever the circuit-breaker has been reset at the power house, the locomotives should be started in succession. Motorman No. 1, for instance, should start as soon as the power is turned on, as indicated by the illumination of the headlights on his locomotive; No. 2 should follow, in say 15 seconds; No. 3, in 30 seconds, and so on until the entire haulage system is again in operation. This system may

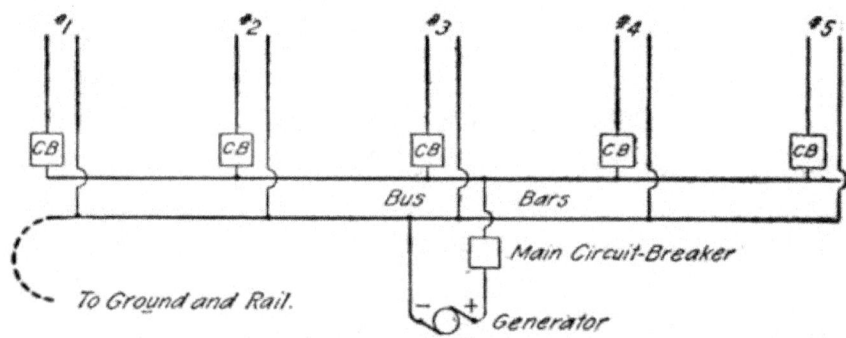

Fig. 32

seem elaborate, but the motormen quickly appreciate its value, and even the last man will prefer to wait 2 or 3 minutes rather than suffer a much greater delay, due to repeated interruption of the power.

3. Each feeder should have a separate circuit-breaker, so that unusual demand for current, or a short circuit, due to a fall of roof, for instance, may be localized and the interruption confined to the circuit on which the trouble has occurred. This suggestion is illustrated by the diagram in Fig. 32, where the locations of the generator and its circuit-breaker, the bus-bars, and the feeder circuits *1, 2, 3*, etc. each equipped with a circuit-breaker CB, are indicated. The diagram is intended to show the location of only the circuit-breakers; ammeters, voltmeters, etc. not being shown.

90. Schedule.—The schedule is as important a feature of a mine haulage system as of a street railway and has an important bearing on the size of the generator and engine. Assuming two locomotives, it is obvious that a smaller generator would be required if while one locomotive is hauling out a loaded trip, the other is hauling in a trip of empties, than if both were hauling loaded trips at the same time. Where several locomotives are operating simultaneously from the same generator with the schedule arranged to the best advantage, the load factor is from 50 to 70 per cent.; that is, the average load and capacity of a generator is from 50 per cent. to 70 per cent. of the aggregate full-load-kilowatt capacity of all the locomotives, assuming the line loss to be not more than 15 per cent. The generator and the locomotive both possess large overload capacities, but it is well to be liberal in generator capacity to provide for future extensions of the haulage system and other uses of the current that are sure to follow, such as driving coal cutters, drills, pumps, hoists, etc.

91. Care of Locomotives.—The maintenance account is one of the principal evidences of good management, and eternal watchfulness is the only means by which it may be reduced to its minimum. Mine-locomotive service is even more severe than ordinary street-railway work, and the proper care of the apparatus is of correspondingly greater importance. The locomotive should be kept in a dry and secure place at night and covered with tarpaulin; neglect of this precaution leads to grounded circuits, or burnt insulation and charged motors. A chain block hung from the roof should be provided for changing armatures and other parts. Each locomotive should be run over a pit and should be carefully inspected and cleaned for the next day's work; while an extra-thorough cleaning should be given the locomotive at least once a month. In this way, loose nuts and bolts, worn-out gears, thin bearing linings, and a number of other minor defects will be discovered and remedied before they become the source of more serious trouble. The efficiency of the inspection system may be greatly increased by carrying in stock in the repair shop at the mines, a liberal supply of those parts of the locomotive that experience has shown to be subject to repair or renewal. The stock need not be large, but it should be maintained by placing frequent orders. A spare armature or two should always be carried in stock, and each of the armatures should be regularly renewed and then thoroughly cleaned and repaired and given a coat of insulating paint.

92. The electric mine locomotive has been developed within a few years and the present type is the result of a

> Motormen must carefully inspect their motors after each day's run, and report their condition on this form. They will be held responsible for every defect not reported. No attention will be given to verbal reports or reports not signed by Motorman.

> **West Virginia Central and Pittsburg Railway Co.**
> MINING DEPARTMENT (Thomas Plant).
> *REPORT OF CONDITION OF MOTOR_____ 190_*
>
> *Motor No._____ was carefully inspected by me this P. M. and found to be in need of the following repairs, viz.:_____*
>
> _____ *Motorman.*
>
> NOTE.—If no repairs are needed, return this blank so marked.

series of changes and improvements in design and construction. Those features of the older types of locomotives that constitute the supply parts, become obsolete at the factory, and for that reason a long time is often required to fill orders for the renewal of such parts. The work of inspection should devolve on a careful and painstaking man of some mechanical ability, and if there are many locomotives, he, with his assistants, should be employed as an all-night force. The inspector should also be required to make a daily report on suitable blanks similar to that shown above, covering the condition of each locomotive. This may seem to be an unnecessary refinement, but it interests the inspector in his work, places before the management the result of his labors, and constitutes a powerful incentive to faithful service.

93. Summary.—The following suggestions may be offered in regard to an electric mine haulage plant.

1. Purchase locomotives of reliable manufacturers, who can offer machines that are the result of their experience either with mine locomotives or in street-railway work, where in some respects analogous conditions exist.

2. Purchase locomotives of ample capacity and do not attempt to operate them at continual overloads.

3. Be sure that the controlling apparatus is of proper design and so arranged that the locomotive may be started slowly and operated for at least two speeds with no resistance in circuit.

4. Equip each locomotive and feeder circuit with an individual circuit-breaker.

5. Lay heavy rails, especially in the main entries, and use as little sand as possible.

6. Employ intelligent and reliable motormen and pay them sufficiently high wages to make it an inducement for them to interest themselves in their work.

7. Last and most important, establish a rigid system of inspection and carry a liberal stock of repair parts.

94. Cost of Electric Haulage.—The cost of operating an electric haulage system depends very largely on local conditions. The example on page 70 taken from actual practice is given merely to show the items usually considered as determining the cost, and is not intended as an average test of such haulage. The haulage is charged, on the pay roll, with one-half the electrician's services, one-fourth the engineer's, and one-third the firemen's, as their duties also include the care of other machinery than the generators, locomotives, and the haulage system. The electric haulage operates only on day shift, while the air and other electric machines, such as electric pumps, coal cutters, etc., work both day and night, hence one-third the cost of fuel is charged to haulage account. The average length of haul is about

8,000 feet and three 12-ton locomotives are employed. For the particular month taken, the output was 38,000 tons of coal. Cost per month:

$\frac{1}{2}$ electrician's salary, at $90	$ 45
$\frac{1}{2}$ day engineer's wages, at $70	35
$\frac{1}{3}$ firemen's wages (3, at $50)	50
3 motormen, at $50	150
3 trip runners, at $45	135
Supplies (oil and repairs)	35
Depreciation and interest (12 per cent. per year)	250
Total cost per month	$700

$$\frac{\text{Cost}}{\text{Tonnage}} = \text{cost per ton} = \frac{700}{38,000} = 1.842 \text{ cents per ton}$$

Fuel cost per ton . . . . 1.5 cents
$\frac{1}{3}$ fuel cost $+ 1.842$ cents . . . . 2.342 cents per ton

95. For gathering to the motor haulage, twenty-eight mules and drivers are employed. The drivers are employed 20 days at $2.10 each, total $42 each, total for the month $1,176. The cost per mule per day, covering depreciation, feed, harness, and attendance is approximately 50 cents each, or $28 \times 50 \times 30 = \$420$ per month; making a total cost of $1,596, or 4.2 cents per ton for gathering, or a total from room to tipple of 6.54 cents per ton. The repairs on this plant, which has been in constant operation for nearly 5 years, have been very slight, probably about $\frac{1}{10}$ cent per ton of coal handled. In no case, with a plant properly designed and installed, should this item exceed $\frac{1}{5}$ cent per ton.

HAULAGE
(PART 4)

ROPE HAULAGE

INTRODUCTION

1. Definition.—A rope-haulage system, applied to mining, is one in which the cars are attached to a rope, and both are moved by means of an engine, motor, or by gravity. Usually, rope haulage is applied to nearly horizontal tracks; but if the haulage is on an inclined plane, the gravity pull of the descending load will greatly assist the motive power and in some cases will be sufficient to move the entire system.

In the first rope-haulage plants, the cars were attached to hemp or manila ropes, or in some cases to iron chains; however, wire ropes are now used almost universally, as they are more economical and serviceable.

2. Rope-Haulage Systems.—There are four systems of rope haulage:

1. *Endless-rope haulage*, in which the rope travels continuously in the same direction, taking empty cars into the mine and bringing loaded cars out.

2. *Tail-rope haulage*, in which there are two ropes, a main rope and a tail rope, attached, respectively, to the outby and the inby ends of a trip of cars. The tail rope is used for drawing the empty cars into the mine and the main rope for bringing the loaded cars out, the ropes thus traveling together alternately in opposite directions.

Copyrighted by International Textbook Company. Entered at Stationers' Hall, London

3. *Gravity planes*, or *self-acting inclines*, arranged so that a loaded car in descending the plane will pull up an empty car.

4. *Engine planes* are of such an inclination that cars will descend the plane by gravity and pull the rope after them, but an engine must be used to haul the cars up the plane.

ENDLESS-ROPE HAULAGE

3. Distinguishing Features.—An endless-rope haulage system is one in which the rope travels continuously in the same direction, passing from the haulage engine drums to a deflecting sheave at the other end of the system and back again to the drums. The ends of the rope are fastened together and it thus forms an endless belt to which the cars are attached by suitable couplings.

As usually installed, the system has two tracks arranged side by side in the main entry; but in case the roof is bad, it may require that one track be placed in each of two parallel entries, one for the loaded cars and one for the empty cars.

4. Description of Endless-Rope System.—Fig. 1 shows, in plan and elevation, an endless-rope haulage system. The rope, in traveling its circuit, passes back of the drums a and b to the balance car c. At this point, it is given a half turn about sheave d and returning past the drums, is guided into and through the mine by means of suitable deflecting sheaves and rollers to the tail-sheave e. The rope is given a half turn about the tail-sheave e, which, in this case, is also mounted on a balance car f and is then returned, by the aid of rollers and deflecting sheaves, to the drums, thus completing the circuit. The object of placing the sheaves d and e on movable balance cars c and f is to take up any slack rope due to stretching. If the sheaves were fixed and the rope should stretch, it would not grip the drum sufficiently tight to move the cars. It will be noticed that the rope passes under pulleys g, over pulleys h, and to one side of pulleys i for the purpose of deflecting the rope and giving it alinement. In the center of the mine track, rollers j are placed at intervals to keep the rope off the ground. The

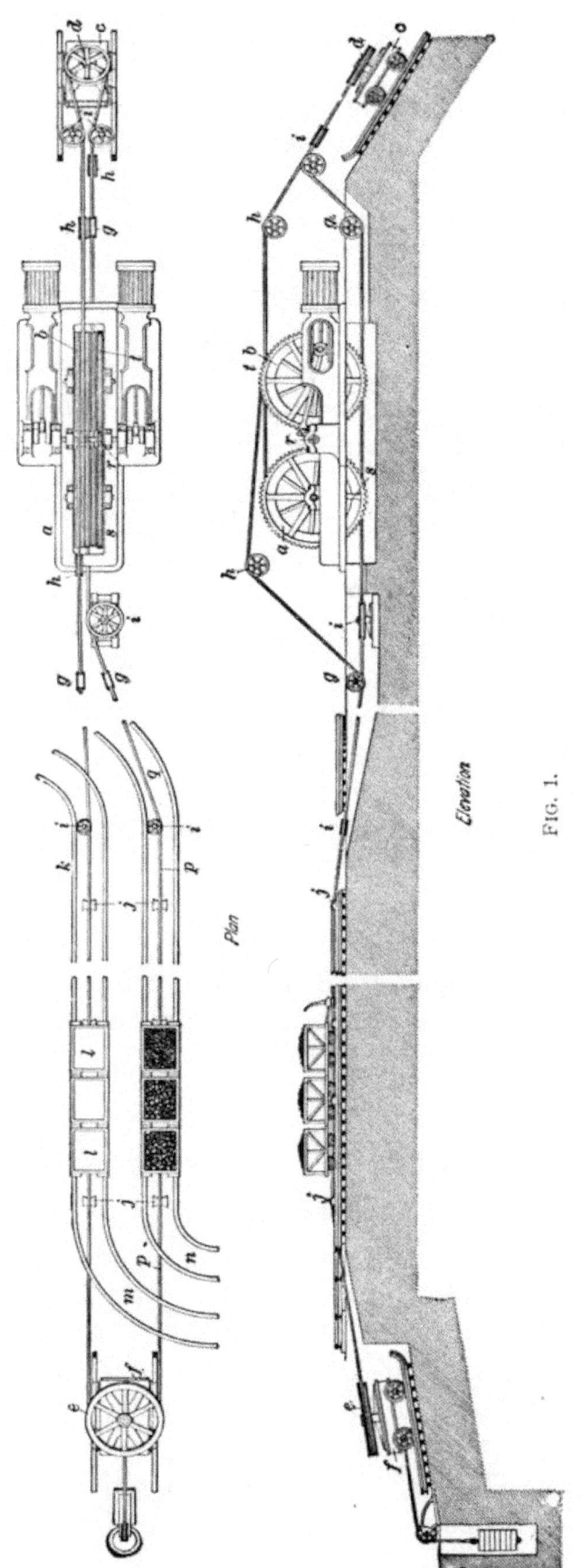

Elevation

FIG. 1.

rope k, in going into the mine, pulls the empty cars l to curve m, where they are unhooked and distributed to the working places. At this same point, but on track n, the loaded cars are attached to the outgoing rope p and hauled to the curve q, where they are unhooked and delivered to the men who see that they are sent to the proper unloading place.

5. Advantages of Uniform Load on Engine.—An endless-rope haulage system is particularly adapted to mines where the main haulage roads are level or have a light uniform grade throughout their length. The cars should be distributed at regular intervals along the entire length of the rope so as to make the load on the engine uniform at all times. There are, however, numerous delays in delivering cars to the haulage system, and it is practically impossible to thus keep the cars regularly spaced, the result being that with a variable grade empty or loaded cars may accumulate on an ascending or a descending grade, thus throwing a variable load on the engine and making it very difficult to regulate it for any given pull.

Where the grades are variable, the rope will lift from the rollers in low places and frequently lash to such an extent as to throw the cars from the track; while on an up grade, the rope bears heavily on the rollers, producing wear and greatly increasing the friction.

POWER FOR DRIVING ENDLESS-ROPE SYSTEM

6. Endless-Rope Engines.—The engines for an endless rope haulage may be of the plain slide-valve or Corliss-valve types, the latter being preferred in case the grades are variable, since they are most economical in the use of steam and are automatic in their action. Second-motion engines are used for this system of haulage. They are run continuously in one direction during working hours and where the loads are variable they should be supplied with flywheels and governors and use steam expansively.

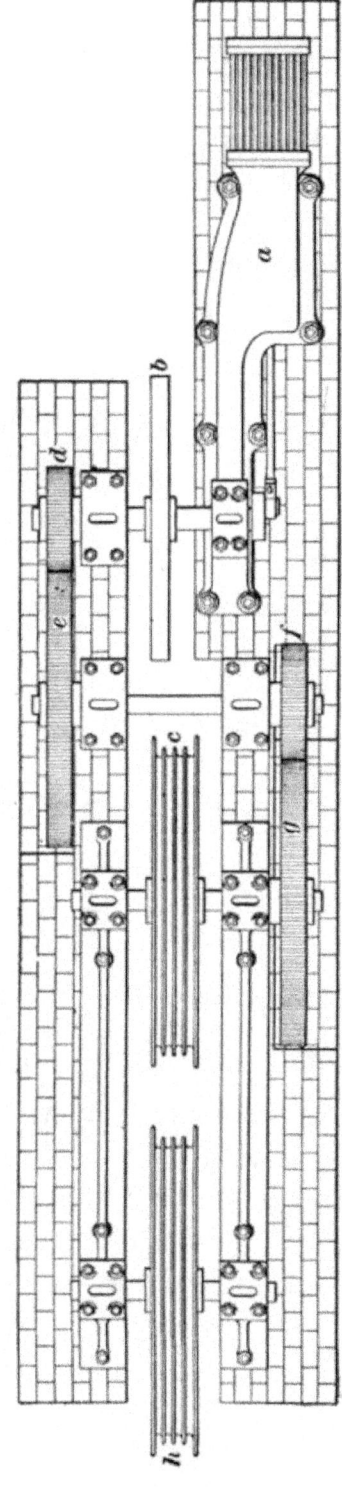
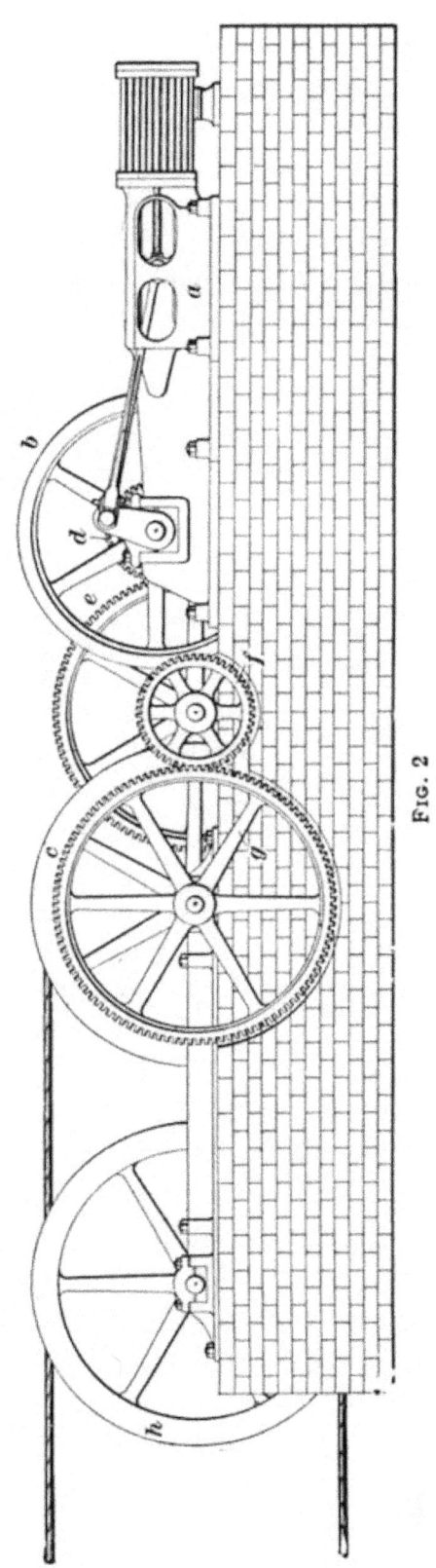

Fig. 2

Fig. 2 shows an engine *a* provided with a flywheel *b* and suitable gear-wheels *d*, *e*, *f*, *g* for driving the rope-haulage drums *c* and *h*. If the gears *d* and *f* are of the same size and *e* and *g* are of the same size, and the diameter of *e* is four times that of *d*, the engine shaft must revolve sixteen times in order to revolve the drum *c* once, thus reducing the speed of the haulage rope, which in the endless-rope system varies from 100 to 400 feet per minute. The drum *h* is a follower;

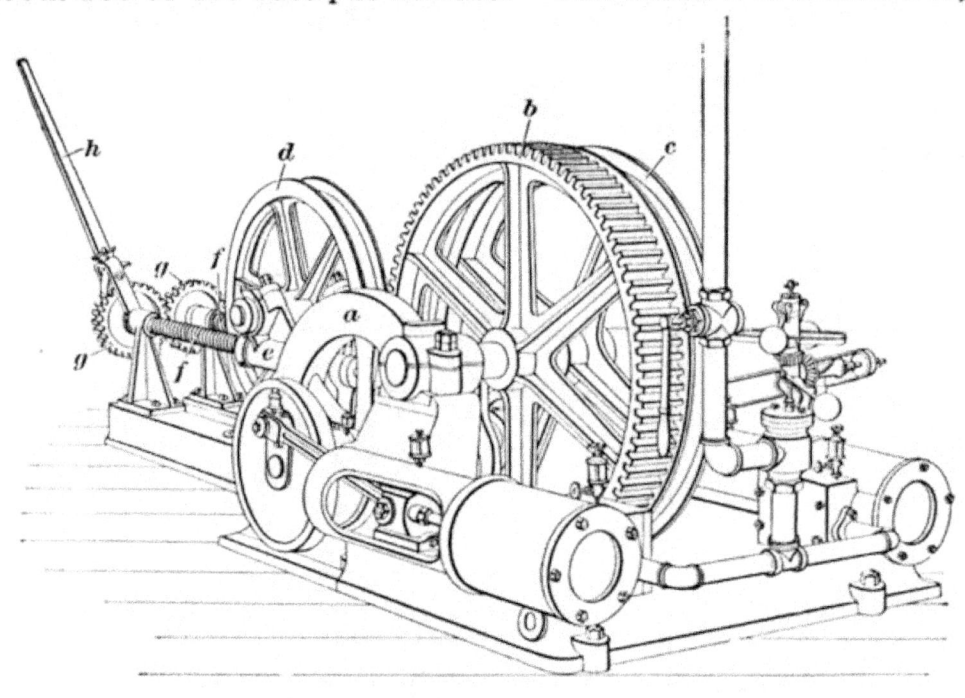

Fig. 3

that is, it is revolved by the rope unwinding from the driven drum *c* and not by gearing.

The engine shown in Fig. 1 has both drums driven by a pinion *r* on the crank-shaft placed centrally so that it will mesh with the two gear-wheels *s* and *t*. By thus placing one of the drums back of the engine shaft, a compact engine is obtained; but, when possible, it is advisable to allow more space for the drums.

Fig. 3 shows a pair of endless-rope haulage engines, *self-contained;* that is, mounted on one bedplate. There is one flywheel *a* for both engines and on the engine shaft there is a pinion, not seen, that drives the large gear *b* and the drum *c*.

The small drum or follower d is set on movable pillow-blocks e that can be moved back and forth along the feed-screws f by means of the ratchet wheels g and the dog h. The follower drum d is a tightener that puts the necessary tension on the rope to insure a firm grip of the rope on the drums. In this case the follower takes up some of the slack due to the stretching of the rope.

7. Endless-Rope Drums.—There are usually two drums on an endless-rope engine, arranged tandem on separate shafts. They vary in diameter according to the size of the rope, size of the engine, and length of haul. The drums and sheaves should be proportioned to the diameter of the rope, particularly where sharp bends in the rope are required. This is oftentimes neglected, resulting in greater wear of the rope.

The driven drum, or follower, is sometimes independent of the engine and is made smaller than the driving drum; it is also sometimes permitted to run loose on its shaft so that the rope will lead properly from one groove to another. Drums with a concave rim are sometimes used, but there is with such a drum considerable surging and jerking of the rope, and grooved drums are preferable. The best results are obtained when the drums are placed with their centers 12 feet apart, in order that there may be a slight sag to the rope and a better bite on the drums.

8. Grip and Tension.—When ropes are coiled around wheels, the hold, or what is called the **grip** of the rope increases directly as the square of the number of coils. The same law holds true if there are two wheels placed tandem; for example, when there are two half coils on each wheel, they are equal to two coils on one wheel and the grip is four times greater than the grip of one coil on one wheel or one-half coil on two wheels. Tension is stress caused by pulling.

9. Balance or Tension Cars.—A new wire rope may stretch as much as 1 to $1\frac{1}{2}$ per cent. of its length, according to the load hauled. The rope for an endless-rope haulage system 1 mile long would be 2 miles in length, and assuming the stretch to be $1\frac{1}{2}$ per cent. this would amount to $5,280 \times 2$

× .015 = 158.4 feet. There may also be a slight contraction or expansion of the rope due to changes in temperature. Were no provision made for this increase due to stretching and temperature, or a decrease due to temperature, the rope would be slack at one time and tight at another, and if the rope were laid around a fixed tail-sheave and the rope should expand, it would fall off the sheave. The stretch of the rope does not take place at once but is gradual, and both it and the variation due to temperature should be provided for by means of balance cars, or weights, that act automatically to take up slack so that at no time is the rope too slack.

The balance car c shown in Fig. 1 compensates for the shortening and lengthening of the rope due to changes in temperature and to the stretching due to the load. If it is not practicable to make the track on which the car runs long enough to provide for all the stretch, a piece can be cut off the rope when the rope stretches too much.

10. Instead of the balance car being on an inclined track,

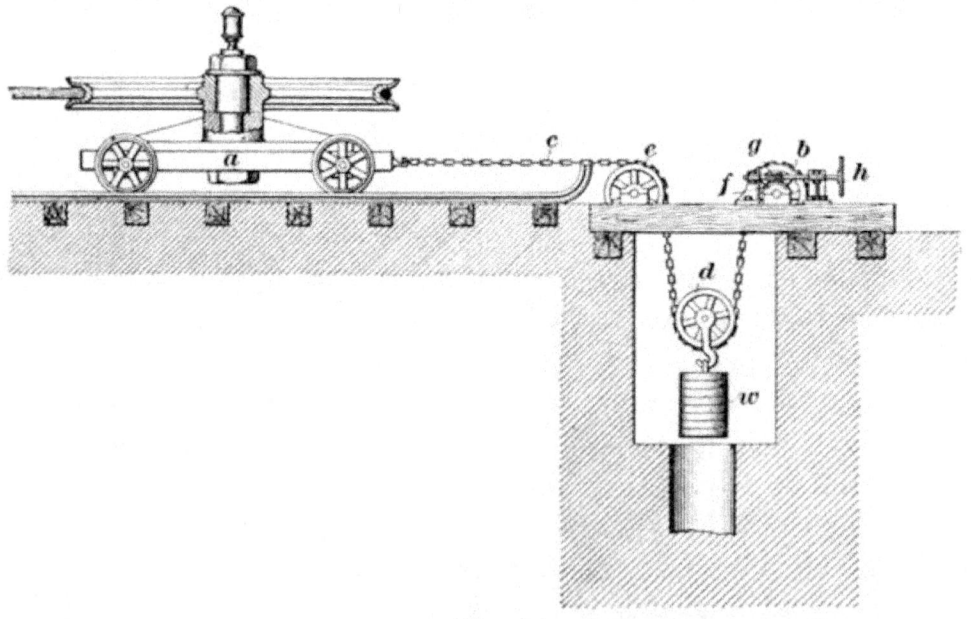

Fig. 4

it may be on a level track, as in Fig. 4. Any variation of the rope in this case is compensated by means of a weight w. The sheave is fastened to a movable carriage a, consisting of a frame mounted on wheels. To the carriage is fastened one end of the chain c, the other end being fastened to a small drum b. In a pit near the balance car is suspended a weight w, by which a tension is maintained in the wire rope, and which always tends to move the car toward the right. The weight is hooked to a wheel d that turns on the chain c, the latter being led over the deflecting sheave e. On the shaft carrying the drum b is rigidly fastened a worm-wheel f meshing with a worm g. This worm-gearing is operated by a hand wheel h. After the haulage rope has stretched enough to cause the weight w to hang low in the pit, the weight is raised by turning the hand wheel h, which causes the drum to revolve through the intervention of the worm

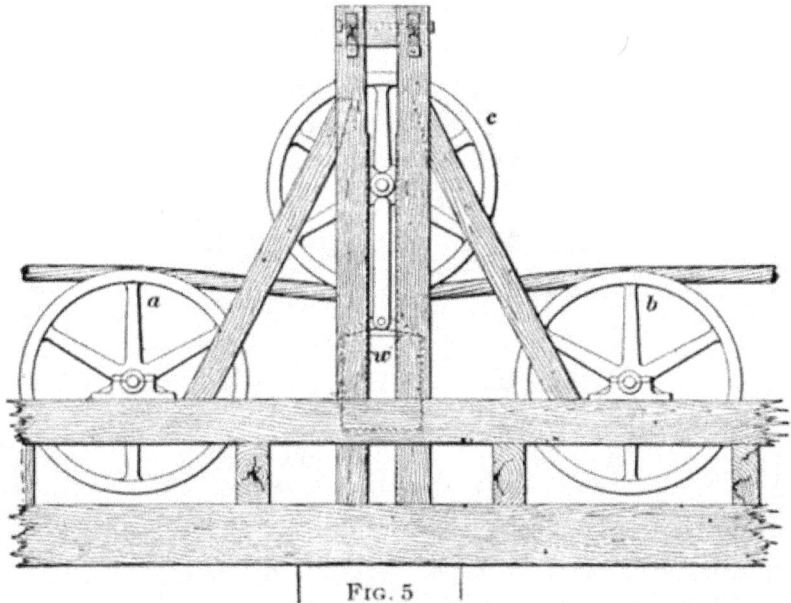

Fig. 5

and the worm-wheel, thus shortening the chain c. The weight w in this case has a movement of only one-half that of the carriage, but must be twice as heavy as one that simply hangs at the end of the chain. The rails on which the balance car travels should be turned up at their ends, to form horns beyond which the car cannot pass. It is well to have a second rope or chain lying loose, having one end fastened to the car and the other to the drum b, so that in case the first breaks the second may be quickly substituted.

11. Compensating Sheaves.—When it is required to provide compensators for the expansion and contraction of the rope only, arrangements as shown in Figs. 5 and 6 are used, and the permanent stretch of the rope is taken up by other means. In Fig. 5, the rope is led over sheaves a, b and firmly held by suitable supports. Between these sheaves is a timber guide frame so constructed that a sheave c can work up and down. The sheave c has a weight w suspended from it that keeps the rope taut. When the rope expands, the sheave c and weight w descend; and when it contracts, they ascend.

12. In Fig. 6, a somewhat different arrangement is shown. Here, cast-iron standards having suitable bearings are provided for the sheaves a and b. At the upper extremity of the larger standard is located a lever c, carrying a sheave d at one end and a counterpoise w at the other. The counterpoise is so constructed that it can be moved and the rope kept taut.

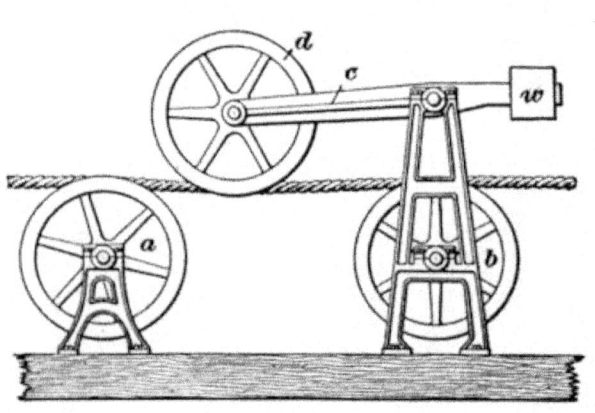

Fig. 6

13. Fig. 7 shows a take-up device used by the Pittsburg Coal Company at its Banning Mine No. 2. The pulling rope enters the engine room, passes over and around the two drums and back over the sheave wheel a on the carriage b and returns to the mine. The carriage b moves forwards or backwards on the track c. At the rear of the carriage is a horizontal pulley d around which the tension rope e passes, and both ends of this rope pass under pulleys f, then up over pulleys g at the top of the tension tower down to the pulleys h attached to the weight w, up over pulleys i, and then to the winch j. The tension tower is of such a height that the weight w can be raised or lowered about 20 feet, thus allowing a considerable travel to the carriage b. The weight w is sufficient to take up and hold the slack of the haulage rope. If the rope stretches much, so as to let the weight

Fig. 7

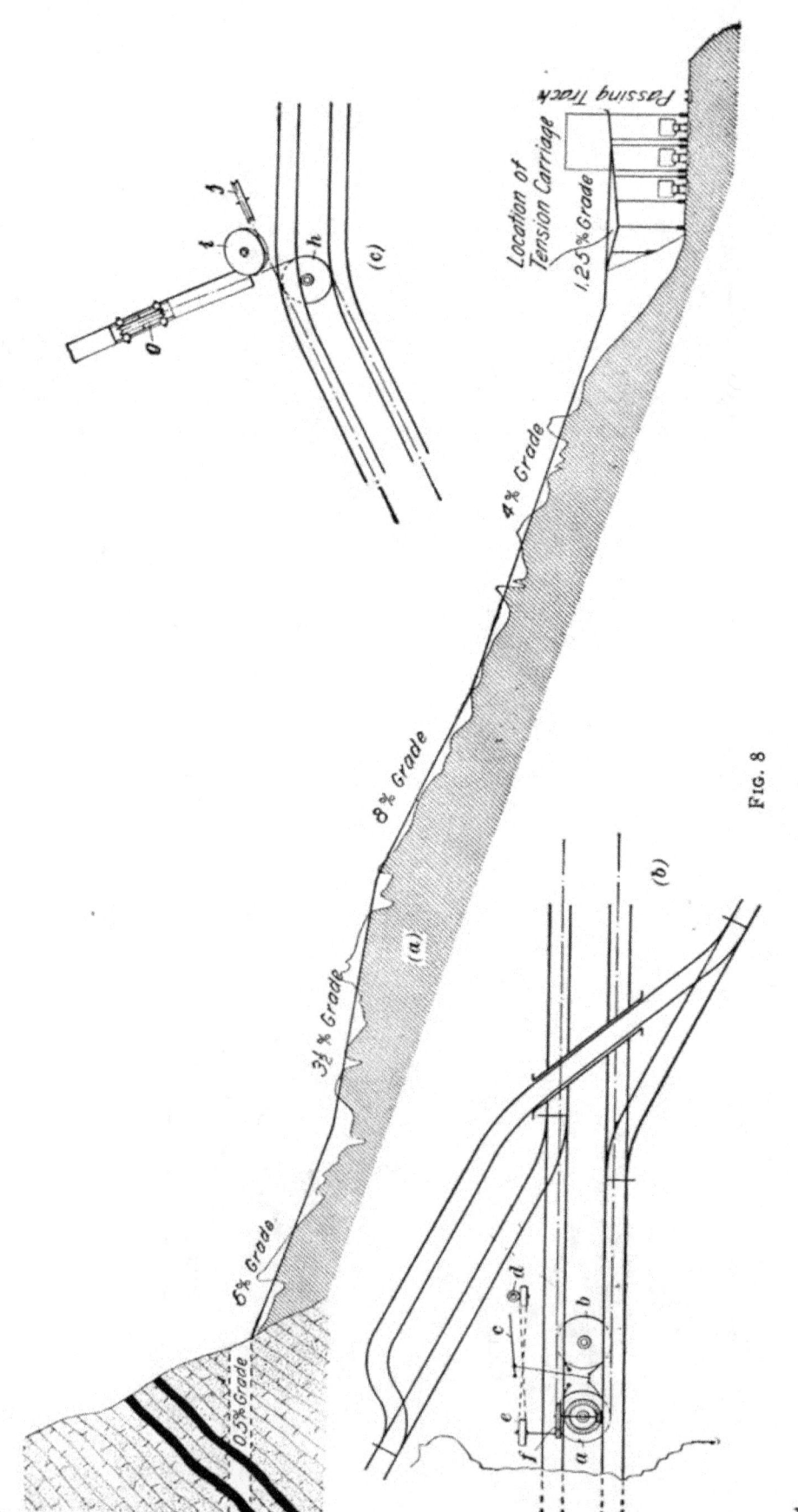

Fig. 8

fall too low, it can be shortened and the weight raised to the proper height by means of the crab winch j operated by the handle k through two sets of gearing l and m.

COMBINED ENDLESS ROPE AND GRAVITY PLANE

14. Fig. 8 shows the application of an endless-rope haulage system to a gravity plane as adopted at the Cuatro Mine, Colorado. A profile of the road is shown in (a) and the terminal arrangements in (b) and (c). At the head of the plane, the rope passes about two sheave drums a and b 72 inches in diameter, which are controlled by brake bands operated by the brake lever c. The drums are operated by gears f attached to the shaft of the pulley wheel e that is driven by a belt from a small engine d. The inclination of the plane is such that only a very small amount of power is required to maintain a speed of 3 to 4 miles per hour, and if the empty and loaded cars are properly spaced the system becomes self-acting. At the bottom of the plane, Fig. 8 (c), the rope passes about the deflecting sheaves j and i, then to a sheave g that is mounted on a tension carriage, then back to a sheave h that guides the rope to the empty track. The cars are attached to the rope by **S**-shaped grapple hooks, termed goosenecks, fastened to the tops of the cars. The rope is supported on ordinary rollers between the tracks; but when the cars are spaced for the full capacity of the mine, the rope does not sag between the cars so as to allow the rope to touch the rollers. With the rope running 3 miles per hour and with 2-ton cars spaced 300 feet apart on the rope, the capacity of the system is a car every $1\frac{1}{8}$ minutes, or about 1,000 tons per day.

DISTRICT HAULAGE

15. While in many cases endless-rope haulage is applied only on the main entries, there are cases where it is extended to cross-entries. This may be accomplished, as shown in Fig. 9, by placing a deflecting sheave m at the entrance to the cross-entry and carrying the rope down this entry to a terminal sheave at the inby end of the parting where trips are made up, and thence back to the main entry, passing over the deflecting sheaves b in the center of the track. While the system is more or less successful, it wears the rope, and the uncoupling and the coupling of cars that must pass the entrance of the side entry is unsatisfactory work,

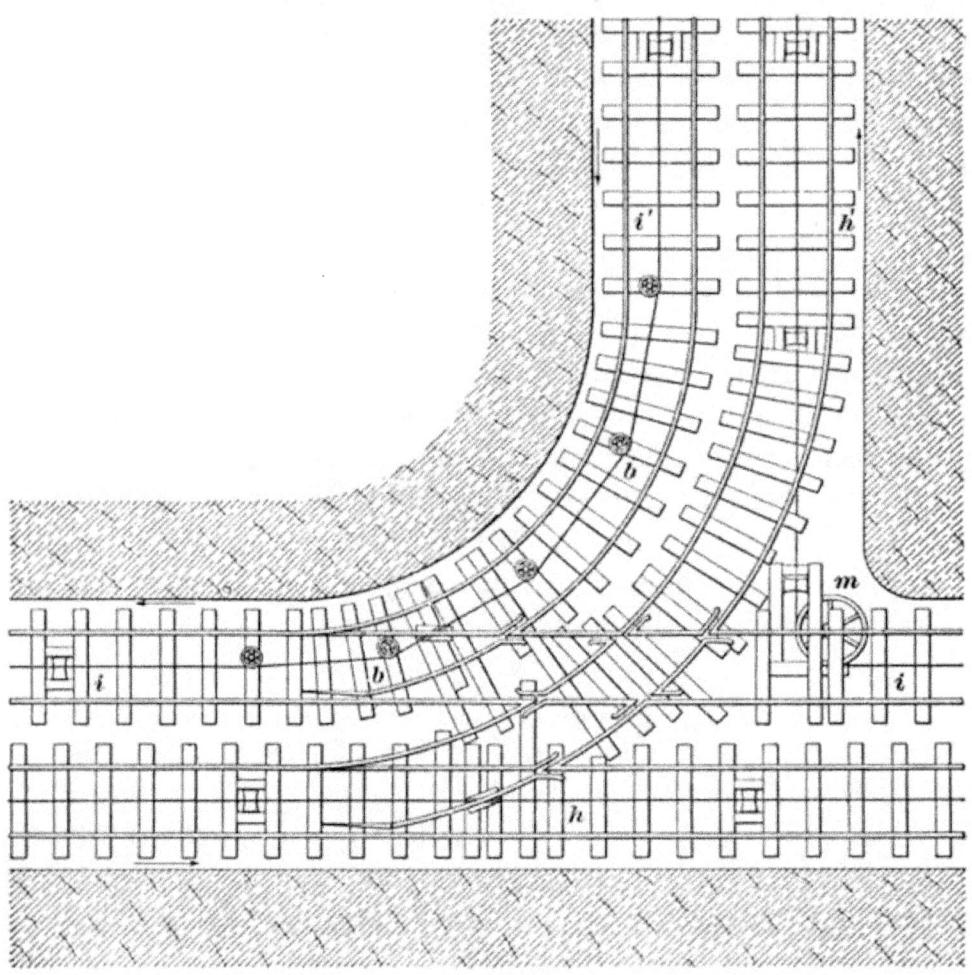

FIG. 9

and besides interferes with the work on the main haulage road. Any accident to cars or apparatus on the cross-entries stops the entire system. The arrows in the figure show the direction in which the rope runs, thus h is the empty track and i is the loaded track on the main entry while h' is the empty track and i' the loaded track on the cross-entry.

A somewhat similar system has been used in Illinois, but instead of a double track on the cross-entry the rope passes up one cross-entry around sheaves placed at a suitable cross-cut between the two entries, then back on the other entry to the main entry. Two stations are made at the end of the rope haul in the cross-entries, one for empty and one for loaded cars. Mules haul the cars to and from the stations and the working faces. The rope, after passing out of the cross-entry, travels along the main entry to the next cross-entry and is deflected so as to pass into that entry and out of the parallel entry. A space on the main entry between any pair of cross-entries with this system has no rope; consequently, the track must be graded so that the loaded cars from farther in the mine can be run by gravity across this space in order to again grip the rope at the next entry. Probably this system is more unsatisfactory than the former, since it will interfere with ventilation and be more expensive than a double-track entry. Besides the objectionable features noted, the variable tension on the rope causes the cars to travel with a jerking motion and the longer the rope is made, the heavier must the rope be for increased haulage, and jerks become more pronounced as the rope becomes heavier. It may also not be practicable or economical to grade the track so that the trips will run by gravity along the entry where there is no rope. For these reasons, the use of a continuous rope for main and district haulage is not to be recommended.

16. The haulage system shown in Fig. 10 (a) and (b) has the advantage that the rope on each pair of cross-entries can be operated as needed. In Fig. 10 (a), H and I are the main-entry tracks. H_1 and I_1 are the cross-entry tracks. The empty cars for the cross-entry are switched from track H to track H_1 and the loaded cars from track I_1 to track I. At m on the main loaded track, there is a grip wheel around which the loaded rope makes two turns for the purpose of giving this wheel sufficient power to work an endless-rope system on the cross-entry M.

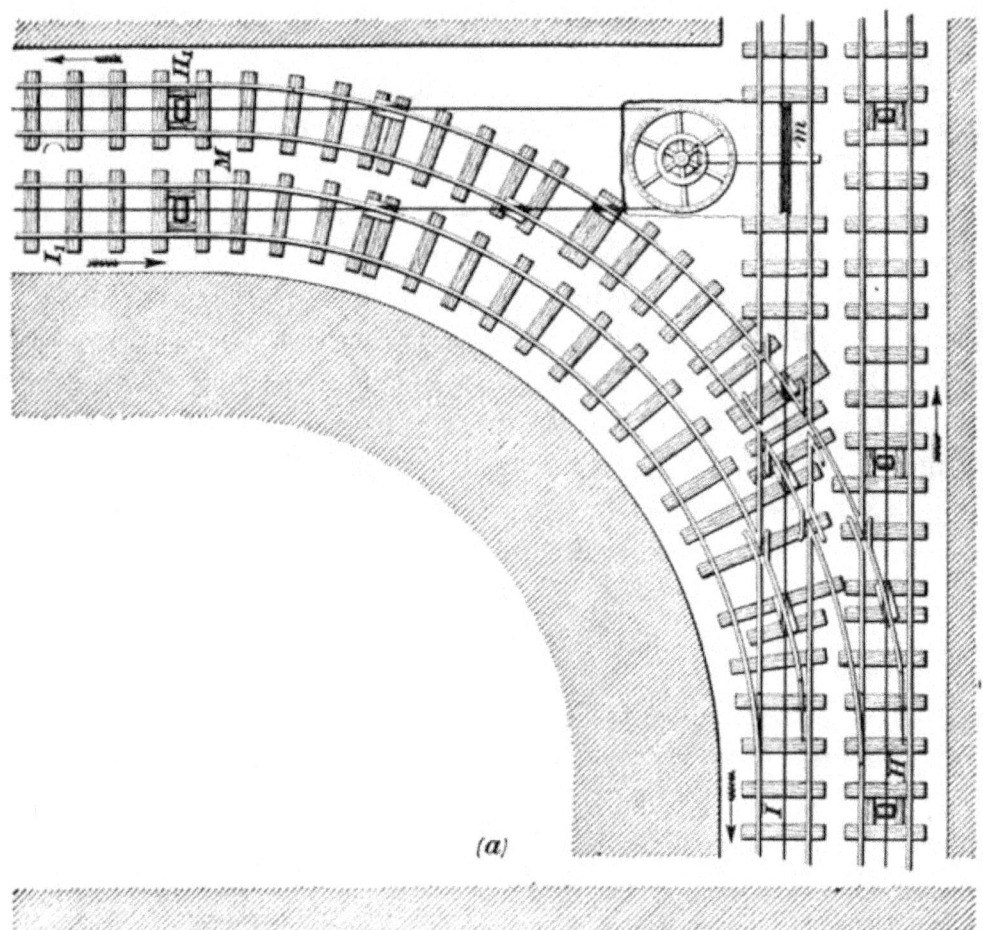

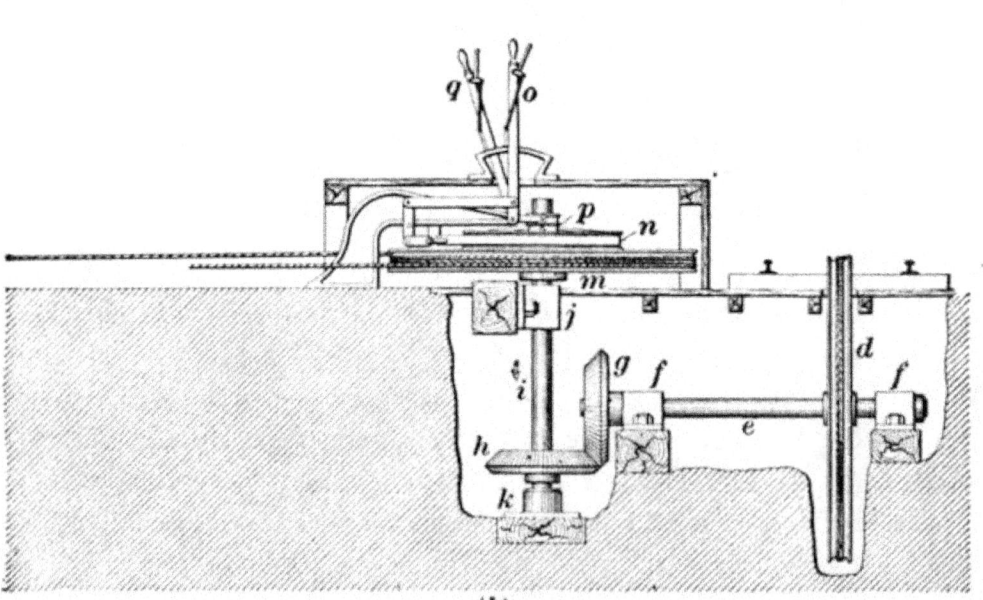

Fig. 10

An elevation of the mechanism necessary to work the rope on the cross-entry is shown in Fig. 10 (*b*), to consist of a grip wheel *d* on a shaft *e* provided with suitable journal-boxes *f*. On one end of the shaft *e* there is a bevel gear *g* that meshes with a gear *h* on the shaft *i*, which also has journal-boxes *j* and *k*. To the shaft *i* is keyed a grip wheel *m*, a brake wheel *n* worked by the lever *o*, and a friction clutch *p* worked by the lever *q*. The grip wheel *d* moves continually, while the grip wheel *m* moves only when the clutch *p* is thrown into gear. If this were not done it would be necessary to keep the rope on the side entry moving all the time. The pull on the main haulage rope can be reduced by this method, for it seldom happens that all the ropes will be needed on the cross-entries at the same time. Since the main entry receives cars from several cross-entries, there is more haulage on the main than on the cross-entry; it is necessary, therefore, for the main-entry rope to travel with more speed than the cross-entry rope, and the reduction in speed is effected by gears *g* and *h*.

17. Since the whole traffic of the mine depends on the main haulage rope, this rope should be large and strong, or otherwise, after it has been in use a short time, it is liable to break and cause delays, which are always a serious matter in mine haulage. Lighter ropes may be used on cross-entries, which will lessen the stress on the main haulage rope, but it is customary to use on side entries those parts of main haulage ropes that are worn but are sufficiently strong for the work on those entries though not for the main entries. This is doubtful economy in material, since the heavy rope on the cross-entries will increase the load on the main haulage rope. The size of the main haulage ropes may be decreased if independent motors operated by compressed air or electricity are used for moving the ropes on side entries. The advantages to be secured by this latter system are: energy is not wasted when no work is done on cross-entries; the haulage may be done quickly or slowly as the case demands. The loss in transmitting energy for separate motors is compensated by the saving in energy required to bend the stiff main haulage rope around the grip wheels when the side-entry ropes are run by power from the main rope.

18. Grips.—The cars in an endless-rope haulage system are attached to the rope by short chains, tongs, grips, bars, or hooks. There are several kinds of **grips** of which a

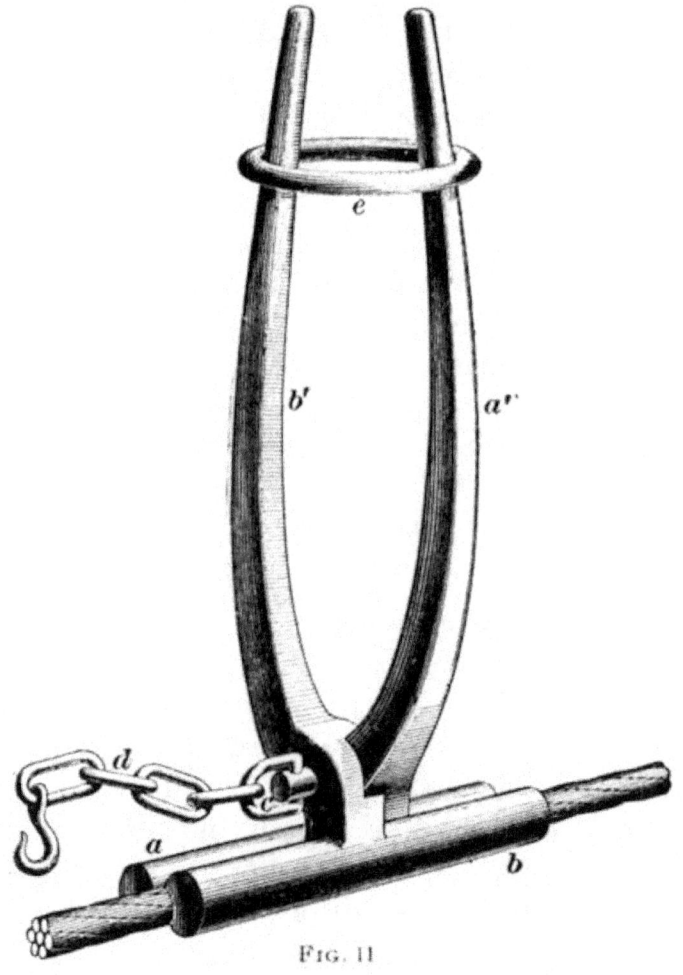

Fig. 11

common form is shown in Fig. 11. It consists of two jaws a and b, having handles a' and b', respectively. The jaws are held together by a bolt c, having a nut on its other end. That part of the bolt that passes through the handles is of smaller diameter than the part to which the chain d is fastened; a shoulder, therefore, is formed between which and the nut the jaws are firmly held. The grip is fastened to the car by the chain d, the hook being attached to the drawbar of the car. Each jaw is made concave inside so that the rope can be held between them. The friction between the jaws of the grip and the rope being greater than the resistance of the car, the latter is moved along with the speed of the rope. To hold the jaws so as to grip the rope and haul the load, a link e is wedged over the handles. Notwithstanding that this kind of grip is very simple in construction, it is cumbersome and can only be used for hauling light loads.

19. Where heavy loads are to be hauled, it is necessary that the construction be such that a great pressure can be

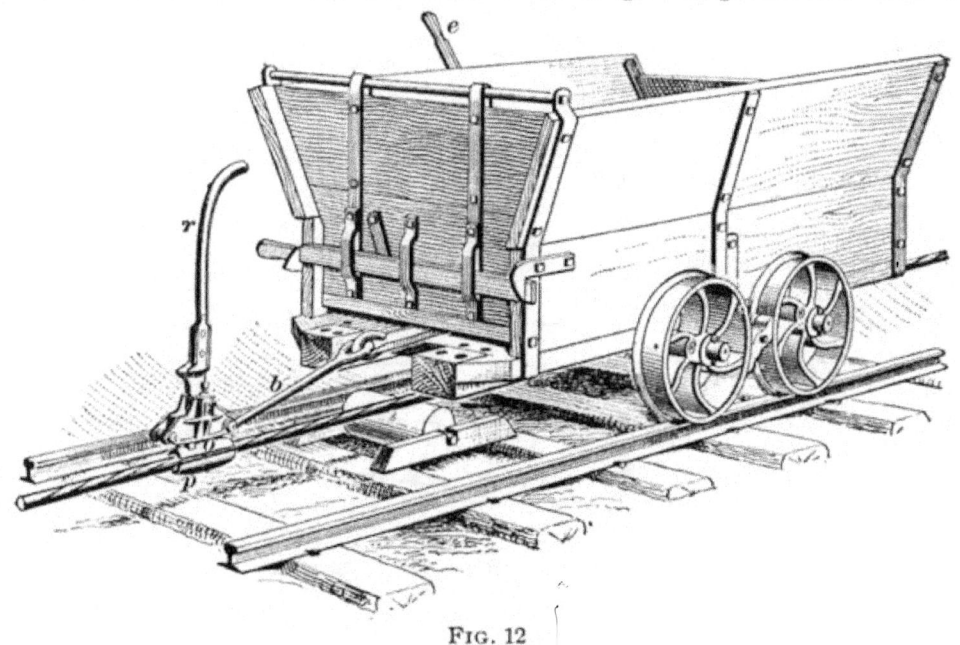

Fig. 12

exerted on the cable by the jaws of the grip. A grip that will meet this requirement and one that is extensively used, is shown in Fig. 12. The car is attached to the rope by means of the grip and the drawbar b. The lever r forming part of the mechanism of the grip is moved backwards to open the jaws $o p$ and raised to the position shown to close them about the rope. In order to stop the cars, when they are uncoupled from the rope, it may be necessary either to use the brake lever e or to insert sprags in the wheels. The details of this grip are shown in Fig. 13 to consist of two plates *m* and *n* bolted together, so as to form a recess in which the shank *b* of the upper jaw *o* can slide up and down. The lower jaw *p* is part of the plate *m*, and does not move. Between the upper extremities of the plates there is a lever *r* that operates the movable jaw *o*; this lever has trunnions *s, s'* that work in bearings *t, t'* fitted to the plates. That end or part of the lever that works between the plates is of such a shape that, when the lever *r* is raised upright, it forces the jaw *o* downwards and causes it to grip the rope firmly. After the jaws *o* and *p* are worn, so that the pressure required to haul the load cannot be secured by raising the lever *r*, the

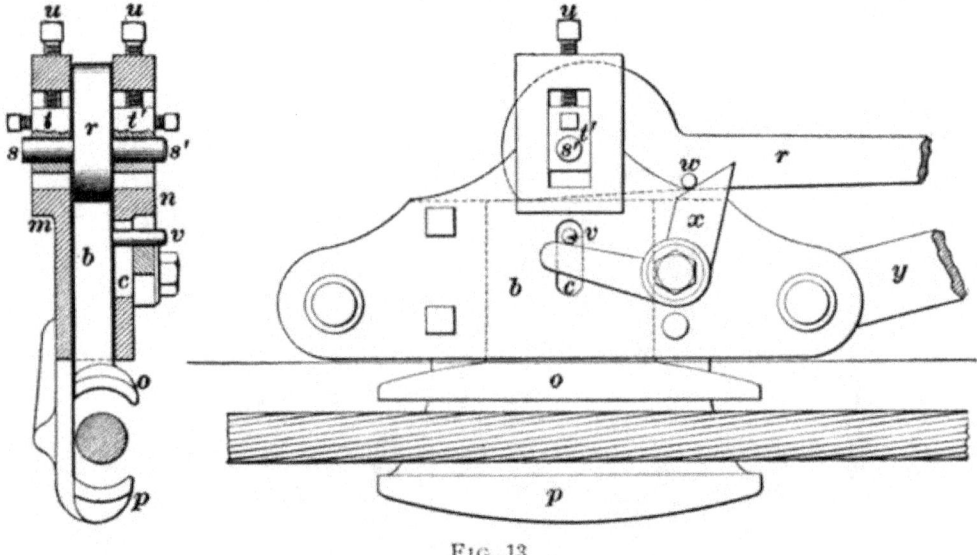

Fig. 13

end of the lever between the plates is lowered by forcing the bearings *t, t'* downwards by means of the setscrews *u*, after which the cable can again be gripped firmly. To cause the jaw *o* to clear the cable after the grip has been released, a pin *v* that projects through an elongated hole *c* in the plate is fastened to the shank *b* of the jaw *o*. Firmly secured to the lever *r* is another pin *w*, and to the plate *n* is pivoted a bell-crank *x*, having the upper arm beveled at the end. As the lever *r* is raised to a vertical position, the jaw *o* is moved downwards, also the pin *v*, thereby causing the bell-crank *x* to swing on its center. In returning the lever *r* to its original position, the pin *w* strikes the nearly perpendicular arm of the bell-crank *x*, causing it to take the position shown in the figure, thereby raising the jaw o by the pin v. The grip is made of steel, except the jaws o and p, which are lined with soft metal so as not to injure the rope; these, when worn out, can easily be replaced. This grip is fastened to the car by the drawbar y having a hook at one end.

20. Fig. 14 shows a simple form of grip that can be used when the rope runs either under or over the car. It consists of an arm a having an eye b, with coupling links c, at one end, the other end being coiled into a spiral of suitable diameter to enclose the rope without much play. The spiral consists of one turn only, and enough space is left between the parts e and e' to permit the rope to pass sidewise between them. The gripper is connected to the rope

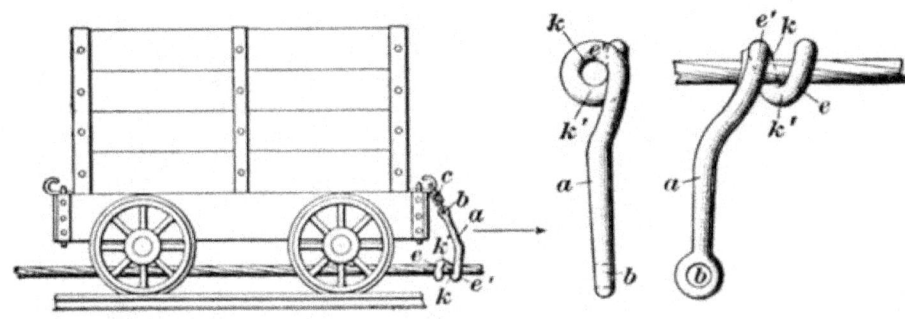

Fig. 14

by merely entering the rope into the coil, and drawing the end b in the opposite direction to that in which the rope is moving. The parts k and k' then bear on the rope on opposite sides, gripping the rope between them; and the greater the resistance of the car to which it is attached the firmer the grip will be.

21. Grip Cars.—When the cars are run singly, each car must be provided with a grip; but if they are run in trains, a powerful grip may be attached to the first car, or the grip may be mounted on a special car, to which the whole train of cars may be attached. When the grip is mounted on a special car, it is applied to the rope by either a combination of levers or a hand wheel. In Fig. 15 is shown a grip car, which consists of a strong platform mounted on wheels. The grip is fixed in the center of the car in such a manner that no longitudinal movement can take place; it may, however be moved sidewise automatically, in order to clear the guiding sheaves when passing around curves. The car is provided with brake shoes b on brake beams and with a brake lever d, for stopping the car, and to prevent any backward movement of the train on slight inclines when the

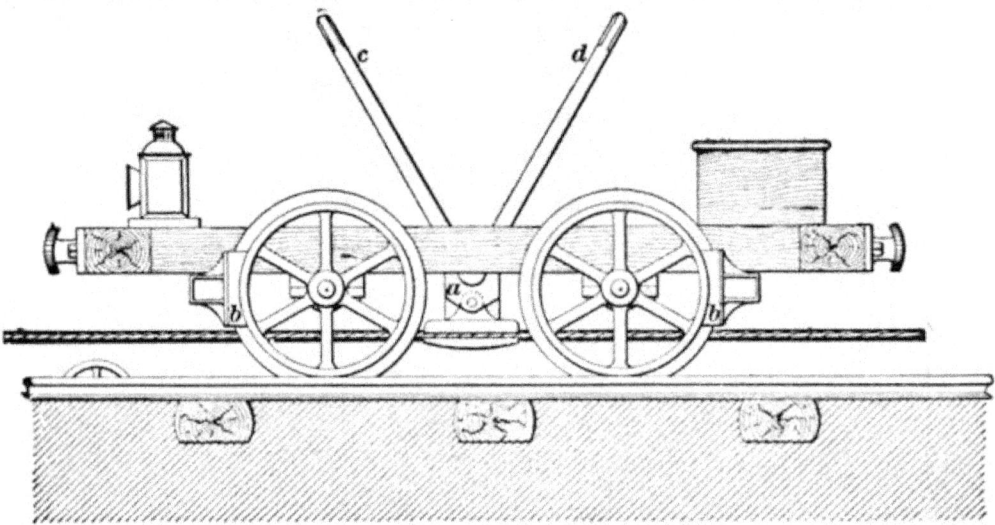

Fig. 15

grip is released from the rope. The grip is placed sufficiently high to clear the track rollers, and is operated by the lever c.

OVERHEAD ENDLESS-ROPE HAULAGE SYSTEM

22. In the **overhead endless-rope haulage system**, the rope is carried above the cars and rests in hooks or carriers fastened in the sides or ends of the cars. The cars are attached to the rope by simply raising the rope and allowing the cars to run underneath it by gravity, the rope being guided into a **V**-shaped automatic clutch grip, or *grab*, attached to the end of the car. The empty cars are taken off the rope by simply elevating the rope to a pulley fixed overhead in the roof. As the cars approach this pulley, the rope gradually and automatically lifts out of the grab, thus releasing the car, which passes off the main line on to the switches or side tracks. The grabs or clutches a, shown in detail in Fig. 16, fit into two sockets b on the rear end of the car. The friction due to the motion of the rope in the grab causes it to turn slightly sidewise and grip the rope. The tension of the grip on the rope will depend on the grade of the track and weight of the loaded car; that is, the heavier the grade or loaded car, the firmer will the grab take hold of the rope. These grabs are free to turn in the sockets in either direction so that they either pull the car when going up grade or hold it back when going down grade. When the cars reach their destination, they are automatically

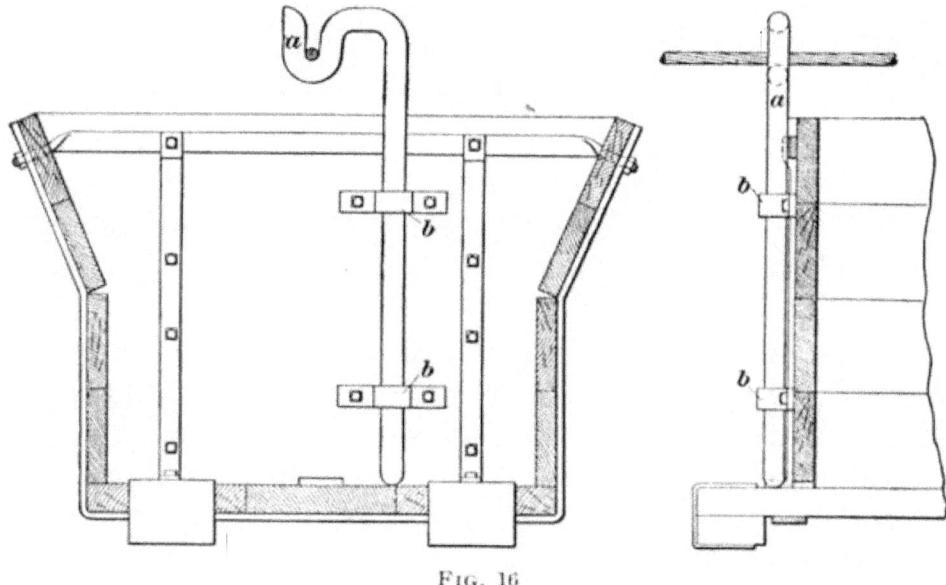

Fig. 16

released by a reversal of the grade together with a gradual ascent of the rope to an overhead sheave, which lifts the rope from the grab.

The rope when not carried by the cars runs on rollers placed in the center of the track for that purpose, but practically all the time the rope is suspended by the cars at points 100 to 200 feet apart, and does not sag enough to allow it to drag on the track or run on the rollers.

In an endless-rope system, where the rope is carried on rollers continuously underneath the cars, there is considerable wear and tear on the rope and a constant renewal of the rollers is necessary. With the overhead system just described, there is practically no wear on the rope, except at either end of the system, where it passes around, in one case, the drums of the driving mechanism or, in the otner case, the sheave at the terminus of the haulage system in the mine. The life of the cars and rope is very much increased by this system. No attendants ride with the cars as they are being hauled and the only labor required is that necessary to attach the empty cars as they are sent into the mines, and the loaded cars as they start out. The cars take care of themselves while on the journey, and as they usually run quite slowly, there is no danger to persons crossing the haulage roads.

TRACK ROLLERS AND CARRYING SHEAVES

23. To prevent the rope of a haulage system from wearing on the roadbed, or, when the roadway is uneven to prevent the rope from coming in contact with the roof, **rollers** and **carrying sheaves,** made of wood, iron, or steel, are used. The ropes attached to the cars lash more or less, and, owing to their movement and construction cut into ties wherever they hit them, besides wearing themselves and sometimes damaging the track. On account of these tendencies, the method of placing rollers and the proper construction of rollers to meet the requirements are important factors in rope haulage.

24. Wooden Rollers.—The hardwoods, such as white birch, gum, cherry, beech, and oak, are made into cylinders

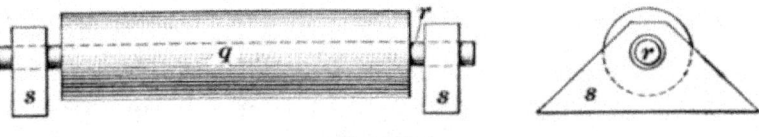

Fig. 17

from 6 to 8 inches in diameter and from 16 to 24 inches long, as shown in Fig. 17, for use as track rollers. Through the center of the roller q a hole is bored into which is inserted an iron rod r that projects beyond each end of the rollers so as to form journals that run in the bearings s. The bearings s are made of wood and are of such length that they can be nailed to two cross-ties.

The centers of the rollers should not be located directly in the center line of the track; then, if the rope wears one end of the roller more than the other, the roller can be changed end for end.

This is a cheaply constructed roller, but not likely to prove satisfactory, since it will almost continually need repairs. The long rod r will probably bend, and will wear the bearings unless they are supplied with thimbles. The rope is very likely to work in between the end of q and the bearing s and break the bearing.

25. Because of these annoying features, more satisfactory rollers can be constructed by using cast-iron gudgeons a, Fig. 18, at the ends of the rollers and small cast-iron babbitted boxes for the journals. The gudgeons have wings b to fit in

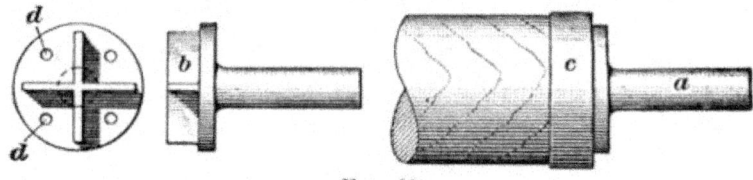

Fig. 18

grooves cut in the end of the roller. To hold the gudgeons fast, irons bands c are shrunk on the ends of the rollers, or the gudgeons are constructed with holes d and spiked to the ends of the rollers.

26. Cast-Iron Rollers.—Fig. 19 shows a cast-iron roller that consists of a cylindrical shell about $\frac{3}{4}$ inch thick, having a flange at each end, as shown, to prevent the rope from being shifted off the

Fig. 19

rollers sidewise. The iron journal bearings a are provided with caps that can be taken off when the rollers are to be removed. The journals are smaller in diameter than the shaft proper, so as to form shoulders on it, in order to prevent the rollers from moving sidewise. These rollers are generally made about 6 inches in diameter.

Fig. 20 shows another form of cast-iron roller. Here, the cylindrical shell of the roller Q is somewhat tapered, and has a small groove in its center, in which the rope runs. The roller is tapered, so that the rope will always slide down to the groove. The shell Q of the roller is fastened to the shaft R by means of the arms and hub S of the roller. The journals t of the shaft are smaller in diameter than the remaining part of it, thus forming shoulders, and preventing the rollers from moving sidewise. The bearings U are of cast iron, of the form shown in the figure, having such a width that they may rest on two cross-ties, to which they are generally bolted. The bearings are provided with a receptacle v, into which oil is poured through the top (by removing the

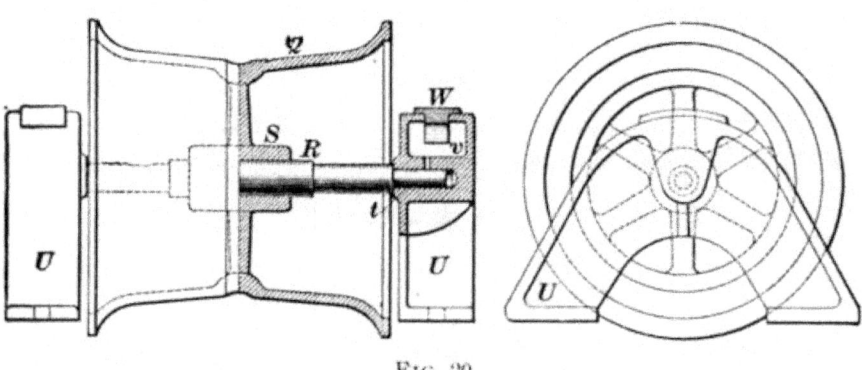

Fig. 20

cap W) for lubricating the journals. These rollers, or carrying sheaves, are generally made 10 or 12 inches in diameter.

27. Roller Journal-Box.—A journal-box that has proved satisfactory on inclined planes is shown in plan and in section in Fig. 21 (a), (b), (c), and (d). Fig. 21 (a) is a plan of the box with a part of the top removed to show the journal bearing a $1\frac{3}{4}$ inches in diameter. There is an oil hole in this top for lubricating the bearing and roller journal. To close the hole and so prevent dirt from working into the bearing a bolt with a solid head is used. The thread for the bolt is shown in Fig. 21 (c). From the oil hole to the journal-box collar and cap b, there is a groove c for the circulation of the oil to the cotton waste packed about the journal in a recess of the collar. The cap b of the collar is made in

§ 57 HAULAGE 27

the shape shown in Fig. 21 (*b*) and (*c*), and is held in place by the bolts *d* having solid heads.

Fig. 21 (*b*) is a section of the journal-box taken through the collar on the line xy. In the recess *e* about the journal, cotton waste thoroughly saturated with lubricant is packed and the waste held down by the cap *b*. The groove *c* carries oil to this waste and prevents it from hardening.

Fig. 21 (*c*) is a section of the journal-box through the

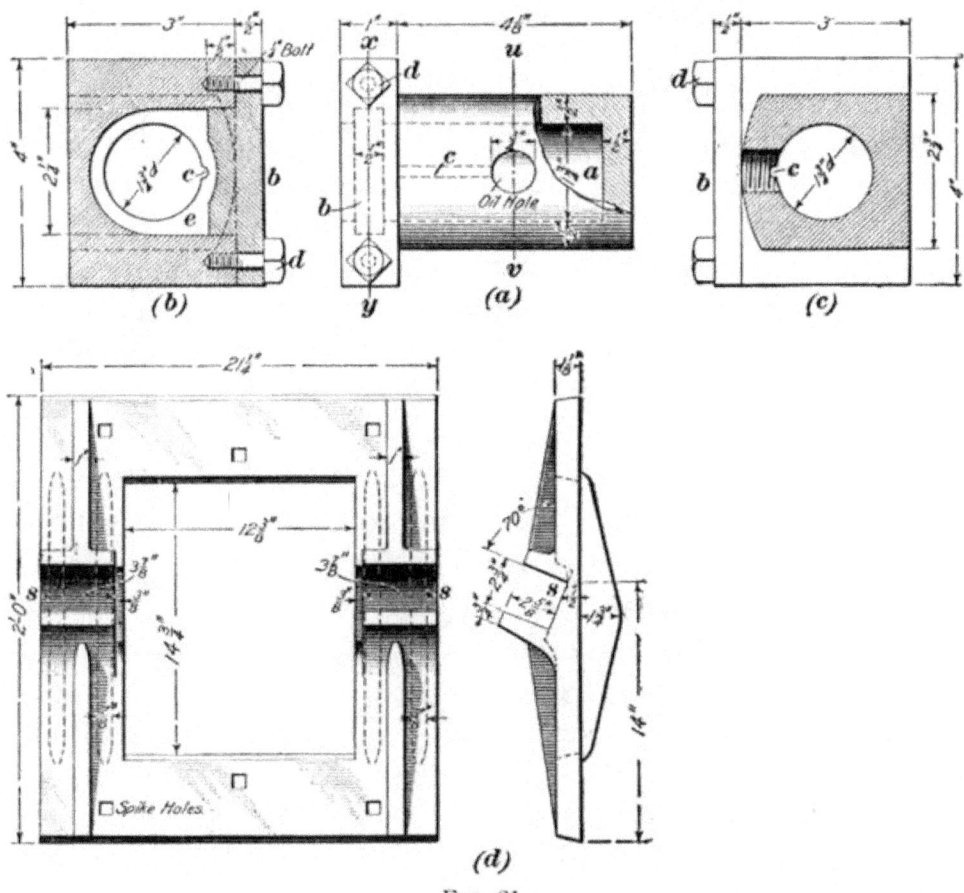

FIG. 21

line uv. Fig. 21 (*d*) is a cast-iron plate in plan and section for the two journal-boxes necessary for each roller. This plate is spiked to ties at each end, and on each side has seats *s* cast a sufficient size and of proper form to take and securely hold the journal-boxes. The seat *s* is made at an angle to conform with the incline of the slope in order to resist the tendency of the rope to move the box out of the seat.

28. Roof Rope Rollers.—It frequently happens on underground inclines, where the grade of the roadway is irregular and there is a sudden change from a steep to a slight dip, that rollers have to be placed above the track near the roof to prevent the rope scraping against the roof. In such cases, iron rollers are generally used. They may be any of the kinds described or merely 2-inch gas pipes plugged with wood and having iron rods passing lengthwise through their centers and projecting at each end to act as journals.

29. Care of Rope Rollers.—The rapid wear of the track rollers is a source of constant expense in the operation of any wire-rope system, if they are not kept in good condition. The irregular surface of ordinary kinds of rope is the chief cause of this rapid wear of track rollers and it is always greater when the rope is new. As the spaces between the rope strands become filled with dirt, tar, or grease and the outer wires become worn down, the wear of the rollers becomes less.

Generally, not enough attention is paid to the matter of supports for the rollers. The gudgeons are often bent, battered, and out of center before they are put in the journal-boxes. The bearing brackets or boxes are sometimes carelessly set, throwing the rollers out of line with the track; or they may be set so close as to cause the rollers to bind; and when this happens the roller refuses to turn, hence the rope slips over it and friction is induced, which wears both the rope and the roller. Often the rollers are spaced too far apart and not enough attention is paid to keeping them clean, lubricated, and in good repair. The enormous friction of rollers, even when taken care of, is not fully appreciated and is often ignored. In actual tests made by the engineering department of the Frick Coke Company in running empty cars down grades varying from 4 to 6 per cent., the empty cars pulling the rope, the resistance due to the rope varied from 6 per cent. to 15 per cent. of the weight of the rope. These tests were made over tracks and rollers in fair shape, and confirm the finding of the Rope Haulage Commission of the North of England Institute of Mining Engineers, in whose report the resistance is given as 10 per cent. of the weight of the rope. There have been cases in which, owing to undulating grades and poor rollers, the power was unable to pull the empty rope on long rope hauls.

30. Spacing Rope Rollers.—Rollers should be placed from 12 to 18 feet apart, averaging about 15 feet; and to insure steadiness of pull, the spacing should not be regular, so as to avoid giving a vibrating undulating movement to the rope. By this whipping motion, very great stresses are brought on the haulage system.

CURVES

31. The track of a haulage road is made straight whenever possible and curves are avoided, since they are likely to give rise to trouble owing to derailment of the cars; they also cause excessive wear on the ropes. When, however, it is necessary to introduce a curve, the inside rail is sometimes made higher than the outside, since the pull of the rope is toward the center of the curve. This is the reverse of the practice on a locomotive road, where the outside rail is higher to prevent a car going off the track on account of the centrifugal force. Very often, the rails are laid at the same height, or the outer rail may even be higher, as in locomotive practice. The relative elevation of the two rails depends on the grade of the road, the amount of the curve, and the speed of haulage.

32. Whenever a curve occurs in a rope-haulage system,

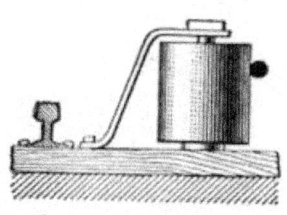
Fig. 22

Fig. 23

the rope must be guided along the curve by rollers, sheave wheels, or drums, in order to keep it away from the side of the entry and to prevent its pulling the cars from the track. Fig. 22 shows a roller sometimes used at the commencement of a curve and placed in the center of the track. Fig. 23 represents a roller sometimes used near the center of a curve either outside or inside the track.

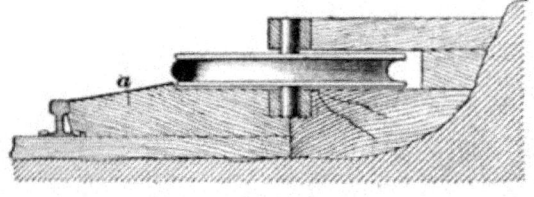

Fig. 24

This roller has a wooden step extending to the rail to guide the rope to the roller; it has also a flange at its upper end to prevent the rope running off the roller. Figs. 24 and 25 illus-

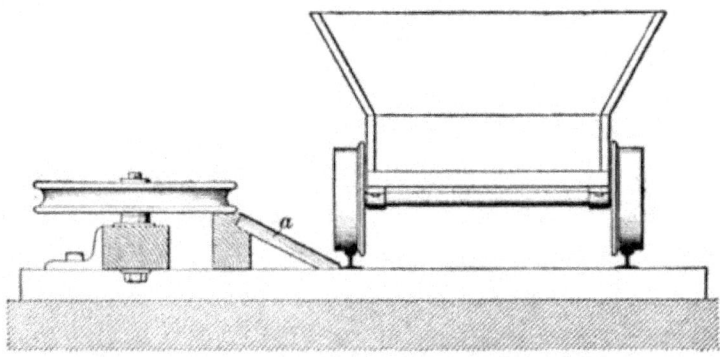

Fig. 25

trate sheave wheels that are often placed on the curves outside the inner rail and near the center of the curve. The rope is guided to the grooves of the wheels by inclined planks a sometimes shod with plate iron.

Fig. 26 shows a small sheave such as is placed between the tracks at the beginning of a curve or when there is a slight curve. The pulley is placed at an angle to allow the rope to slip out of the groove readily, as the car approaches the place.

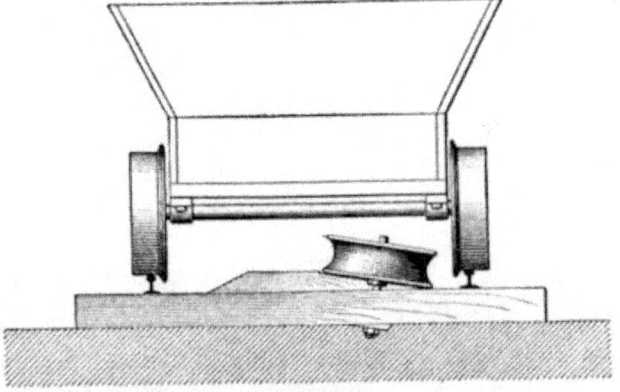

Fig. 26

Fig. 27 represents a cheap and serviceable roller, manufactured from two worn-out car wheels, placed as shown on

§ 57 HAULAGE 31

an old car-wheel axle for a journal. The wheels are clamped together and held in place by a cotter in the axle. The axle may be braced and placed in an upright position, as in the illustration, or used horizontally.

33. In Fig. 28, the rollers or sheaves are placed in the center of the track. The diameter of the sheaves is limited in this case by the gauge of the track, and usually rollers such as were shown in Figs. 23, 26, and 27 are employed.

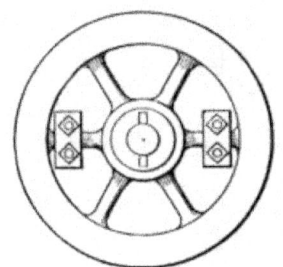

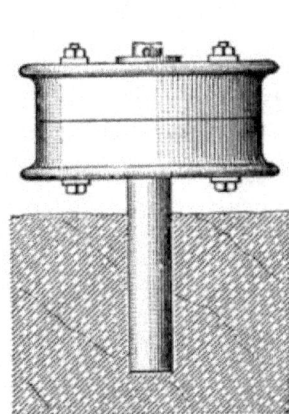

Fig. 27

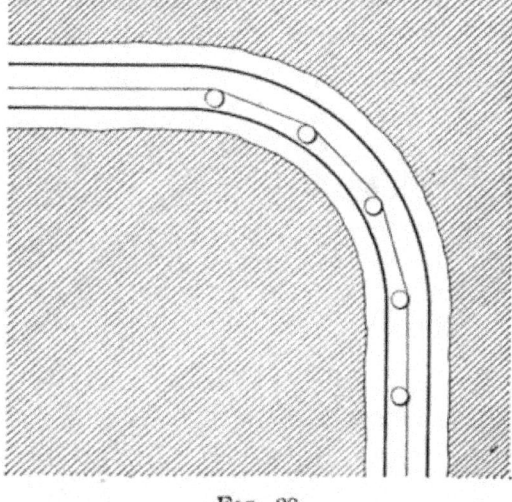

Fig. 28

34. Fig. 29 shows sheaves or drums a placed outside the

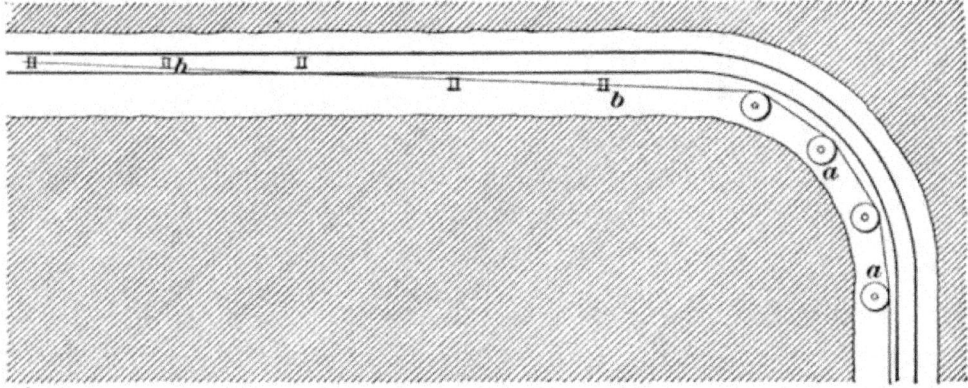

Fig. 29

track and rollers b leading to the straight track beyond the curve. Sheaves or drums placed outside the track, as in

Figs. 29 and 30, may be larger in diameter than those placed inside the track; in fact, they should be from 36 to 48 inches

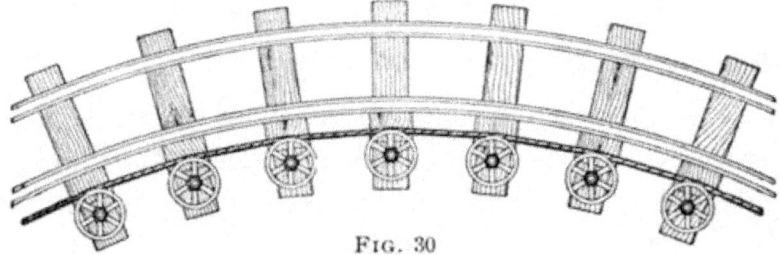

Fig. 30

in diameter in order to prolong the life of the rope and reduce friction.

35. Rope Guides.—Where rollers are placed outside the track, it becomes necessary to guide the rope to the roller and keep it from the rail. This is sometimes accomplished

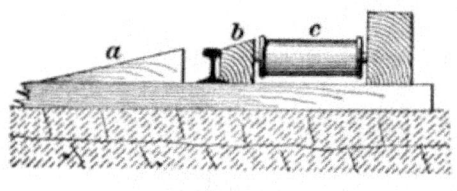

Fig. 31

as illustrated in Fig. 31, where the sheet-iron sheathed block a raises the rope so that it lifts over the rail and then by block b is guided to the roller c.

In rounding a curve the rope, instead of slipping over the rail and taking the roller or sheave, is apt to catch under the flange of the rail and thus wear both the rail and the rope. To avoid this, the device shown in Fig. 32 (a) and (b) may be used. It consists of a finger a pivoted at b and constructed with the lower end heavier than the top end so that it will stand upright when

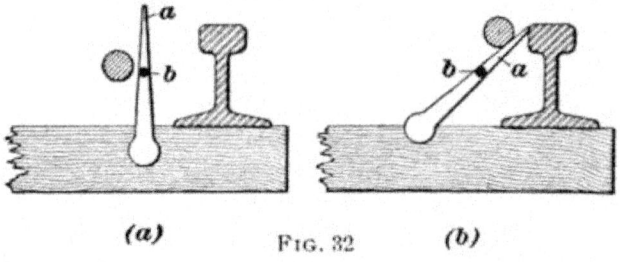

(a) Fig. 32 (b)

at rest. When the rope strikes the finger, it leans until it rests on the rail and thus guides the rope over the rail, as shown in Fig. 32 (b).

36. Reverse Curves.—Where reverse curves occur, the rope is usually guided by small sheave wheels arranged near each other in the center of the track, as shown in Fig. 33. In such cases, unless the rope is very slack, it will not leave the sheave wheels until the car approaches close to them.

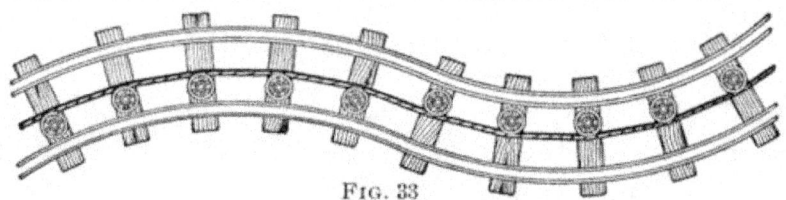

Fig. 33

Rollers such as were shown in Fig. 23 could be used on reverse curves.

ENDLESS-ROPE HAULAGE CALCULATIONS

37. To determine the pull on the rope in an endless-rope haulage system,

Let L_w = working load, that is, the load on the engine or pull on the rope, in pounds;

o = output, in pounds per minute;

v = speed at which rope travels, in feet per minute;

l = length of rope haulage road, in feet;

W_l = weight of a loaded car, in pounds;

W_e = weight of an empty car, in pounds;

W_r = weight of rope, in pounds = $(2\,w\,l)$;

w = weight of rope per foot, in pounds;

f = coefficient of friction (usually taken as $\frac{1}{40}$);

N = number of cars on either side of rope at one time.

Then, on a level road, the pull on the rope is the friction pull alone, and

$$L_w = f N (W_l + W_e) + f W_r \quad (1)$$

The horsepower of an engine to move the system is represented by the equation,

$$HP = \frac{L_w v}{33{,}000}$$

Substituting the value of L_w as found in formula **1**, in the preceding equation,

$$HP = \frac{f v N (W_l + W_e) + f v W_r}{33{,}000}$$

$$HP = (N W_l + N W_e + W_r) \frac{f v}{33{,}000} \quad (2)$$

The value of N is determined as follows:

Let $\dfrac{l}{v}$ = time taken to make one trip in or out, in minutes;

$\dfrac{lo}{v}$ = amount of coal on the rope at one time.

The number of loaded cars on the rope at one time equals the amount of coal on the rope at one time divided by the weight of coal on one car $(W_l - W_e)$;

then, $\qquad N = \dfrac{\dfrac{lo}{v}}{W_l - W_e} = \dfrac{lo}{v(W_l - W_e)} \qquad (3)$

If this value for N be substituted in formula **1**,

$$L_w = \dfrac{flo}{v(W_l - W_e)}(W_l + W_e) + fW_r$$

$$= f\left[\dfrac{lo}{v}\left(\dfrac{W_l + W_e}{W_l - W_e}\right) + W_r\right] \qquad (4)$$

then, $\qquad HP = \dfrac{fv}{33,000}\left[\dfrac{lo}{v}\left(\dfrac{W_l + W_e}{W_l - W_e}\right) + W_r\right],$

$$HP = \dfrac{f}{33,000}\left[lo\left(\dfrac{W_l + W_e}{W_l - W_e}\right) + vW_r\right] \qquad (5)$$

EXAMPLE.—Find the horsepower required to operate on a level an endless-rope haulage system that delivers 1,000 tons of coal per day of 10 hours, when the length of haulage road is 5,000 feet, the loaded car weighs 3,200 pounds; the empty car weighs 1,200 pounds; the rope travels 4 miles per hour, and a 1-inch rope weighing 1.58 pounds per foot is used, assuming the coefficient of friction to be $\frac{1}{40}$.

SOLUTION.— $W_l = 3,200$ lb., $W_e = 1,200$ lb., $W_r = 2 \times 5,000 \times 1.58 = 15,800$ lb.; $l = 5,000$ ft.; $v = \dfrac{5,280 \times 4}{60} = 352$ ft. per min.; $f = \frac{1}{40}$; $o = \dfrac{1,000 \times 2,000}{10 \times 60} = 3,333$ lb. per min. Substituting these values in formula **5**,

$$HP = \dfrac{1}{40 \times 33,000}\left[5,000 \times 3,333\left(\dfrac{3,200 + 1,200}{3,200 - 1,200}\right) + 352 \times 15,800\right] = 31.99$$

Ans.

38. Endless-Rope Haulage on Inclines.—When the track of an endless-rope haulage system is inclined, the calculation of the pull on the rope and of the power required to operate the system takes account not only of the frictional resistance, as in formulas given in Art. **37**, but also of the gravity resistance due to lifting the load through a certain height, which depends on the grade of the track. If the grade is variable, though not very steep at any place, it is customary to assume a mean grade for the whole length of track. If, however, there is an excessive grade at any part of the system, the pull on the rope and power necessary to haul the maximum load that may come on that portion of the road at any time must be calculated. Since an endless-rope haulage is not well adapted to steep variable grades, it is not necessary to give formulas for this unusual case, which can be calculated as a self-acting incline or an engine plane, as explained in *Haulage*, Part 5. Since the weights of the cars and the rope are supposed to be balanced, the weight of coal only has to be considered in calculating the gravity resistance. If, then, the total difference in elevation, in feet, between the ends of a haulage road is represented by h and the grade is assumed to be uniform, the extra amount of work done by the engine, in foot-pounds per minute, is equal to the weight o of coal delivered per minute multiplied by the height h, in feet, through which it is raised—that is, $h\,o$—therefore, the total horsepower of the engine is expressed by the formula

$$HP = \frac{f}{33{,}000}\left[l\,o\left(\frac{W_l + W_e}{W_l - W_e}\right) + v\,W_r\right] + \frac{h\,o}{33{,}000} \qquad (1)$$

when the loaded cars are pulled up the grade, and

$$HP = \frac{f}{33{,}000}\left[l\,o\left(\frac{W_l + W_e}{W_l - W_e}\right) + v\,W_r\right] - \frac{h\,o}{33{,}000} \qquad (2)$$

when the loaded cars are pulled down the grade.

If $\dfrac{h\,o}{33{,}000}$ is greater than $\dfrac{f}{33{,}000}\left[l\,o\left(\dfrac{W_l + W_e}{W_l - W_e}\right) + v\,W_r\right]$, the system is self-acting and no power need be applied by the engine.

Example 1.—An endless-rope haulage road is 1 mile in length. The rope hauls out 800 long tons of coal in 10 hours. The cars each carry 1 long ton of coal, and an empty car weighs 1,760 pounds. The velocity of the rope is 2 miles an hour, and the size of the rope is such that it weighs 2 pounds per foot of length. What is the pull on the rope and the horsepower of the hauling engine, supposing the mean grade of the road to be 2° downwards toward the opening, and assuming a coefficient of friction of $\frac{1}{40}$?

SOLUTION.—In this case, $o = \dfrac{800 \times 2{,}240}{10 \times 60} = 2{,}986\tfrac{2}{3}$, or (2,987 lb.); $v = \dfrac{5{,}280 \times 2}{60} = 176$ ft. per min.; $l = 5{,}280$ ft.; $W_l = 1{,}760 + 2{,}240 = 4{,}000$ lb.; $W_e = 1{,}760$ lb.; $W_r = 5{,}280 \times 2 \times 2 = 21{,}120$ lb.; $f = \tfrac{1}{40}$; $h = 5{,}280 \times \tan 2° = 5{,}280 \times .0349 = 184+$ ft. Substituting in formula **2**,

$$HP = \dfrac{1}{40 \times 33{,}000}\left[5{,}280 \times 2{,}987\left(\dfrac{4{,}000 + 1{,}760}{4{,}000 - 1{,}760}\right) + 176 \times 21{,}120\right]$$
$$- \dfrac{2{,}987 \times 184}{33{,}000} = 133.54 - 16.65 = 16.89. \quad \text{Ans.}$$

EXAMPLE 2.—Assume all the conditions as in the preceding example except that the grade is away from the opening and against the loaded cars.

SOLUTION.—In this case, the values will be the same, but, using formula **1**,

$$HP = 33.54 + 16.65 = 50.19. \quad \text{Ans.}$$

OPERATION OF ENDLESS-ROPE SYSTEM

39. Speed of Operation.—The best speed at which to operate an endless-rope system should be carefully considered. Suppose, for instance, that a given output is secured by having twenty loaded cars on the rope at one time and with the rope running at 2 miles an hour; the same output can be obtained with ten loaded cars on the rope at one time with the rope running 4 miles an hour. When endless-rope haulage is carried on with a velocity of even 4 miles an hour, the ropes become kinked, flattened, and permanently injured by the grips much sooner than with a speed of 2 to 3 miles an hour. To start a heavy car from a state of rest, and give it a velocity of 4 miles an hour, may strain a light rope that is sufficiently heavy to haul the load; therefore, for this velocity, ropes must be made larger than those required for the stress due to ordinary pull. This stress varies as the square of the velocity and the stress due to starting a car in motion at 4 miles an hour, as contrasted with the stress due to starting one with a velocity of 3 miles an hour, is as $4^2 : 3^2$ or $16 : 9$. If an endless-rope haulage be run at 6 miles an hour the number of cars on the rope in a given distance will be $\frac{3}{6}$ the number required to be put on a rope moving with a velocity of 3 miles an hour to deliver the same output. Then, the diameter of the rope necessary to haul half the former load will be reduced to $\sqrt{\frac{3}{6}} = .707$ of its original diameter. On the other hand, the stress due to setting the cars suddenly in motion at a 6-mile velocity, as contrasted with a 3-mile velocity, will be as the squares of the velocities and the diameter of ropes required will increase as the velocities or as $\frac{6}{3}$; consequently, if the diameter of a rope for a 3-mile velocity were 1 inch, that for a 6-mile velocity would be $\sqrt{\frac{3}{6} \times \frac{6}{3}} = 1.414$ inches.

In addition to the increased stress on the rope due to a high speed, there are also other disadvantages. It is not safe for a person who is inspecting the road to cross a double track of this character. Again, the cars leave the rope at too high a velocity, and are, therefore, not sufficiently under control at the period when they are detached. The best results of the endless-rope haulage are probably obtained with a maximum velocity not exceeding 2 to 3 miles an hour; for then persons doing work on the tracks, such as oiling the rollers, inspecting the roof, sides, and timber, and making the necessary inspection and repairs of rails, ties, etc., are able to take care of themselves. This is an important matter, for great delay and expense are saved by proper inspection and repairs.

TAIL-ROPE HAULAGE

40. Distinguishing Features.—In a **tail-rope haulage,** two ropes are used; one, called the *main rope*, or haulage rope, pulls the loaded cars out of the mine; while the other, the *tail-rope*, is attached to the rear of the loaded cars and is dragged behind them out of the mine. The connecting link between the ropes is therefore the train of cars. To take the empty cars into the mine, the tail-rope is attached to the front of the trip and pulls it in, while the main rope is attached to the rear of the trip and is dragged in just as the

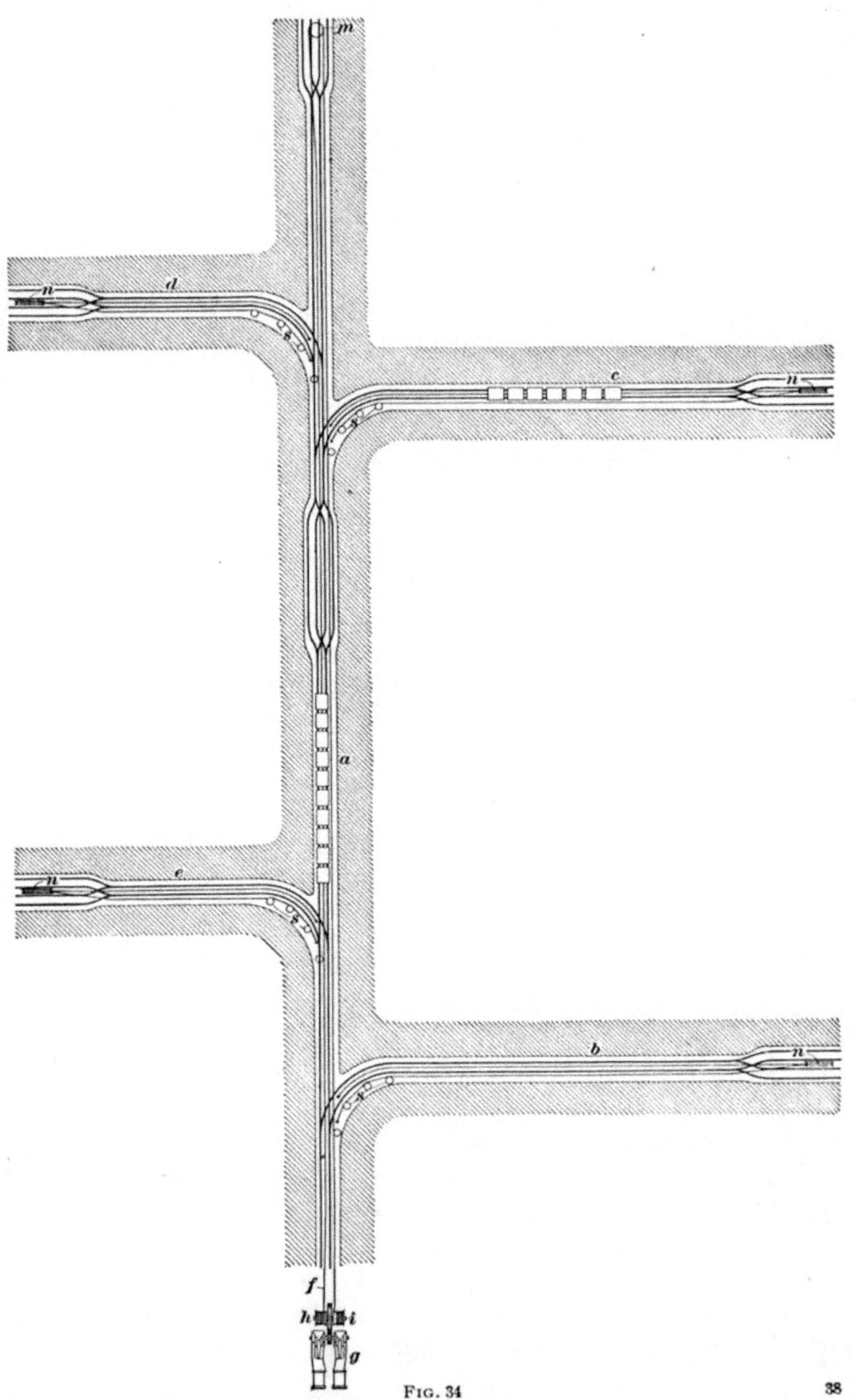

Fig. 34

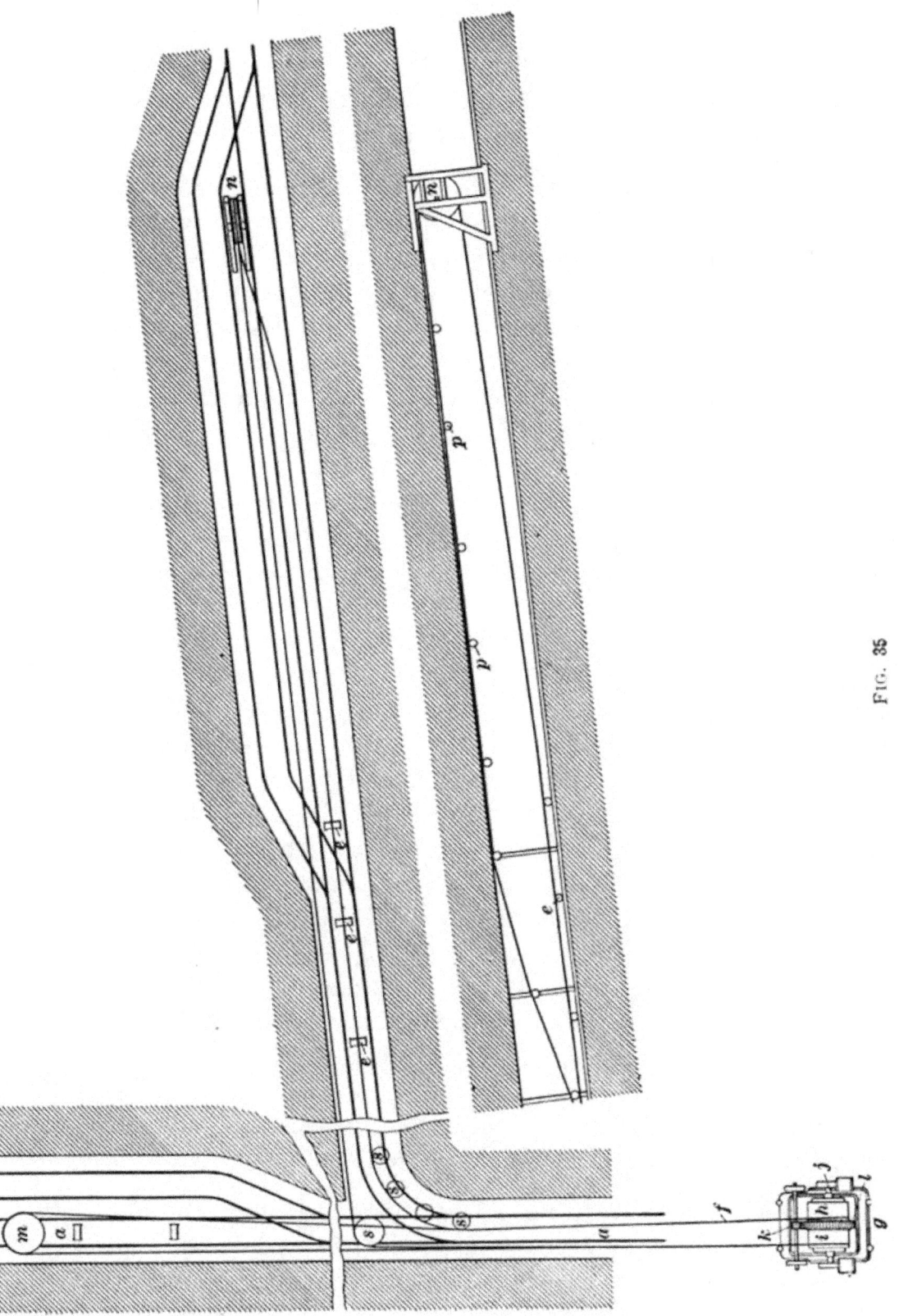

Fig. 35

tail-rope was dragged out. The tail-rope is therefore twice the length of the main rope, since it must extend from the drum to the tail-sheave and back again to the drum.

41. Description of Tail-Rope Haulage.—Fig. 34 is a diagrammatic plan of a tail-rope haulage system, without regard to the actual mine workings and ventilating system, while Fig. 35 shows details of the arrangements of tail-sheaves, and sheaves at curves leading into the side entries, etc. The main entry a, Fig. 34, leads to the shaft bottom or to any other opening through which cars are taken from the mine. The side entries b, c, d, and e are laid off at regular intervals along the main entry and each has its independent haulage system operated from the system on the main entry, as will be explained later. The main rope f is operated by a duplex second-motion engine g, which may be located either on the surface or underground. The engine has two separate drums on one of which h, Fig. 35, the main rope winds and on the other i the tail-rope winds. These drums are loose on the shaft j and are operated by clutches on this shaft. At the end of the main entry a, Fig. 34, the tail-rope passes around a sheave m, which may be horizontal, and under one of the tracks as shown; or it may be vertical and placed between the tracks, as shown at n at the ends of the side entries. Deflecting sheaves s are placed at the mouths of the side entries or on curves to deflect the rope in order to prevent the rope wearing and to permit it to make the curves without undue friction. In the center of the track, rollers e, Fig. 35, are placed at suitable intervals to prevent the rope wearing on the floor. Detailed dimensions for the rollers e are given in Fig. 36; these rollers do not differ in construction from those previously described for endless-rope haulage, and any suitable form of rope roller may be used.

42. Spacing of Rollers.—In a tail-rope system, the rollers are generally not so closely spaced as in an endless-rope haulage system. There is no fixed rule for the spacing between rollers, although they are generally placed so that the ropes will be on them and not drag on the ground. The distances between rollers will depend on the grade of the track and other similar conditions. However, they should not be spaced the same distances apart; for instance, if the average spacing be 15 feet, rollers should be spaced 14 and 16 feet alternately. The reason for spacing the rollers unevenly is to prevent the whipping of the rope, which may occur if they are uniformly spaced.

The rollers for the return tail-rope are sometimes placed over the track near the roof, as shown at a, Fig. 36, but more often on one side of the entry, and either near the roof or

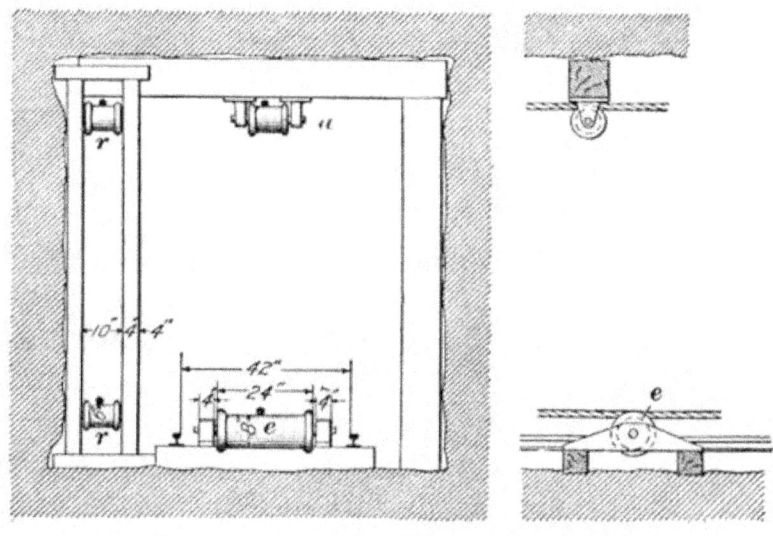

FIG. 36

near the floor, as shown at r, r, Fig. 36. Of the two positions, that one near the roof offers less chance for accidents than that near the floor.

43. Method of Operation.—The method of operating a tail-rope system is as follows: If an empty trip is ready to be hauled into the mine, the tail-rope is coupled on to the front end of the trip and the main rope on to the rear. When the signal is given to the engineer that all is ready, the tail-rope drum is thrown into gear, and the main-rope drum allowed to run slack. The trip is then pulled in to the desired parting by the tail-rope, while the main rope is allowed to drag on behind. At the parting, the ropes are uncoupled and the main rope is then coupled on to the front of the loaded trip and the tail-rope on to the rear of it. When the engineer receives the signal that all is ready, he throws into gear the main-rope drum and throws out of gear the tail-rope drum. The main rope thus hauls out the loaded trip while the tail-rope drags on behind and controls the speed of the cars on down grades.

In order to obtain a large output, since one track only is used, it is necessary to move the cars rapidly and in good-sized trains. The legal speed limit in some states is 6 miles per hour, or $\dfrac{5{,}280 \times 6}{60} = 528$ feet per minute; in most cases, the speed should be from 500 to 600 feet per minute, though it is often exceeded.

DISTRICT TAIL-ROPE HAULAGE

44. When it is desired to extend the tail-rope system of haulage to branch roads and side entries, they are provided with suitable rollers, ropes, and sheaves, and the system is termed **district tail-rope haulage**. Fig. 37 shows the plan of a mine having four gathering stations a, b, d, and e on side entries and one gathering station c on the main entry. Each district is provided with a rope reaching from the entrance to the district to a tail-sheave at the gathering station and back to the main entry. Both the main-entry and district ropes are provided with couplings at their ends. In the plan, assume that the trip is empty and is about to be sent into district b. The main tail-rope n is uncoupled from the trip and the rope f is coupled in its place. The main tail-rope t is also uncoupled at h and the branch rope g coupled to it. The tail-rope can now pull the empty cars and main haulage rope m to w. At this station, the loaded cars are substituted for the empties and the ropes coupled as before to the front and rear of the trip. The signal is now given to the engineer to haul out to the main entry, where the district rope g is uncoupled from the main tail-rope t, which is again coupled and the tail-rope n attached to the rear end of the trip. The signal is then given the engineer and the trip is hauled to the mine opening.

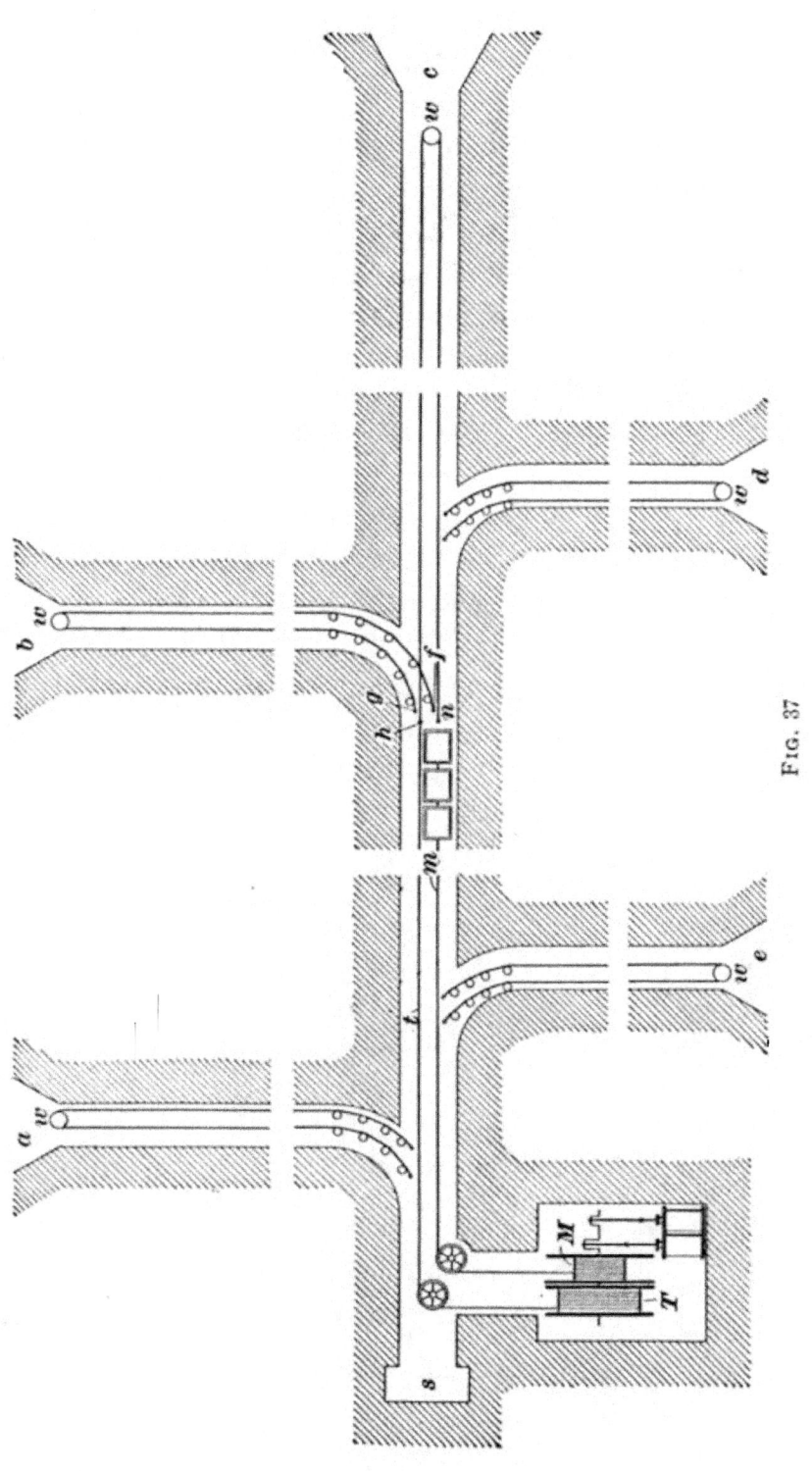

Fig. 37

In the illustration, the engines are of the compound class and situated below ground—that is, in the mine—since *s* represents the shaft bottom. The drums *M* and *T* are of the same size, although *T*, at first sight, looks the larger owing to its being the tail-rope drum and having more rope wound on it than the main haulage drum *M*.

In practice, instead of waiting until the trip reaches the entrance of the district to couple to the district rope, it is customary to couple them to the main-entry tail-ropes at the same time that the empty trip is being coupled to the ropes at the shaft. The rope is then continuous from the shaft to the end of the side entry and the trip is hauled without stopping from the shaft to the gathering-up station. This method economizes in time and in engine power, since the run is continuous and the wear due to starting and stopping the engine one or more times is obviated.

In this connection, it is well to remember that to economically work a tail-rope haulage system, there must be enough cars gathered at each station to make a trip that can be pulled to the mine opening directly without having to stop and make up the trip from other stations or side entries.

45. Fig. 38 shows a main entry *A*, a cross-entry *B*, and a

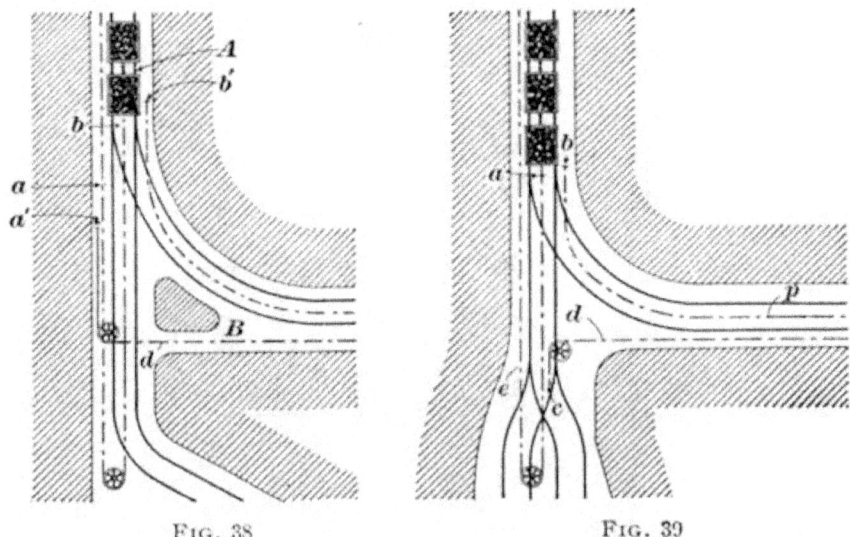

Fig. 38   Fig. 39

loaded trip of cars that have evidently just been hauled from the cross-entry by the rope ends *a' b'*. The main tail-rope is coupled at *a* and *b* so that this system has a direct run to the shaft from the main tail-sheave, thereby reducing the power that would be necessary to overcome the friction and weight of the side-entry rope. The friction of the rope a' around the sheave in order to take the direction *d*, and of the rope b' about the curve may be so great that the method of changing ropes may be found economical in engine power, even if it becomes necessary to stop and start the engine.

46. Another plan for running the tail-rope into a district is shown in Fig. 39. The main tail-rope is uncoupled at *a* and the district tail-rope *b* coupled to the trip. At the same time, the main tail-rope *a* is uncoupled at *c* and coupled to the district rope *d*. In pulling cars either into or out of district *p* by this method, the main tail-rope *e* passes about two sheaves besides the tail sheave at the end of the cross-entry not shown. In Fig. 38, there is but one sheave. In each of these methods, although the main haulage rope is not uncoupled from the trip until the district parting is reached, the engine must be stopped at the entrance to the district in order to uncouple the tail-ropes.

CROOKED-ENTRY HAULAGE

47. The simplest system of tail-rope haulage is where the main and side entries are straight and the tracks are assumed to have a uniform grade, as illustrated by Figs. 34 and 35. Although similar conditions are desirable, they are not necessary; in fact, one of the chief advantages of the tail-rope system is that it can be used on curved and crooked entries and when the grades are variable.

Fig. 40 shows a tail-rope haulage system in a crooked entry. The tail-rope *d* in such cases is carried on sheave wheels *a*, rather than on rollers, in order to keep it in line with the deflecting sheave wheels *b* placed at the curves. The main rope *c* is guided around curves by a number of small sheave wheels *e* placed opposite the large sheave wheels *b* at the curves. This arrangement affords a gradual deflection of the rope at this place, the number of sheaves

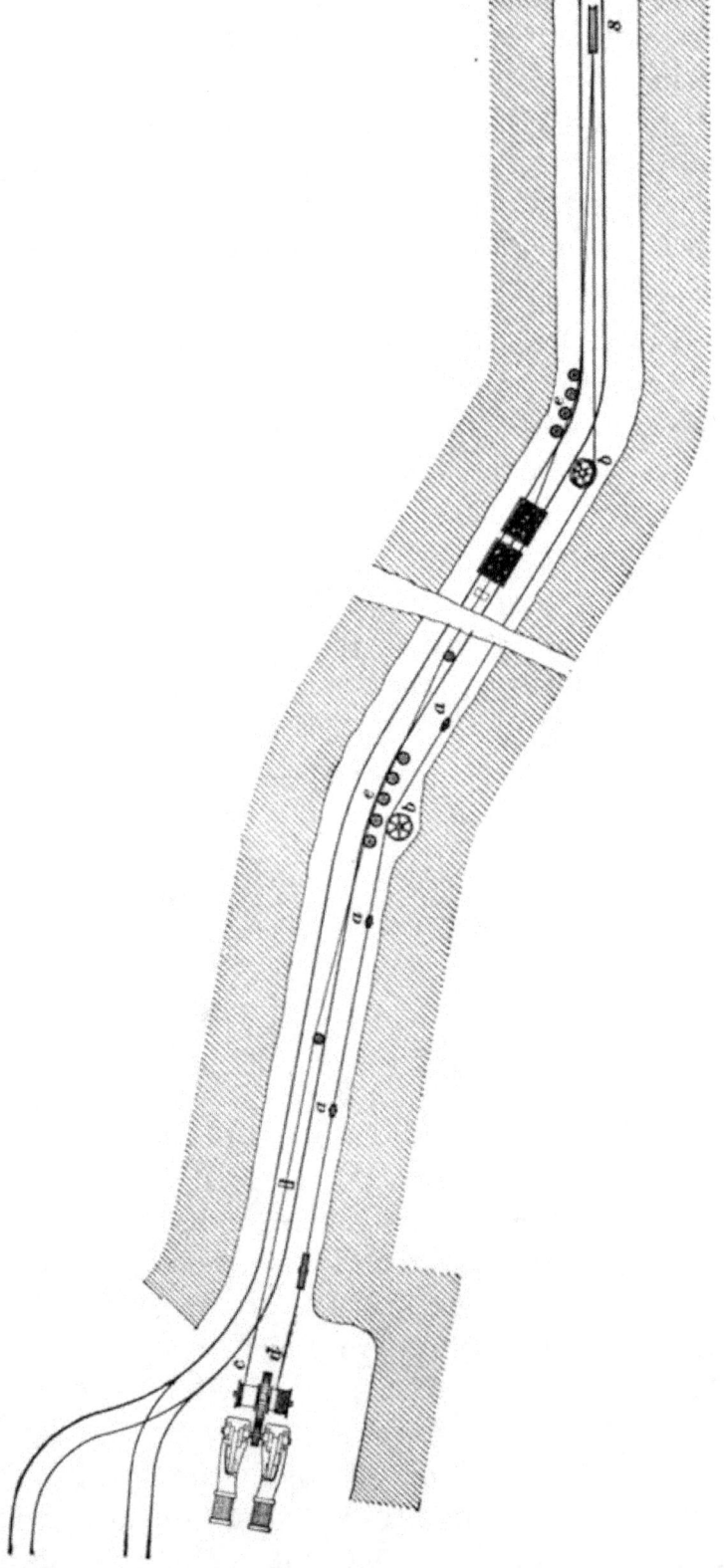

Fig. 40

or rollers used depending on the degree and length of the curve. Fig. 41 shows the usual method of lining the sheaves a to carry the rope to deflecting sheaves.

48. The possibilities of the tail-rope system are shown at a mine near Scranton, Pennsylvania, where the haulage rope makes a 90° curve, then ascends a grade of 18° for 500 feet, then down a grade of 8° for several hundred feet and around another 90° curve in a reverse direction from the first. In order to bring

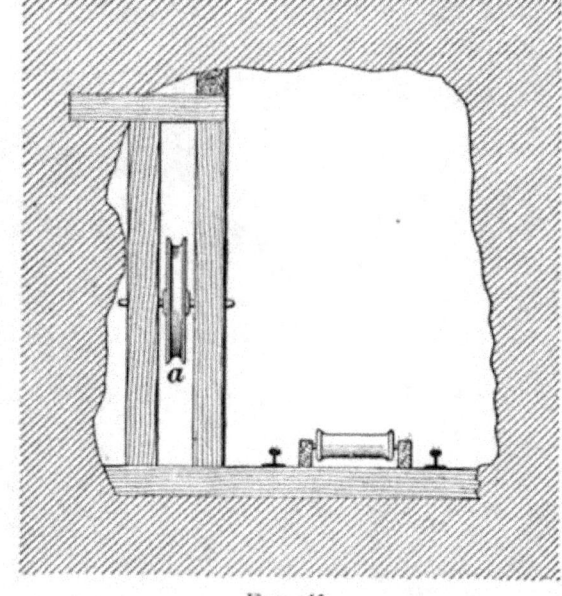

Fig. 41

a loaded trip down the grade of 18°, timbers are placed in the center of the track to such a height that the cars are raised from the track and slide on their axles on the timber.

49. Fig. 42 illustrates the method of rounding curves in one mine where tail-rope haulage was adopted. The large horizontal pulley wheels a are for deflecting the main haulage rope, because that is generally larger in diameter and has more pull on it than the tail-rope. The arrangements b are for guiding the rope to the groove in the wheels. It will be noticed that the end of the haulage rope terminates in a chain that is coupled to or uncoupled from the cars by the trip rider. This chain is more readily uncoupled than the end of a rope, and is not so easily injured by the twisting and bending to which it is subjected. Small sheaves c for the tail-rope are placed near the rib where they will be away from the main haulage-rope pulleys.

50. In many haulage plants, large wood-lagged drums are used on curves for the haulage rope. In Fig. 43, the wood-lagged drums, one of which is marked a, show grooves worn in them by the rope. When the grooves become too

Fig. 42

Fig. 43

§ 57 HAULAGE 49

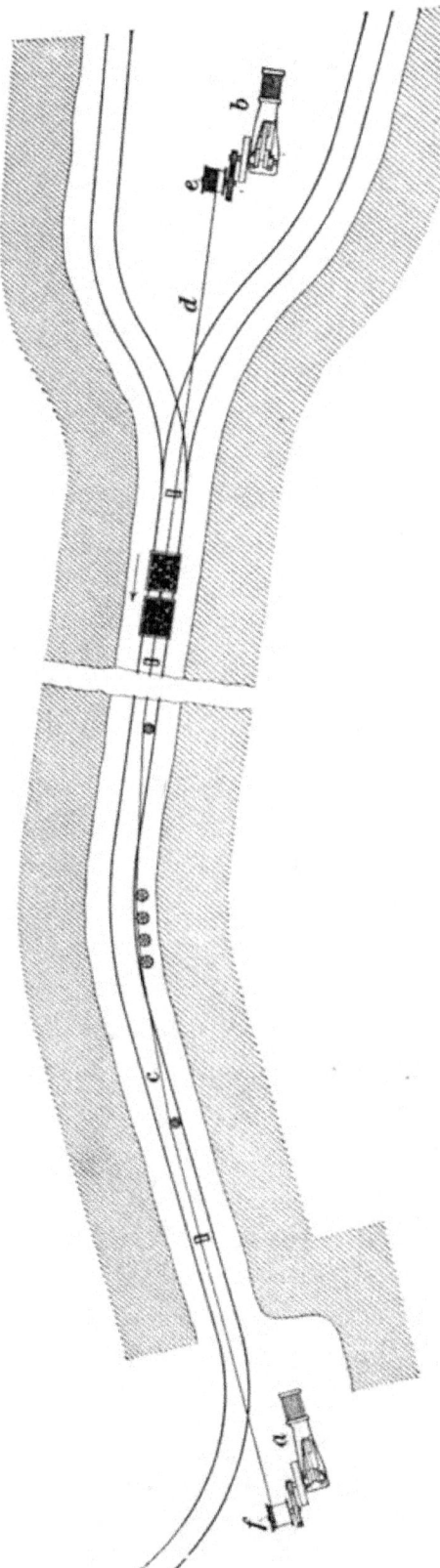

Fig. 44

deep, the drum is reversed end for end, and it is only when the lagging wears through that the drum is relagged. The tail-rope at this mine is carried on sheave wheels b near the roof. To prevent the cars leaving the track on the curve, the guard c is constructed and the inside track rail elevated.

51. Fig. 44 shows a peculiar arrangement of tail-rope haulage. There are two haulage engines, one a placed near the entrance to the mine, the other b placed inside the mine at the extreme end of the system. The main rope c pulls the cars out from the mine and at the same time unwinds the tail-rope d from the drum e. On the return trip, the tail-rope d is wound on the drum e and this pulls in the empty cars. The main rope c, being fastened to the empty trip, is unwound from drum f and pulled into the mine.

It might be possible to use this arrangement where the haulage road is level and the distance between engines long, but the system is not so economical as either the return tail-rope or the endless-rope haulage systems, and is seldom used.

COMPARISON OF ENDLESS- AND TAIL-ROPE HAULAGE

52. Tail-rope haulage differs from endless-rope haulage in that the haulage is done by two ropes, the loaded and the empty trip, traveling alternately in opposite directions on the same track, while the endless rope travels constantly in one direction, hauling the loaded and empty cars in and out of the mine on separate tracks at the same time. In tail-rope haulage, the cars are coupled in trains; while in endless-rope haulage the cars are usually, though not always, hauled singly or in pairs. Endless-rope haulage requires two mine tracks for its operation, while the tail-rope system requires but one mine track; this is an advantage in favor of the tail-rope system where the entry roof is weak and requires support, for, in addition to the extra cost of driving and supporting a wide opening for two tracks, the chances for stoppage of the system due to falls of roof are much greater with a wide than with a narrow entry. As the tail-rope system is run at a much higher velocity than the endless-rope system, often from 10 to 12 miles or more per hour, the chances for a train becoming derailed and knocking out the timbers are much greater; this would probably mean a fall of roof and delay. The tail-rope is better suited to haulage in crooked entries, with heavy grades against the loads, and for side entries or branches that leave the main entry at sharp angles than the endless-rope system. To obtain a large output with a tail-rope plant having average-sized engines, the trips must be run at high rates of speed, and, as the workings are extended, the speed of haulage must be increased. A tail-rope haulage does not deliver cars as regularly as an endless-rope haulage; however, the latter requires more cars. Usually, the wear on roadways, ropes, and engines is greater with tail-rope than with endless-rope haulage.

The endless rope, as ordinarily employed (on main entries alone), requires one-third less rope than the tail-rope system, but the rope must be of good quality and of the same size throughout its length, while second-hand or old ropes of different sizes can be used in tail-rope haulage. As endless ropes travel only from 2 to 4 miles per hour, less power is required to haul the load and rope with endless-rope than with tail-rope haulage. Endless-rope haulage is generally considered not so well adapted to haulage from side entries as tail-rope haulage, but the former is more easily extended. Where grip cars are used to attach the trips to the rope, the endless-rope system costs more for labor than the tail-rope system; on the other hand, when grip cars are not employed, endless-rope haulage costs less for labor than tail-rope haulage.

53. Cost of Operation.—The cost of operating an endless- or tail-rope haulage system is about the same and is estimated by the Trenton Iron Company to average from 2 to 3 cents per ton per mile. This cost is divided up among the items in the percentages given in Table I.

TABLE I
COST OF OPERATING WIRE-ROPE HAULAGE SYSTEMS

System	Rope Per Cent.	Maintenance Per Cent.	Power Per Cent.	Labor Per Cent.
Tail-rope	15	24	30	31
Endless-rope	13	26	12	49

54. Endless Tail-Rope System.—The continuous slow-speed plan of endless-rope haulage is well adapted to mines where the workings are not too distant from the tipple or shaft, say within a mile. Although endless-rope haulages longer than these are in operation, nevertheless, when the workings become extensive and remote, an excessive supply of mine cars is necessary, by reason of so many of them being engaged with the rope, going and coming. This large number of cars, both loaded and empty on the rope at one time, also considerably increases the strain on the rope and engine.

To avoid these features in extensive haulages, the plan adopted by some is to run the rope at high speed from 8 to 10 miles per hour, hauling trains containing from 50 to 100 mine cars. In this plan, it is common to use a reversing engine, alternating the direction of the trains. This plan may be called the **endless tail-rope system**, since it combines features of both these systems.

POWER FOR OPERATING TAIL-ROPE

55. Tail-Rope Haulage Engines.—First-motion engines, or direct-acting engines, are not so well suited to tail-rope haulage as second-motion or geared engines, as a

Fig. 45

first-motion engine works to better advantage when the load is steady. It is usually stated that long hauls and large trains favor the use of first-motion engines; however, regular grades are seldom found in mines and irregular grades cause variable loads on the engine. While engine builders will, of course, quote prices on such engines for tail-rope haulage, they seldom advise their use for this purpose. The drums on first-motion haulage engines must be small or the speed of the train will become too great.

§ 57 HAULAGE 53

A second-motion engine constructed for tail-rope haulage is shown in Fig. 45. The drums are both mounted on one heavy shaft, on which also are two large spur wheels. The drums are loose on the shaft, and are supplied with wide flanges *a* and brake bands *b*. The drum brakes are worked through the medium of wheels *c* and the engine brakes by foot-lever *f*, the latter being for the purpose of bringing the engine to rest at any given place. On the other end of the foot-lever is a block brake, which works on the engine shaft and thus stops the engine. When either drum is to be set in motion for hauling, the friction clutch for that drum is brought into action by a series of compound levers worked by one of the levers *c*. There is a dowel-pin, not shown but moved by the levers *d*, that, by pressing against the drum key *g*, slides the drum along the shaft to the friction wheel on the spur wheel. The contact between the two causes the drum to turn. A heavy spring is compressed when the drum is moved into the friction wheel; and when the leverage is removed, this spring reacts and slides the drum away from the friction wheel. The throttle-valve stem for such engines is within easy reach of the engineer and the valve is usually attached above his head at *h*.

If the main-rope drum is to perform the haulage, the tail-rope drum is permitted to run loose; and if the tail-rope drum is to perform the haulage of empty cars, the main-rope drum is permitted to run loose. No reversing gears are needed for these engines as they run continuously in one direction.

56. It requires special attention on the part of the engineer to brake the drums so as to prevent damage to the rope. The natural friction caused by one rope dragging after a trip will keep the rope paying out uniformly, but it is evident that, if unchecked, the drum running loose may pay out rope faster than the other drum can wind it up and thus permit the rope to kink. Under such conditions, the engineer must apply sufficient brake power to the loose drum to prevent this taking place. A certain amount of momentum is attained by the engine and drums so that, when the steam is cut off and the brakes applied to the drum the engine must be brought to rest by the foot-brake. The subject of braking is again important in that the ropes should be slack enough to be unhooked from the trip but not so slack that they will have a tendency to kink and perhaps unhook themselves.

57. Fig. 46 shows a self-contained second-motion engine

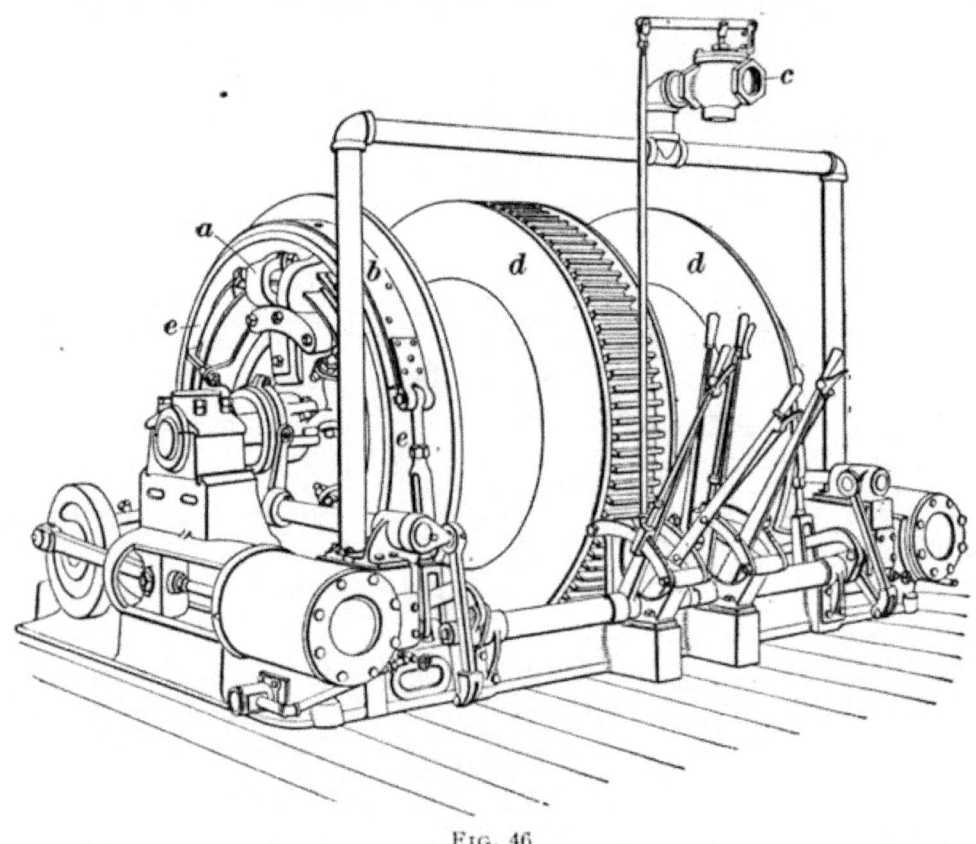

Fig. 46

constructed for tail-rope haulage. There is but one large gear-wheel on the drum shaft and one smaller pinion wheel on the engine shaft, the latter not shown in the figure. The friction clutches a, the brake bands b, and the throttle valve c are worked by hand levers centrally located within easy reach of the engineer. The drums are designed for long rope haulage, as can be seen by inspection of the wide flanges d. Both drums are loose on the shaft, each having separate levers for clutches and brake bands. Each drum has a polished friction ring e, either cast on or fitted to the drum, for the clutch to bite when thrown into friction.

SURFACE TAIL-ROPE SYSTEM OF HAULAGE

58. In order to avoid two haulage ropes in the shaft, when the haulage engine is on the surface, and also to avoid the annoyance of carrying the return tail-rope in the haulage way, the following plan may be adopted: At a point determined by a survey to be immediately above the terminal sheave wheel in the mine, a drill hole is bored from the surface, and the necessary sheaves placed in position on the surface and in the mine. Sheaves for the rope at the surface are supported on low frames erected for the purpose, consisting of two upright posts a, Fig. 47, held together by bolting strong ties b to them. The sheaves are keyed to shafts that run in babbitted boxes. The cost of erecting such frames is very much less than the cost of erecting frames for sheaves in the mines. If there is any tendency on the part of these frames to lean or move out of alinement it is readily noticed and is then remedied by braces.

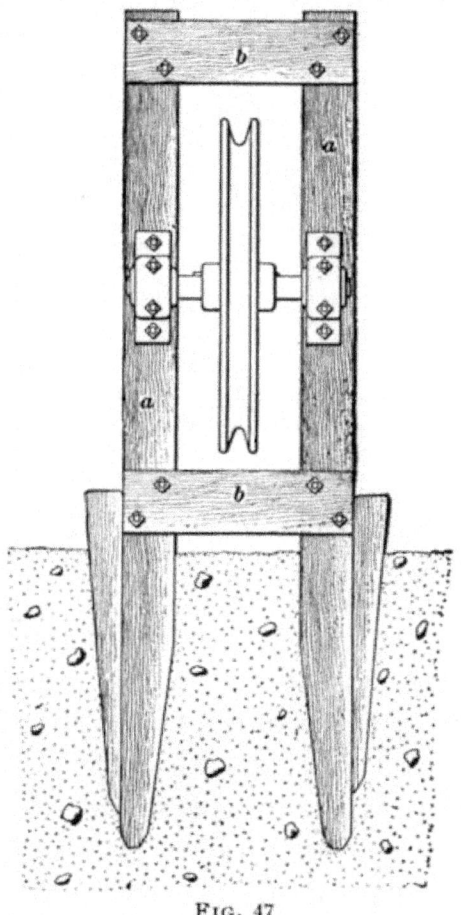

Fig. 47

When rollers are used on the surface, similar frames must be constructed, and as rollers continually wear out and need replacing, the smaller sheaves will be found more economical.

The **return sheave** in the mine is situated so as to line up with the bore hole. It may be supported on a suitably strong frame arranged as in Fig. 48, if it is located at the end of the inside parting. The sheave frame is braced at the roof and floor, as at times it must withstand the pull of the entire loaded trip, should it be necessary to stop an outgoing loaded train and pull the cars back up grade. The tail-rope must be of a suitable strength to withstand the strain thrown on it under the conditions just named.

The timber d is inclined at an angle such that the resultant of the horizontal and vertical pulls is nearly in line with it. Strong braces c are used to keep the timbers b and d properly

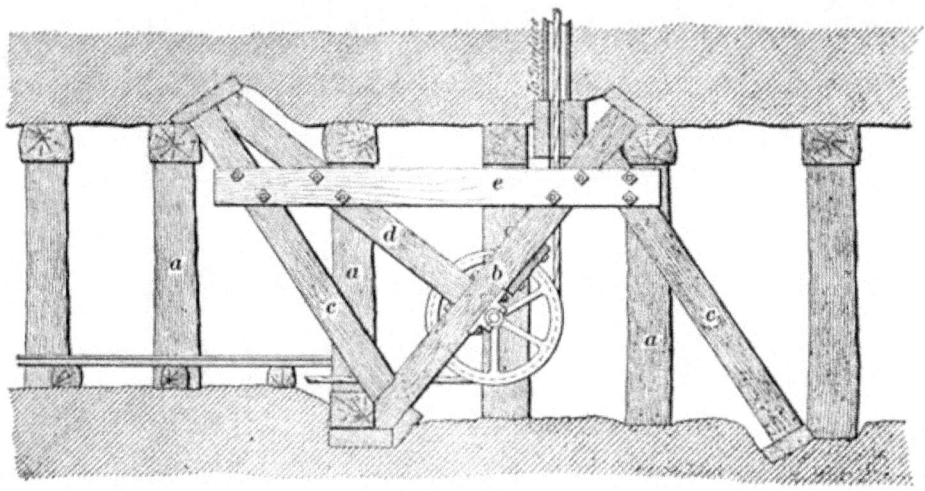

Fig. 48

keyed in their positions. The inclined timbers and the braces are made secure at the roof and floor, by inserting the ends in hitches, which provide an additional safeguard against movements due to the pull on the sheave wheel. As an additional precaution against movement, the tie e is bolted to the timbers b, c, and d.

Timber sets a are erected to guard against a roof fall that would permit the framing to move or would injure the sheave wheel.

SIGNALING

59. It is often necessary to stop the haulage engines in any system of haulage, but especially so in endless-rope and tail-rope systems. These stops are rendered necessary by reason of falls of rock on roads, derailed cars, broken rails, or poor roadbed, broken or displaced timbers, etc. and since the trip must be stopped quickly at any place along the system, a series of signals to the engineer must be established.

A very common and convenient method of signaling the engineer from any part of the haulage road consists of two bare copper wires, strung along on props or timbers for the entire length of the haulage road. These two wires are separated about 8 inches and the signals are given from any point by pressing the wires together or connecting them by a piece of metal laid across the wires. When the wires are connected, a small bell rings in the engine room. The electric batteries and other apparatus are located in the engine room. Sometimes one wire of the signaling apparatus is utilized for telephone purposes, the telephones being located in the engine room and at several points along the haulage way. It might be thought that by locating the engines at the surface and conducting the ropes through the shafts or bore holes, a difficulty would arise with the signals between the gathering-up stations and the engineer, and between the gathering-up stations and the "make-and-break" boys at the entrances to the district roads. Experience, however, has disapproved this and the signals are found to be as perfectly given and received as they would be if given from the gathering-up stations to the engineer in the mine. The necessity of accuracy in signaling can be readily appreciated when it is stated that many of the roads are undulating, and require, on the part of the engineer, great care to prevent the trains overrunning the main rope on the one hand and the tail-rope on the other, when the trains are running down grade. With a proper code of signals, and with the engineer duly informed of the characteristic down grades of the several districts, the trains are kept well under control.

60. Indicators.—In underground rope haulage, the train cannot be seen by the engineer and for safety some mechanical contrivance must be installed that will show him the position of the train. Such an apparatus is particularly important on roads having unequal grades, for the engineer must know when to supply steam to the engine and when to shut off steam, also when to apply brakes, and the exact position in which to stop a train when the different stations and haulage roads are reached.

Any of the indicators used in connection with hoisting engines to show the position of the cage in the shaft can be attached to a tail- or endless-rope haulage engine and arranged to show the position of the trip merely by arranging suitable gearing between the drum shaft and the indicator shaft.

TAIL OR RETURN SHEAVES

61. Horizontal Sheaves.—The tail-sheave that is placed at the end of a tail-rope haulage system for reversing the direction of the rope and starting it back to the engine drum may be supported in a number of ways. Fig. 49 shows a horizontal tail-sheave that may be used for tail-rope systems. The cast-iron spider m is firmly bolted to a timber bedplate and is provided with a steel or a wrought-iron spindle n, around which the sheave o revolves. That part of the spindle that passes through the spider has a smaller diameter than that which passes through the hub of the sheave, so that, by screwing up the nut, the spindle is firmly held in the spider without clamping the hub of the sheave, and, therefore, offers no resistance to the free turning of the sheave. To prevent the sheave from coming off, and to keep dirt from entering the bearing, the spindle is provided with a nut and washer p. For properly lubricating the sheave, the spindle is provided with an oil cup q, the oil from which flows through suitable connecting channels in the spindle, as shown by the dotted lines.

The sheave should be rigidly anchored, so that there is no possibility of its becoming loose under a heavy load.

The figure shows a common method of anchoring a sheave. Here the foundation is built of $12'' \times 12''$ timbers, to which the spider is bolted, the bolts passing through the timbers as shown. At a distance of about 3 or 4 feet from the sheave two timbers a are firmly secured in the ground, at an angle; to these the sheave frame is tied by the rods b. One end of each rod passes through the upper foundation timbers, and

§ 57   HAULAGE   59

the other through a 12″ × 12″ timber c placed on the outside of the timbers a. The ends of the rods are supplied with nuts and cast-iron washers. For preventing the rope from

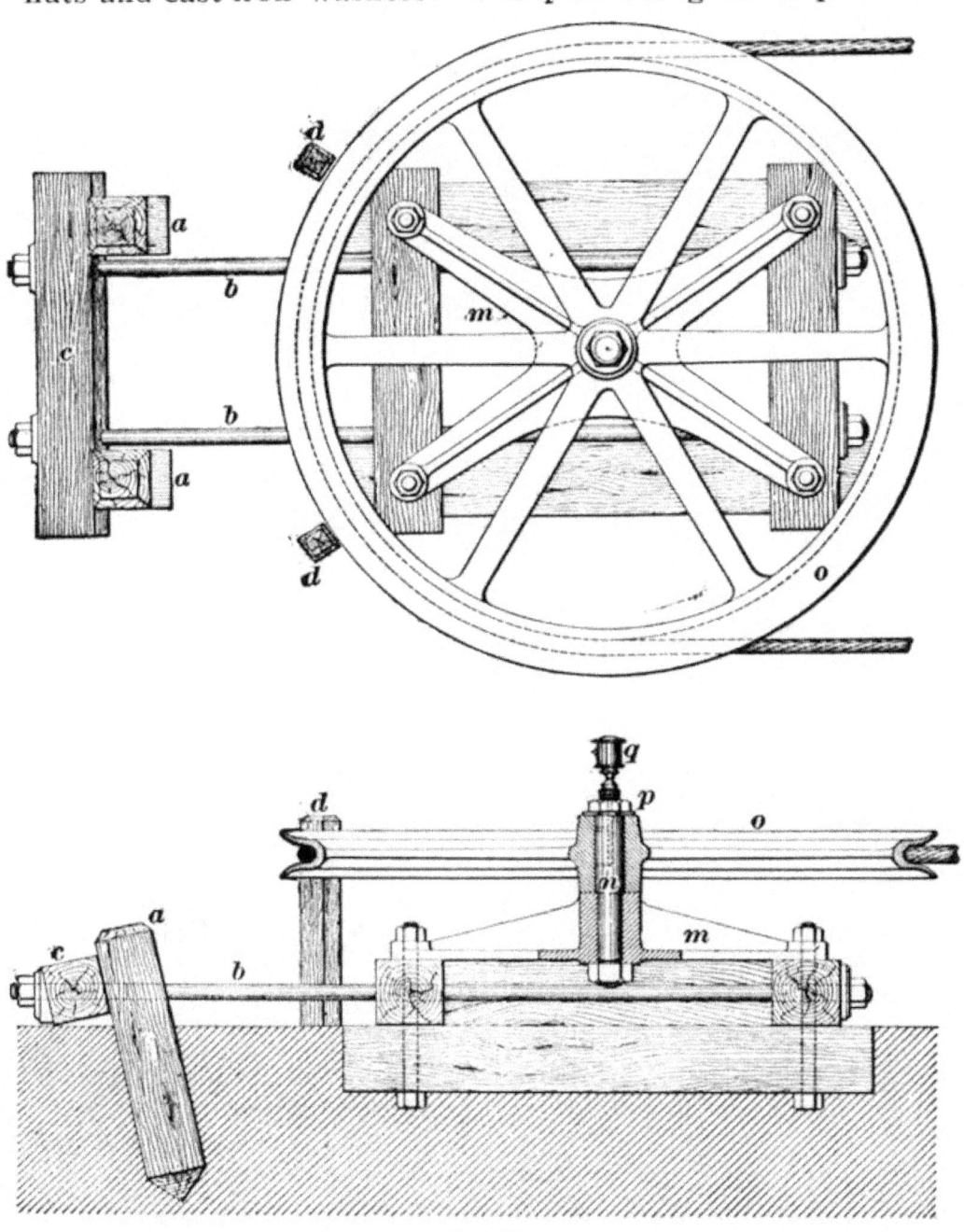

Fig. 49

falling off the sheave when it becomes slack or when the stress is removed, two timbers d are firmly fastened in the ground close to the sheave.

62. When it becomes necessary to place the tail-sheave directly under the track, the arrangement shown in Fig. 50 is generally used. A pit *l* of sufficient depth is provided, in which the tail-sheave *o* is placed. At a short distance from the tail-sheave on each side of the track a deflecting or guide

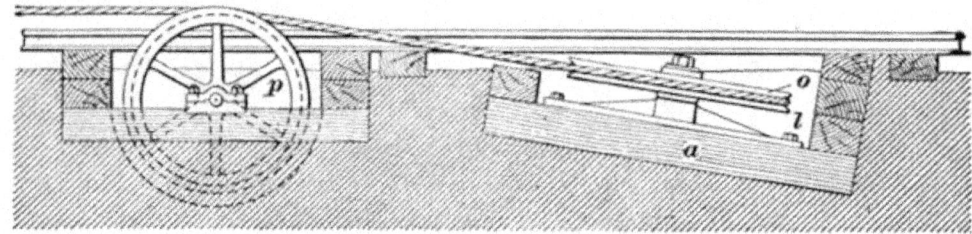

Fig. 50

sheave *p* is placed. The rope, in coming along the track, is deflected to this sheave, and then to the tail-sheave around which it passes to the other sheave.

63. Vertical Sheaves.—Tail-sheaves are sometimes arranged in a vertical position as in Fig. 51. The shaft of the sheave *O* revolves in bearings *a* placed on each side of the wheels, and these bearings are firmly bolted to a strong

Fig. 51

rectangular frame *b* made of heavy timbers. The sheave is elevated a certain height by a wooden structure consisting of posts *c*, rigidly fixed in the floor and roof of the excavation, with the horizontal timber *d* between them. This sheave frame is also held in place by rigidly fastening

§ 57　　　　　　　　HAULAGE　　　　　　　　61

a post e at an angle, and wrapping one end of a chain f several times around it, the other end of the chain being fastened, by means of an eyebolt, to the frame b.

Fig. 52 shows another method of anchoring a vertical tail-sheave, to two heavy timbers a, which are strengthened by braces b. The joint formed by the timbers and braces should be so made that there will be no possibility of the

Fig. 52

braces slipping when under a great stress. This may be accomplished by notching the timbers, as shown in the figure, and by using bolts to hold in place the bearings c, in which the sheave-wheel shaft revolves.

64. Location of Sheaves.—When locating rope-guiding or deflecting sheaves it is best to place them in such a manner that the angle formed by the two parts of the rope as it passes around the sheave is as large as possible, since the greater the angle, the smaller the stress on the bearings in which the sheave revolves. This is illustrated by Fig. 53. The engine is at A, the cars are at B, and three sheaves C, D, and E are used as shown. Assume that the resistance offered by the cars is 5 tons; then, neglecting the friction of the sheaves and the rope, the tension in all parts of the rope will be 5 tons. To determine the load on the spindle of the sheave C, around which it revolves, proceed as follows: To move the cars B, a force of 5 tons must be applied to the part $a\,b$ of the rope in the direction of the arrow c. There is also a force of 5 tons acting in the portion $d\,e$ of the rope in the direction of the arrow f due to the resistance of the cars. The sheave C, owing to these two forces in the rope, tends to move toward J on the line $g\,J$, but is prevented from doing so by the spindle around which it revolves, and by the anchoring. The force that tends to move the sheave C toward J

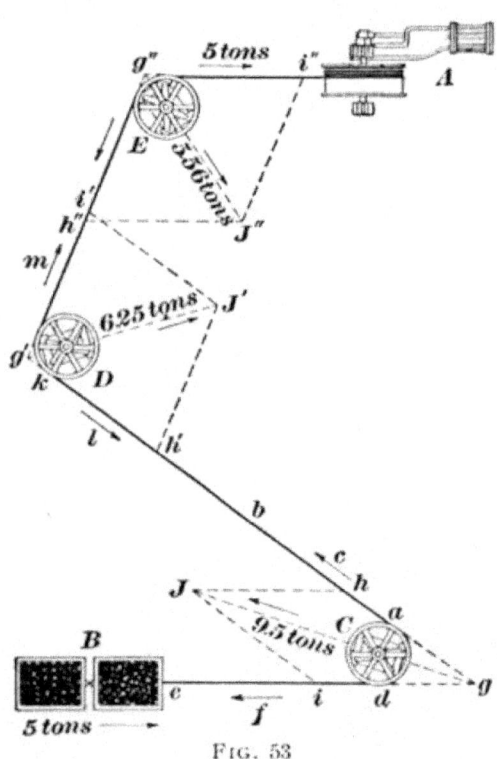

Fig. 53

along the line $g\,J$ may be found by producing the lines of the portions $a\,b$ and $d\,e$ of the ropes until they intersect at g. To any convenient scale, lay off from the point of intersection g the lines $g\,h$ and $g\,i$ on the portions of the rope $a\,b$ and $d\,e$, respectively, each equal to 5 tons. Through the point h draw the line $h\,J$ parallel to the portion of the rope $d\,e$, and through the point i draw the line $i\,J$ parallel to the portion of the rope $a\,b$; then, draw the diagonal $g\,J$ of the parallelogram. This diagonal is the resultant, and represents the greatest force acting on the sheave C. Measuring it to the

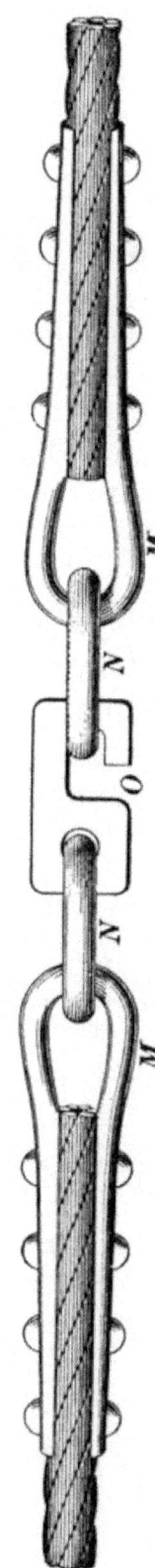

Fig. 54

same scale that has been used to lay off the forces gh and gi, respectively, its value is found to be 9.5 tons. This force of 9.5 tons represents the pressure on the journal and on the bearing of the sheave, and is nearly twice the stress in the rope. The pressure acting on the sheaves D and E is found in a similar manner.

In ascertaining the pressure on the sheave D, the direction of the stress in the portion of the rope kb must be considered as acting from k toward b, as shown by the arrow l; that is, it must be regarded as being produced by the resistance of the cars. The direction of the stress in the portion of the rope $g'g''$ due to the pull of the engine is in the direction of the arrow m. Constructing the parallelogram of forces as above and remembering that the tension in the rope still remains 5 tons, the resultant is $g'J'$, and, measuring it to the same scale as has been used to lay off the forces $g'i'$ and $g'h'$, it is found to scale $6\frac{1}{4}$ tons, a result considerably less than in the case of the sheave C. It will also be noticed that the angle $i'g'h'$, formed by the rope led around the sheave D, is considerably greater than the angle hgi formed by the rope at the sheave C. In a similar manner the resultant stress $g''J''$ on the sheave E is found to be equal to about 5.56 tons. It matters not whether a sheave is placed in a vertical, horizontal, or angular position, the resultant stress may be found in the same manner by the parallelogram of forces.

From this figure it will be seen that the best method for placing spiders to resist stress that comes on the sheave is with the long timbers of the frame parallel to the resultant. Also, it is evident that a sheave may be used satisfactorily in one place while not in another, for instance, the sheave at C which has a stress of 9.5 tons on it must be heavier than the one at E, which has a stress of but 5.56 tons.

COUPLINGS

65. Rope Couplings.—To connect the different ropes quickly and efficiently, tail-rope couplings are required. A number of couplings have been designed whose object has been to make secure connections and at the same time refrain from injuring the rollers, when passing over them, and the rope when wound on the haulage drum. The coupling shown in Fig. 54 is one of the simplest in use, but has two very noticeable defects: first, the flat link O injures the rollers and sheaves, and is likely to injure the rope when wound on the drum; second, while the strap fastenings M are strong, they are not sufficiently flexible for their length to wind on the drum, and they may therefore injure the rope. While the links N are secured to the eyes M for the sake of flexibility, nevertheless the straps must be riveted to a sufficient length of rope in order to make a strong connection.

66. Car and Rope Couplings.—The main and tail-ropes have their ends fastened in sockets to which suitable couplings are attached for coupling them to the cars; Fig. 55

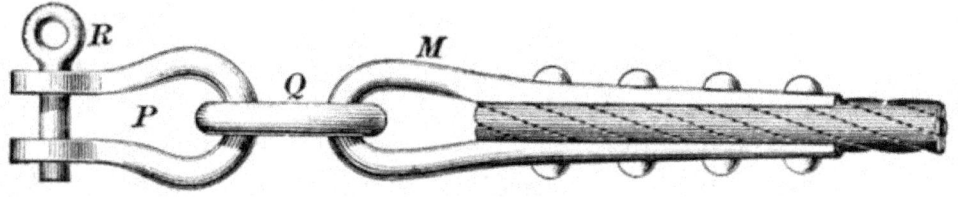

FIG. 55

is an illustration of the simplest coupling of this character. It consists of an eye strap M and a clevis P connected by means of the link Q. This clevis, or shackle, is attached to the drawbar of the car by means of the pin R. An arrangement of this character is attended with serious difficulties in practice, for, in the event of the engine being suddenly stopped, the rope coils under the point of attachment and

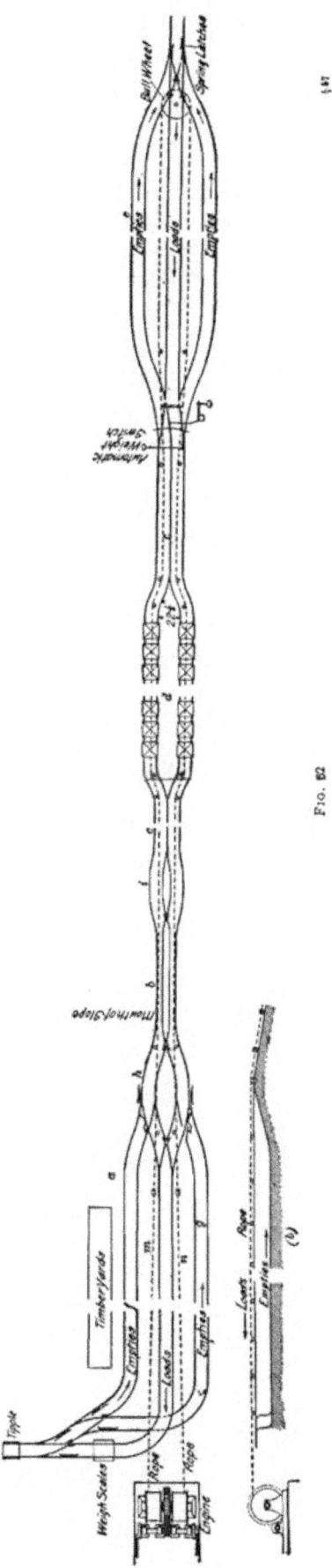

Fig. 52

kinks and becomes permanently injured. To remove the possibility of such an occurrence, a length of 15 or 20 feet of chain connects the socket P, Fig. 56 (b), with a swivel K', the

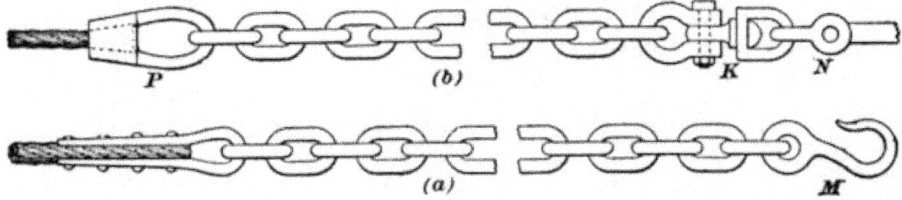

Fig. 56

latter being hooked to the clevis N. Instead of swivel and clevis, a hook M, Fig. 56 (a), may be used.

67. Knock-Off Links or Detaching Hooks.—It is sometimes necessary to detach the rope from the cars when

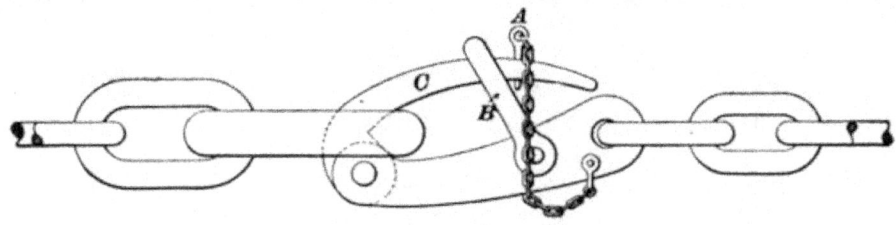

Fig. 57

they are in motion and when the rope is taut. To do this, special devices known as **knock-off hooks** are employed. Fig. 57 is an illustration of a knock-off device, in which a

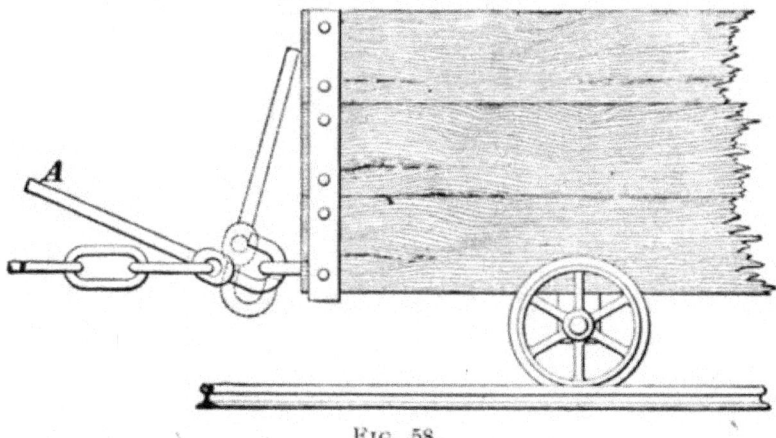

Fig. 58

pin A is set in the bill of the hook to keep the clevis B in position. The moment, however, the pin A is withdrawn, the bill C turns on the hinge underneath the connecting link,

B is forced to the right, and then C in straightening out permits the link held by the hook to slip off and disconnect the cars and rope.

Fig. 58 shows a knock-off hook that can be operated by hand. When a detachment is necessary, the lever A is pulled upwards, and then the hook takes the position shown by the dotted lines and slips from the drawbar.

Fig. 59.

Fig. 59 is an illustration of a hand-detaching arrangement very similar in character to the one shown in Fig. 58. Hooks with levers of this description are dangerous and objection-

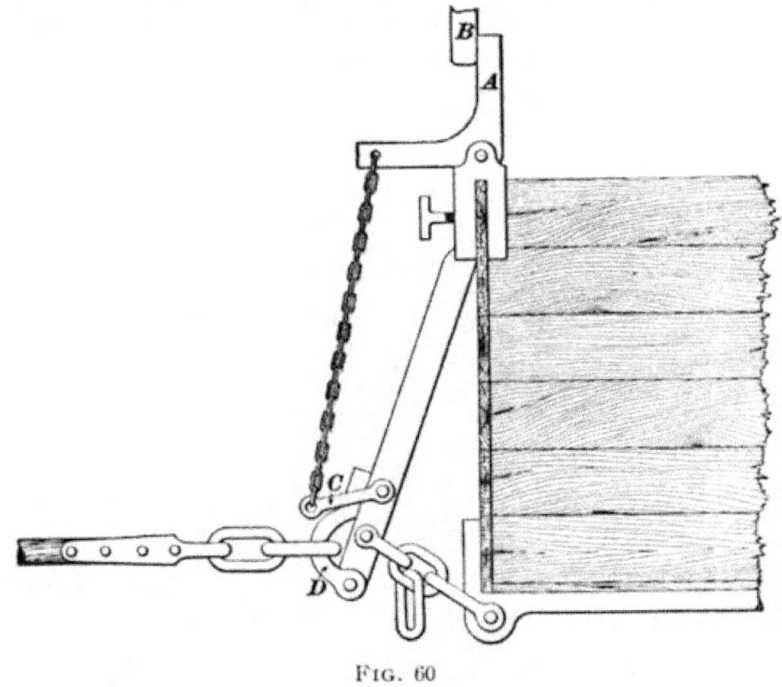

Fig. 60.

able, because they are liable to trip the operator, injure the roadbed, and throw cars off the track.

Fig. 60 shows an automatic detaching hook. There is a knock-off timber B set across the track, and so fixed that

§ 57 HAULAGE 67

when a train passes under it a bell-crank *A* on the car strikes against it, thus raising the clevis *C* and disconnecting the hinge *D* from the rope and its coupling.

Fig. 61 shows another automatic detaching hook, which

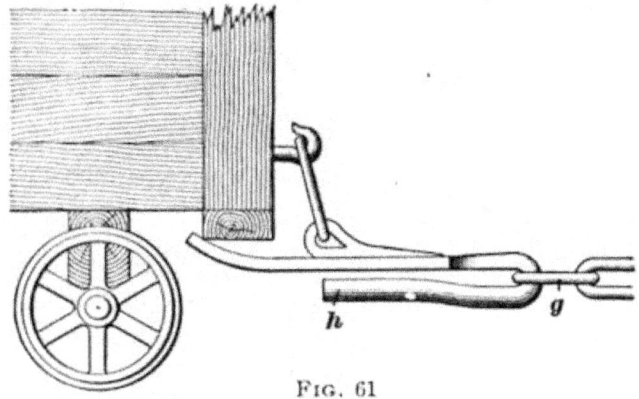

FIG. 61

does not break the connection until the engine stops. The moment the car attempts to overrun the chain, the link *g* slides off the hook at *h* and the rope falls to the ground, leaving the train free to run farther on by its own momentum.

THIRD-RAIL TAIL-ROPE HAULAGE

68. Fig. 62 shows an adaptation of tail-rope haulage to a slope 4,500 feet long with grades varying from 1° to 7° against the load. On account of the grades and local conditions, trips were limited to from sixteen to twenty cars. By means of the three rails and turnouts shown in Fig. 62, it is possible to pull out a loaded trip and run in an empty trip at the same time, thus doubling the output without materially affecting the stress on the rope. In Fig. 62, *a* is an outside landing that contains two tracks for empty cars and one for loaded cars. It was necessary to lap the tracks at *b* on account of the width of the mouth of the slope, which was but 7 feet. This is not a necessary nor advisable part of the system and should be avoided wherever possible. The third-rail part of the track is at *c*; the turnout at *d*, and the gathering station at *e*.

The cars that are loaded are hauled from the center track of the gathering station *e* in the mine to the center track of the receiving station a at the surface. The loaded track a is here elevated so that empty cars going from the tipple to track g may pass under it and so not interfere with cars on the loaded track. Both ropes m and n are above the empty track g, so that they do not interfere with the movement of the cars on this track.

The gathering station at the bottom of this rope-haulage plant is likewise arranged with three tracks, the center one for loads, and the two outside ones for the empties. When a loaded trip is brought out by the rope m, an empty trip must be returned by that rope and the same arrangement is followed with the rope n. Hence, the tracks f and g at the surface are alternately used in lowering empty trips. It is evi-

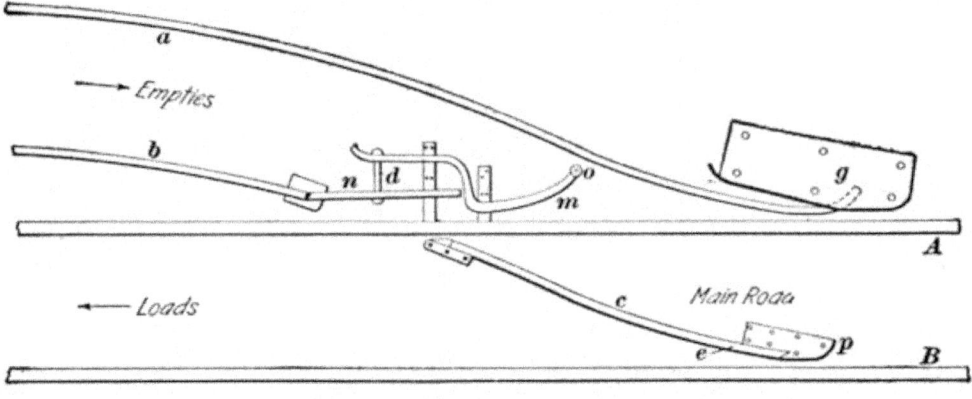

Fig. 63

dent that the track system would be simpler if it were not necessary to have the lapped tracks at b and if the third-rail portion c above the turnout could be continued to the loop h, but since this cannot be done a loop i is needed to direct the cars into the third-rail part c.

Fig. 63 shows a plan of the switch used for directing cars from the loop h to the loaded track at the surface, and from the empty tracks to the loop. When an empty car is going to the loop, it passes over the rail A by the following arrangement: The rail a is brought to within 4 inches of the edge of rail A, then it runs parallel with it for about 14 inches, and finally turns out. The part that is parallel is gradually depressed so that it supports the tread of the wheel until its flange gently touches the top of rail A. A guide

§ 57　　　　　　　　HAULAGE　　　　　　　　69

block g then forces the wheel to take rail A. This guide block is a thick plank spiked rigidly in place and faced with strap iron. A plate p is placed at the point of the rail c, and its end e is depressed so that the flange of the wheel strikes it gently. The wheel finally drops off the plate p when its tread is over rail B; this plate is also spiked rigidly in place.

The latch n pivoted at the end of rail b is connected to the guard m by a bolt d. A short piece of pipe is placed between the latch and guard and the bolt passed through it. This connection is made loosely so that it will not interfere with the swing of the latch. The guard is pivoted at o. As the empty cars advance, the flange of the head-wheel strikes

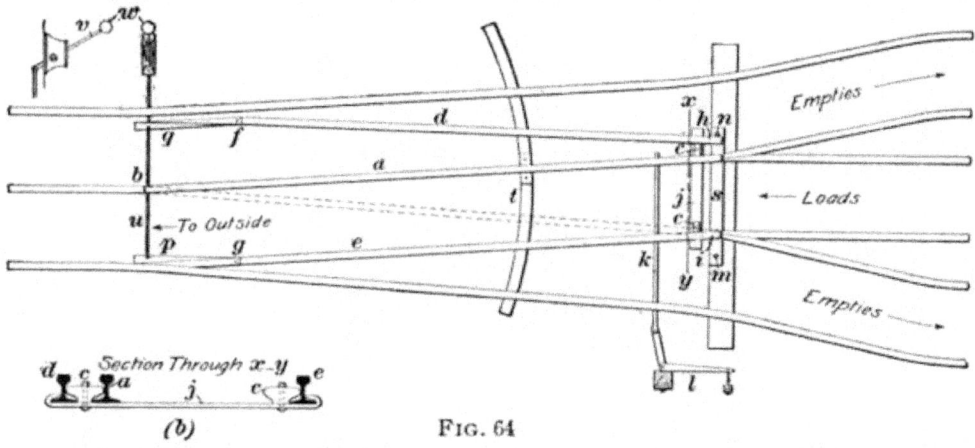

Fig. 64

the latch n and slides the loose end in line with rail c, at the same time causing guard m to slide over rail A.

As the loaded cars move along the main road in the direction of the arrow, the head-wheel strikes guard m and pushes it and the latch n off the rail A to the position shown. This style of switch, it will be noticed, gives an unbroken main track, which is desirable, because the less the jar to the loaded cars, the less damage is done them and there is absolutely no chance of loaded cars getting off the track. As an outside switch, it requires little attention in any kind of weather.

Fig. 64 shows the switch used at the outby end of a gathering station in the mine. The switch is in position for empty cars to pass to the side track x, the loads having previously passed outside. The moving rail a is spiked to the ties at outby end at *b*, leaving about 24 feet free to swing to either side. The normal position of this rail is directly in line with the center line of the middle or loaded track of the siding. A flat iron *t*, bent in the shape of an arc, is attached to this rail and moves with it, working under the other rails. A switch lever *l* is connected to the switch rod *k*, which is connected to the rail *a*; this is used to throw the rail *a* from the position shown to the position indicated by the dotted lines. The rails *d* and *e* are spiked to the ties at their latch ends *f* and *g*, leaving their bridled ends *h* and *i* free, so that they can be easily moved 6 inches. The bridle *j* fits about the flanges of the rails *d* and *e*, Fig. 64 (*b*), and has stops *c* arranged so that the rail *a* cannot be thrown closer than 4 inches to either *d* or *e*. Thus, when the switch is thrown, the end of the rail *a* first sweeps along the slide *s* until it strikes stop *c* of the bridle when all three rails move together

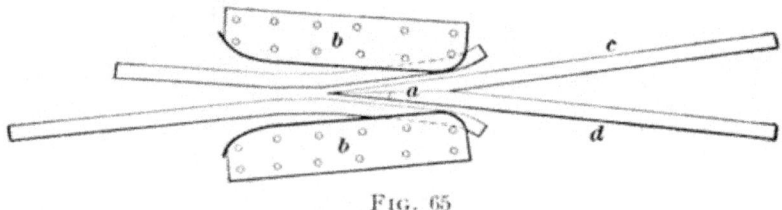

Fig. 65

about 6 inches into place. Stops *n* and *m* prevent the rails from going too far. The latches *p* and *q* are thrown automatically by the loaded cars. Thus, when the loaded cars last left the center track they closed latch *p* and opened latch *q*, making the road clear for incoming empty cars. A lever *v*, with a weight *w* at its top, is so connected to the bridle *u* that whenever the loaded cars strike either latch, the weight is thrown to the opposite side and holds the latches firmly in place. The special advantage of this switch, aside from its simplicity, considering the many ends it meets, is the almost unbroken road it gives for switching empty cars. Fig. 65 shows a frog *a* with guards *b* arranged to prevent loose and badly worn wheels from leaving the track. When cars are passing over a frog rapidly, wheels grooved on their treads by wear will frequently take the wrong side of point of frog and derail the cars. These guards crowd the wheels over on rails c and d. They are simply planks cut as shown and faced with strap iron and well spiked to good ties. The strap iron should be at least $2\frac{1}{2}$ inches wide. All frogs on above-described system are thus guarded, and not one car has left the track at frogs since they have been used.

TAIL-ROPE HAULAGE CALCULATIONS

69. Load on Engine.—The calculation of the load on the engine for a tail-rope system resembles the calculations of the drawbar pull of a locomotive, except that in a tail-rope system the friction of the rope dragging behind a trip increases the pull on the pulling rope. The load on the engine is often considered the same as the pull on the hauling rope. This is true except when the cars are running down a grade steep enough for them to run by gravity and to drag the rope winding off the drum. There will then be no pull on the hauling rope due to the resistance of the trip, but there will be a pull equal to the gravity and frictional resistance of the rope that is winding on the drum.

As already explained, the haulage system for each side entry becomes practically a distinct system when the main ropes and tail-ropes of the main haulage road are connected to the side-entry rope; hence, in calculating the pull on the rope and the load on the engine, it is necessary to calculate for that portion of the mine at which the pull is a maximum, that is, for the longest haul and the steepest grade.

To determine the pull on the rope in a tail-rope haulage system where the grades are not very steep,

Let L_m = load on engine when main rope is winding on the drum, in pounds;

L_t = load on engine when tail-rope is winding on the drum, in pounds;

o = output, in pounds per minute;

v = speed at which rope travels, in feet per minute;

l = length of rope haulage road, in feet;

W_l = weight of a loaded car, in pounds;

W_e = weight of an empty car, in pounds;

W_r = total weight of rope, in pounds;
w = weight of rope per foot in pounds;
f = coefficient of friction (usually taken as $\frac{1}{40}$);
N = number of cars in a trip;
a = grade, in per cent.;
X = angle of inclination.

The weights per linear foot of the main ropes and tail-ropes are sometimes different. The main rope is usually the heavier if it has the loaded cars to pull up grade, while at other times the tail-rope is the heavier if it has to pull empty cars up grade.

In all cases, however, the difference is small, and for that reason their weights per foot will be taken as the same unless otherwise stated. Again, when an empty trip starts into the mine, the length of the moving tail-rope is twice the distance from the hooking-on place to the gathering-up station, and when the trip reaches the gathering-up station, the length of the rope moving is still twice the length of the journey, because then the main rope lies in the middle of the track, and an equal length of the tail-rope lies on the side of the track. Therefore, the total length in motion is never less than twice the length of the track from the outer end to the tail sheave near to which the trip must be hauled. For this reason, the weight of rope due to twice this distance will be taken in each case, that is,

$$W_r = 2wl$$

If the main rope and tail-rope are of different sizes, the weight of each must be calculated and the sum of the two weights gives the total weight of rope. The maximum weight of rope off the drum must be taken; if the tail-rope is lighter per foot than the main rope, the maximum weight is when all the main rope and one-half of the tail-rope is out; if the tail-rope is heavier per foot than the main rope, the maximum weight is when all the tail-rope is off the drum and all the main rope is wound on. When the road is level, the pull on the main rope may be calculated by the formula,

$$L_m = fNW_t + fW_r$$
$$L_m = f(NW_t + W_r) \qquad (1)$$

§ 57   HAULAGE   73

The pull on the tail-rope drawing in an empty trip is given by the formula,
$$L_t = f N W_e + f W_r$$
$$L_t = f(N W_e + W_r) \quad (2)$$

The working load L_w used in calculating the size of engine to be used is the maximum load that may come on the engine and may equal either L_m or L_t, depending on which is the greater.

70. Horsepower of Engine.—The horsepower required for the haulage engine is found by multiplying the maximum load or pull on the rope by the speed, in feet per minute, and dividing the product by 33,000, that is,

$$HP = \frac{L_w v}{33,000}$$

in which

L_w = maximum pull on rope, usually on main rope and equal to L_m.

EXAMPLE.—The greatest length of main- and tail-rope haulage in a mine is 6,500 feet, and the tracks are perfectly level; the weight, per foot, of the main rope is .7 pound, the weight, per foot, of the tail-rope is .6 pound; a full car weighs 5,000 pounds; an empty car weighs 1,800 pounds; the trains consist of twenty cars, and the coefficient of friction is $\frac{1}{40}$. (*a*) What are the pulls on the main rope and tail-rope, respectively? (*b*) If the average speed of the trains is 10 miles an hour, what is the horsepower of the hauling engine, due to the maximum tension of the ropes?

SOLUTION.—(*a*) The total weight of the rope unwound when the trip is at the inside parting is $(.6 + .7) \times 6,500 = 8,450$ lb. Substituting the given values in formula **1**, Art. **69**, $L_m = f(N W_l + W_r)$,
$L_m = \frac{1}{40} \times (20 \times 5,000 + 8,450) = 2,711.25$ lb., the pull on the main rope. **Ans.**

Substituting in formula **2**, Art. **69**, $L_t = f(N W_e + W_r)$,
$L_t = \frac{1}{40} \times (20 \times 1,800 + 8,450) = 1,111.25$ lb., the pull on the tail-rope. **Ans.**

(*b*) According to the conditions of the example the horsepower must be calculated from the maximum pull. The speed of the train is 10 mi. an hr., or $\frac{5,280 \times 10}{60} = 880$ ft. per min. Then, applying the formula for horsepower,

$$HP = \frac{L_w v}{33,000} = \frac{2,711.25 \times 880}{33,000} = 72.3 \text{ H. P.} \quad \textbf{Ans.}$$

It is evident that with a long haul, the power required to overcome the resistance of the rope is considerable.

71. Tail-Rope Haulage on an Incline.—If all or a part of the road is on a grade, the pull on the rope will be increased or decreased, depending on whether the grade is against or in favor of the load. The pull on the rope is then equal to the resistance due to friction plus or minus the gravity pull, depending on whether the grade is against or in favor of the load. The weight of the tail-rope balances the weight of the main rope if the two ropes are of the same size, so that if the gravity pull of one is against the load the gravity pull of the other is in favor of it and the one balances the other.

When hauling out the loaded cars, if the grade is against the load, the pull on the main rope is given by the formula,

$$L_m = f(NW_l + W_r) + aNW_l \qquad (1)$$

If the grade is in favor of the loaded cars, the pull on the main rope is given by the formula,

$$L_m = f(NW_l + W_r) - aNW_l \qquad (2)$$

When returning the empty cars, if the grade is against the cars, the pull on the tail-rope is given by the formula,

$$L_t = f(NW_e + W_r) + aNW_e \qquad (3)$$

If the grade is in favor of the empty cars, the pull on the tail-rope is given by the formula,

$$L_t = f(NW_e + W_r) - aNW_e \qquad (4)$$

EXAMPLE.—On a short portion of a tail-rope haulage system, the main rope hauls a train of thirty loaded cars up a grade of 4 per cent. (a) What is the maximum load on the engine or pull on the main rope when a full car weighs 5,000 pounds, the main rope and the tail-rope weigh each 1 pound per foot, the length of the track being 5,600 feet, and the frictional coefficient is $\frac{1}{40}$? (b) What will be the pull on the tail-rope in returning the empties, if an empty car weighs 1,800 pounds?

SOLUTION.—(a) Since the track is 5,600 ft. long and the ropes weigh 1 lb. per ft., the weight of the rope is $5,600 \times 2 = 11,200$ lb. Substituting these values in formula **1**,

$L_m = \frac{1}{40} \times (30 \times 5,000 + 11,200) + .04 \times 30 \times 5,000 = 10,030$ lb. Ans.

(b) The grade will be in favor of the empty cars, hence, substituting in formula **4**,

$L_t = \frac{1}{40} \times (30 \times 1,800 + 11,200) - .04 \times 30 \times 1,800 = -530$ lb.

that is, the load on the engine is negative and the cars will run down the grade by gravity. Ans.

72. When a haulage road only runs up grade for a short distance, it is the practice to accelerate the speed of the train a little before reaching the rising ground, and then by its inertia the train is carried over with no other loss than that of a reduced velocity, which it soon recovers.

If there is a very steep grade on the haulage road and it is desired to determine the pull very accurately, it may be calculated by the formulas used for an engine plane with the engine at the bottom of the plane, as will be fully explained in *Haulage*, Part 5.

When hauling out the loaded cars, if the grade is against the cars, the load on the engine is given by the formula,

$$L_m = f(NW_l + W_r) \cos X + NW_l \sin X \qquad (1)$$

If the grade is in favor of the loaded cars, the load on the engine is given by the formula,

$$L_m = f(NW_l + W_r) \cos X - NW_l \sin X \qquad (2)$$

When hauling in the empties, if the grade is against the empty cars, the load on the engine is given by the formula,

$$L_t = f(NW_e + W_r) \cos X + NW_e \sin X \qquad (3)$$

If the grade is in favor of the empty cars, the load on the engine is given by the formula,

$$L_t = f(NW_e + W_r) \cos X - NW_e \sin X \qquad (4)$$

It is, however, seldom necessary or advisable to use the formulas given in this article, as those given in Art. **71** give results sufficiently accurate for all practical purposes and are the ones generally used.

73. Speed of Trains.—In tail-rope haulage, it is important that the maximum velocity of the trains should not exceed about 12 miles an hour, for three very important reasons: (1) The tracks are costly to keep in repair when the velocities exceed this speed; (2) cars are derailed more frequently at high speeds than at low speeds; (3) high speeds render the inspection of the tracks, ditches, timbering and roof dangerous during working hours, even though safety holes are provided.

74. Size of Trains.—To estimate the number of cars that should be hauled in a train, the values of four important factors, all of which are variable, or different in different mines must be known. They are: (1) the number of tons of coal each haulage district can produce per day; (2) the lengths of the hauls from each parting; (3) the capacity of a mine car; (4) the number of trains that can be run out of the different districts, that is, the actual time of hauling each day exclusive of delays and time lost in changing ropes during each trip.

The *prospective output* of the mine is found by estimating on the basis of present practice what the probable output will be from each district.

The *lengths of the road* are calculated approximately from lengths measured on the map of the property.

The *size of the mine cars* is made to conform to the dimensions of the hoisting shafts and the heights of the haulage roads, although, in fair-sized beds, the height of the rooms will be a factor.

75. The number of trains is found as follows:

1. Find the mean of all the lengths of the districts, each being measured from the shaft to the making-up stations.

2. Multiply the mean length of the districts by 3, for the following reasons: A journey of full cars out and empty cars in is equal to double the length of a district road, and to compensate for unpreventable and unforeseen delays, another addition must be made to the length of the track, and, therefore, the mean length of the district roads must be multiplied by 3.

3. Find the number of feet a point in the rope will pass through in one working day. Suppose, for example, the mean speed of the rope is 12 miles an hour, and that the time of 1 day is 10 hours; then, $5,280 \times 12 \times 10 = 633,600$ feet, the distance a point on the rope will move through in 10 hours.

4. Divide the distance a point on the rope would move through if kept continually in motion by three times the

mean length of the district roads, and the quotient will be the number of trains that can be hauled out per day.

EXAMPLE.—How many trains can be run out by a main- and tail-rope haulage in one day of 10 hours, the speed of the rope being 12 miles an hour, and the lengths of five districts being as follows: A, 5,012 feet; B, 4,654 feet; C, 3,278 feet; D, 7,101 feet; E, 2,794 feet.

SOLUTION.— $5,012 + 4,654 + 3,278 + 7,101 + 2,794 = 22,839$; $22,839 \div 5 = 4,567.8$, the mean length. Then $\dfrac{5,280 \times 12 \times 10}{4,567.8 \times 3} = 46.24$, or, practically, 47 trains per day. Ans.

76. To find the number of cars in a train, *divide the output, in tons per day, by the number of trains, multiplied by the tons of coal a car will carry; the quotient will be the number of cars in a train.*

EXAMPLE 1.—The output of a mine is 2,000 tons of coal per day; the number of trains to haul out this quantity is 47. If one car carries 2 tons, how many cars must there be in a train to do the work?

SOLUTION.— $\dfrac{2,000}{47 \times 2} = 21.276$, or, as there cannot be a fraction, the number is 22 cars in a train. Ans.

EXAMPLE 2.—A haulage plant consists of six district haulage roads of the following lengths: A, 6,784 feet; B, 4,250 feet; C, 8,276 feet; D, 3,560 feet; E, 5,720 feet; F, 7,963 feet. The mean up grade to the shaft is 2 per cent.; the output is 3,000 long tons of coal in 10 hours; the cars carry $2\frac{1}{2}$ long tons each; the speed of the train is 10 miles an hour; an empty car weighs 1,900 pounds; 1 foot of rope weighs 2 pounds; the coefficient of friction is $\frac{1}{40}$. (*a*) How many trips can be run out each day? (*b*) How many cars will there be in each trip? (*c*) What will be the horsepower of the haulage engine?

SOLUTION.—(*a*) The distance a point in the rope will run in 10 hr. is $5,280 \times 10 \times 10 = 528,000$ ft. To find the number of trains that can be run out in a day, this product is divided by three times the mean length of the haulage roads. The mean length is

$(6,784 + 4,250 + 8,276 + 3,560 + 5,720 + 7,963) \div 6 = 6,092\frac{1}{6}$ ft.

Hence, the number of trains is

$\dfrac{528,000}{6,092\frac{1}{6} \times 3} = 28.89$, or, practically, 29 trains. Ans.

(*b*) To find the number of cars in a train, the output is divided by the number of trains multiplied by the number of tons a car will carry.

Thus, $\dfrac{3,000}{29 \times 2.5} = 41.4$ cars, or, practically, 42 cars in a train. Ans.

(c) Remembering that the output is in long tons, the weight of a loaded car will be $(2{,}240 \times 2.5) + 1{,}900 = 7{,}500$ lb. Therefore, $N = 42$ $W_l = 7{,}500$ lb. Since the engine must be designed for the maximum load, the longest haul C will be taken, and $W_r = 2 \times 8{,}276 \times 2 = 33{,}104$ lb. Substituting these values in formula **1**, Art. **71**, $L_m = f(NW_l + W_r) + aNW_l$,

$L_m = \tfrac{1}{46} \times (42 \times 7{,}500 + 33{,}104) + .02 \times 42 \times 7{,}500 = 15{,}003$ lb.

The velocity of the trips is $\dfrac{5{,}280 \times 10}{60} = 880$ ft. per min. The required horsepower can now be found from the formula in Art. **70**, in which $L_w = L_m$,

$$HP = \frac{L_w v}{33{,}000} = \frac{15{,}003 \times 880}{33{,}000} = 400+. \quad \text{Ans.}$$

HAULAGE
(PART 5)

ROPE HAULAGE—(Continued)

GRAVITY PLANES

1. Definition.—A *gravity plane* is an inclined plane used for transporting material from a higher to a lower level and on which the gravity pull of the descending load is the motive power for raising the ascending load. Such a plane is also sometimes called a *self-acting plane* or incline. Gravity planes may be located either inside or outside the mine, but in either case the principles involved are the same.

2. General Details.—In order to work a gravity plane successfully, good tracks are needed on the plane and at the top and bottom. The tracks must have an inclination of at least 5° for a self-acting incline where the rope is attached directly to the car and at least 10° where a barney is used. Wire rope is generally used to haul the load, and it has been found in practice that lang-lay and locked-wire ropes are more serviceable than ordinary hoisting ropes for gravity planes, because they wear more uniformly as they move over the plane. For winding and unwinding the ropes, drums or wheels of suitable size are placed at the top of the incline; and to prevent the ropes wearing unduly, rollers and, at times, sheaves are required along the plane. Several methods are employed for lowering loaded cars and pulling up empty cars on gravity planes—such as fastening the

Copyrighted by International Textbook Company. Entered at Stationers' Hall, London

hoisting rope to the draw-head of one of the cars; using barneys that lower or push up the cars, the rope being attached to the barneys; counterweights or balance trucks that are sufficiently heavy to pull up an empty car, attached to a rope, and are in turn pulled up the plane by a loaded car attached to a rope.

ARRANGEMENT OF TRACKS

3. Self-acting inclines have tracks laid in any of the following four ways: (1) the incline may be double-tracked from top to bottom, as in Fig. 1; (2) it may have three rails above a parting in the middle of the plane and three rails below this parting, as in Fig. 2; (3) it may have three rails above a parting in the middle of the plane and two rails below this parting, as in Fig. 3; (4) it may be a jig plane with two rails for the car and two for the balance truck, as in Fig. 4.

The method of arranging the tracks adopted on an underground incline depends very largely on the width of the incline and the timbering required. Timbering on inclines is not advisable unless set skin to skin, for a runaway car may knock out timbers that are set here and there to hold up the roof, and let rock down on the tracks. If the roof is tender, it may be necessary to use a system, such as a jig plane, that requires only a single-track throughout in order to avoid the wide partings required by other systems.

4. Double-Tracked Incline.—The best gravity planes are those having two tracks throughout, such as that shown in Fig. 1. The advantages derived from the use of two tracks far exceed in value their increased cost over the other methods, there being less wear and tear on the rope, cars, and rollers. The loaded cars a are descending the plane and hoisting up the empty cars b. At the top of the incline there is an automatic switch c that is thrown by the empty cars so that the next loaded cars will take the track just previously vacated by the empty cars. For instance, the switch was turned for the condition shown so

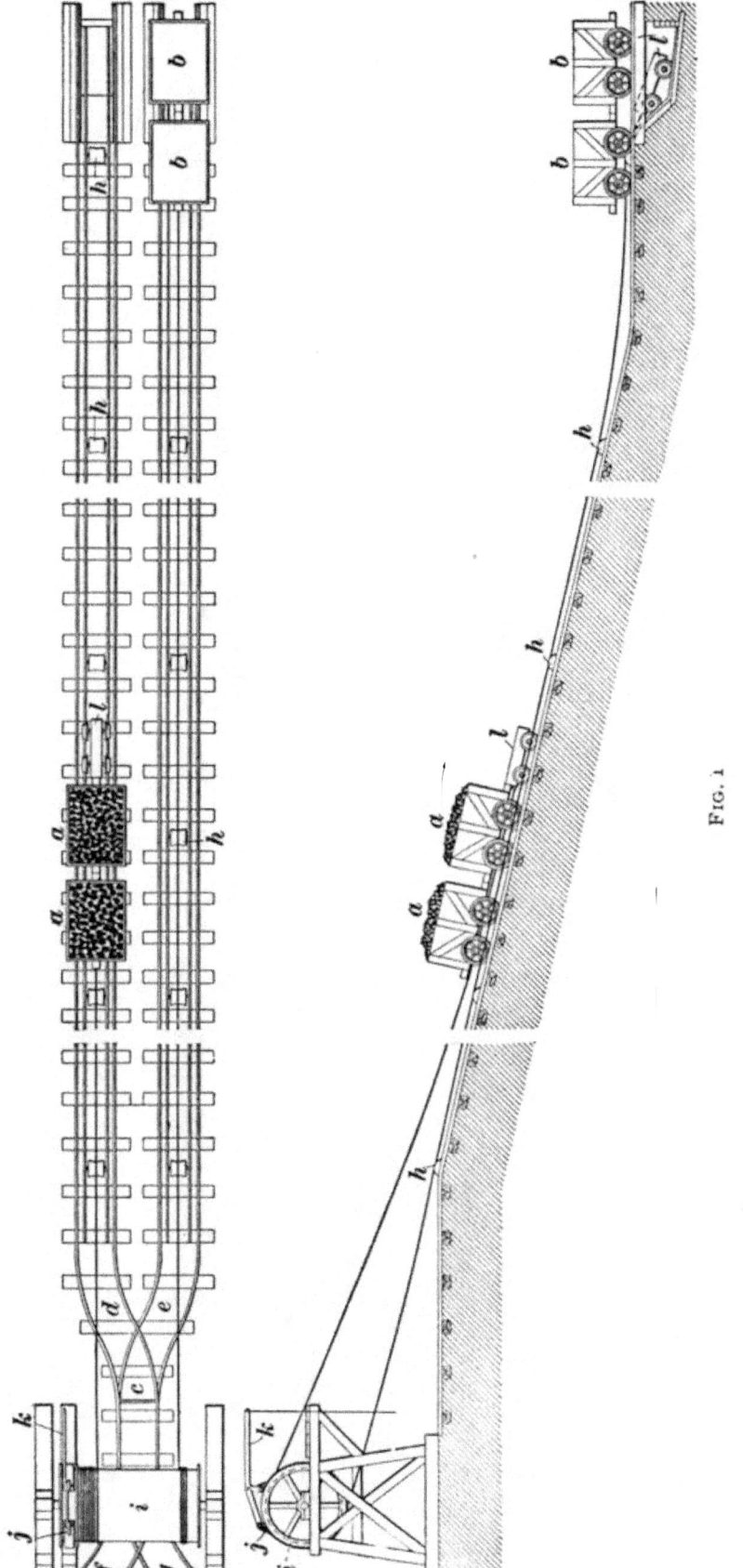

FIG. 1

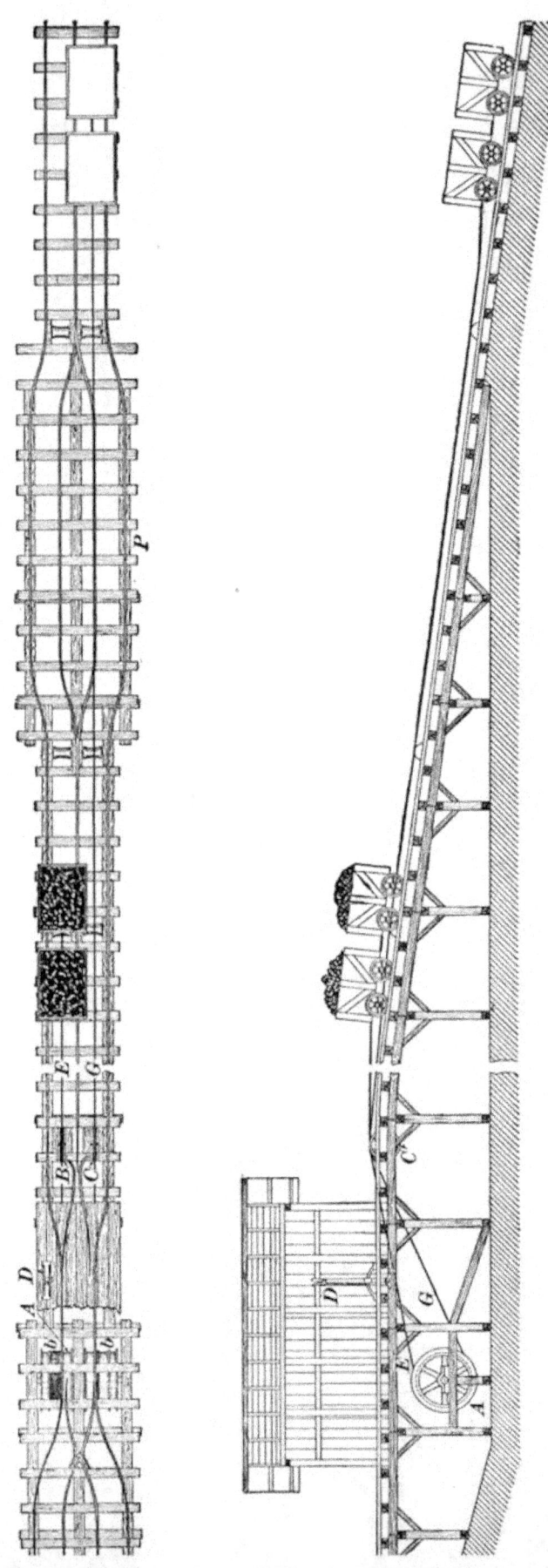

Fig. 2

that cars a took the track d. When cars b reach this switch, they will throw it so that the next loaded trip will take the track c. When the empty cars reach the top of the plane, they are pushed on to track f. At the junction of the tracks g and f, there is a spring switch that is set so the empty cars will take the track f; and when the loaded cars are pushed to the top of the plane, their flanges open the switch, which immediately springs back into place. The rollers h are placed between the tracks to prevent the rope from wearing and also from cutting the ties. At the head of the plane, there is a drum or reel i about which the ropes wind, one winding on the top and the other on the bottom, so that one rope coils on the drum as the other uncoils from it. The drum is controlled by a band brake j operated by a handle k. After the loaded trip has passed the middle point of the plane in descending, the speed will usually increase, and the brake is used to control it.

5. Three-Rail Inclines.—Fig. 2 illustrates a **three-rail gravity plane** with a turnout P in the center of the plane. Since the drum A is below the platform and track, the brake is worked by a lever D. The object of having three rails is to prevent the rope from coming in contact with the ascending or descending car. As it is, the ropes E and G ride the inside parting rails, and are worn by them. The switch arrangements are similar to those described in connection with the double-tracked incline, and at the bottom there is an automatic latch switch that is set by the loaded cars for the empty cars, being the reverse of the action of the switch at the top of the plane. The bent rails at the knuckle form a frog to guide the cars to the proper track. At the knuckle, the ropes bend over the sheaves B and C.

6. Combination Two- and Three-Rail Inclines. In Fig. 3, a three-rail plane a is made from the top to the parting b and a two-rail plane c below the parting. If the plane is long, the descending rope d will strike against the ascending cars, run off the rollers where there are three

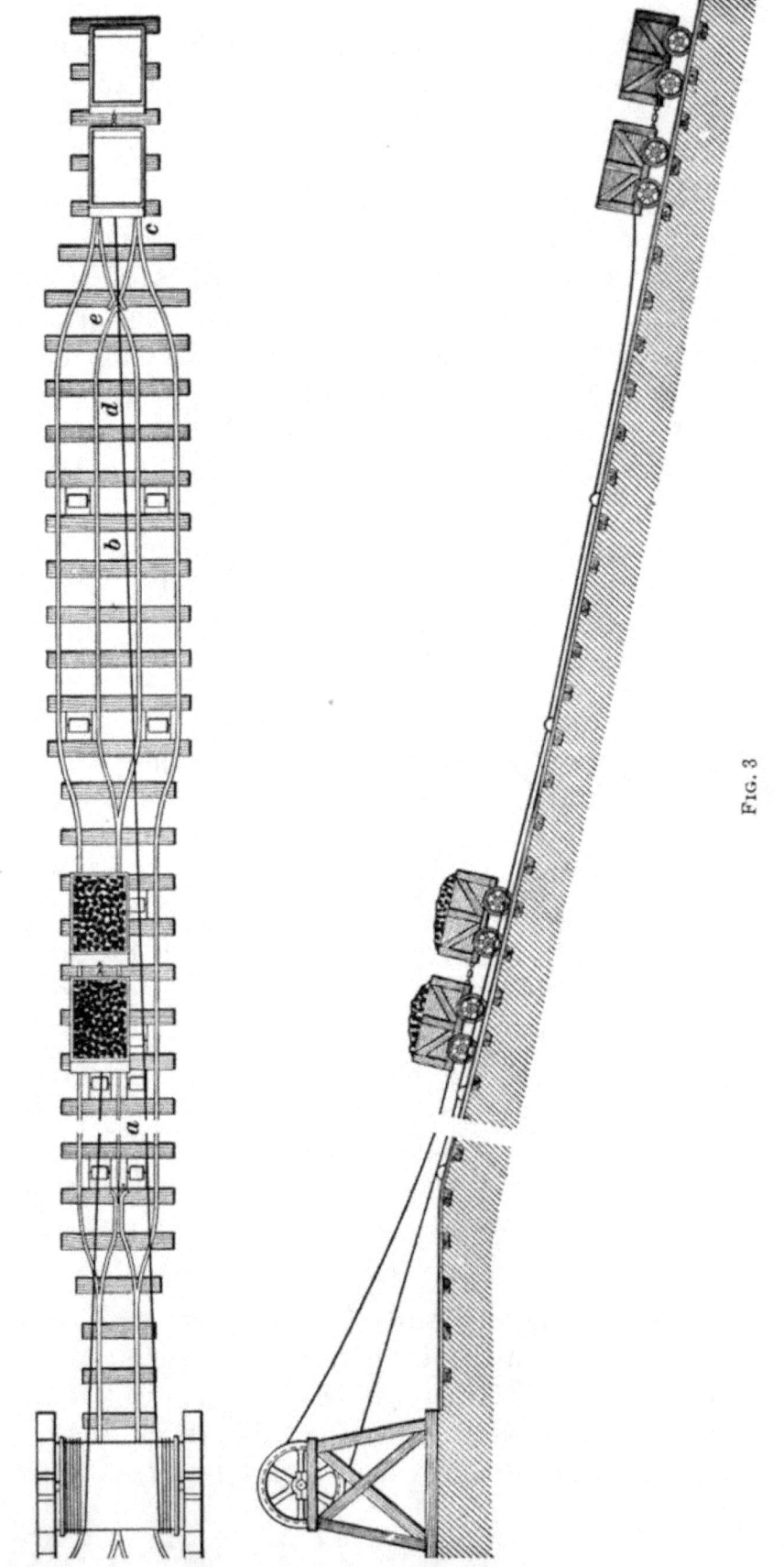

Fig. 3

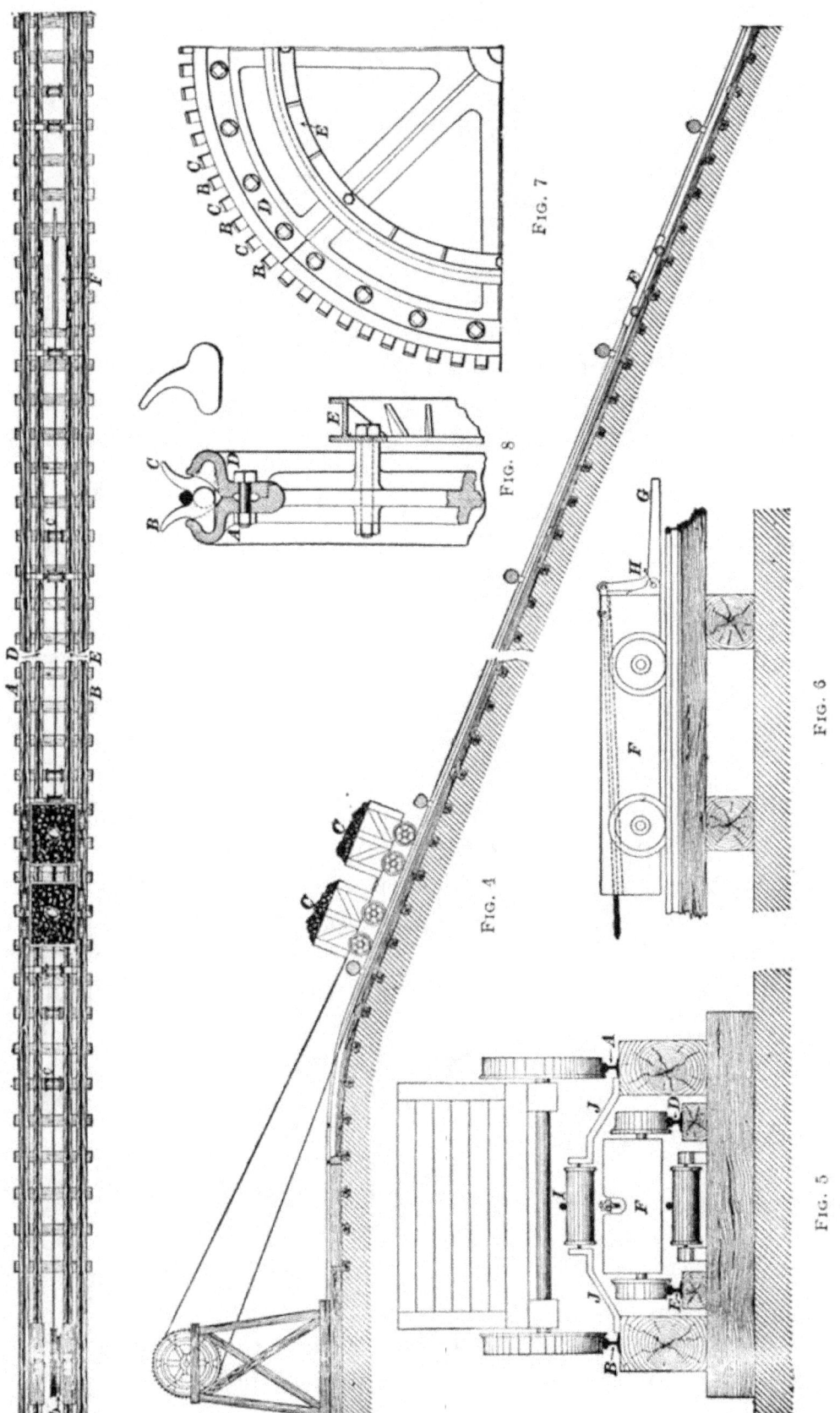

Fig. 4
Fig. 5
Fig. 6
Fig. 7
Fig. 8

tracks, and catch in the switch frog e below the parting. The switch below the parting is automatically thrown by the descending cars; and when the ascending cars reach this point, they take the tracks over which the last descending cars passed. The use of the two- and three-rail plane is not advisable except for short distances.

7. Jig Planes.—The **jig plane** is a gravity plane having a balance truck for returning the empties to the top of the plane. This balance truck runs on a track between and slightly below the rails of the car track. The balance truck is of less weight than the loaded trip, but of greater weight than the empty trip, so that the descending loaded trip raises the truck to the top of the plane and when the empty trip is attached to the rope the truck descends the plane and pulls up the empties.

8. Fig. 4 shows a plan and elevation of a jig plane. The tracks AB are for cars C, while the track DE is for the balance weight F. The grip wheel b takes the place of a drum. The rope leading to the balance truck is supported on a set of rollers c, while the rope attached to the cars is supported by another set of rollers at a higher level. A section through the plane showing the relative positions of the car, balance truck, and rollers is shown in Fig. 5. AB are the car rails and DE the rails on which the balance truck F runs; I is a roller for supporting the rope attached to the cars; and J is a support for this roller. A roller for supporting the balance truck rope is shown beneath the truck. Fig. 6 is a side view of the balance truck F. The rope is attached to a bell-crank at the back end of the balance truck and pivoted at H so that the lever G is elevated to the position shown. In event of the rope breaking, the lever falls and digs into the ground and prevents the truck from running down the plane. Jig planes are arranged usually for light loads and short distances. The detailed plans and elevations of jig planes shown in Figs. 4 to 8 are, however, for heavy loads. Since but one car track and one track for the balance truck are needed, a grip wheel may be used for the rope instead of a drum or multiple-lap wheels. This wheel may be mounted on a frame above or below the tracks so as to line with the center of the tracks. A section and elevation of a grip wheel is shown in Figs. 7 and 8. The grips BC are fixed on the periphery of the wheel like so many teeth and are independent of each other. The rope presses in between the jaws of the grip, which open at the bottom and close at the top until the rope is held as in a vise. If the rope breaks, the grips will hold it and prevent it from running out down the plane. There is a brake wheel E on the side of the grip wheel for the brake band.

9. In pitching seams, where the mineral will not run by gravity, or where it will be broken if allowed to run, the jig is sometimes employed to advantage. In such cases, the balance car runs in between the car rails, and the grip wheel runs vertically; or the balance car runs on another track parallel to the car rails, and the grip wheel is placed horizontally. The weight of the balance truck is made equal to half the sum of the weights of the empty and loaded cars.

10. Fig. 9 shows a good arrangement for a jig wheel and

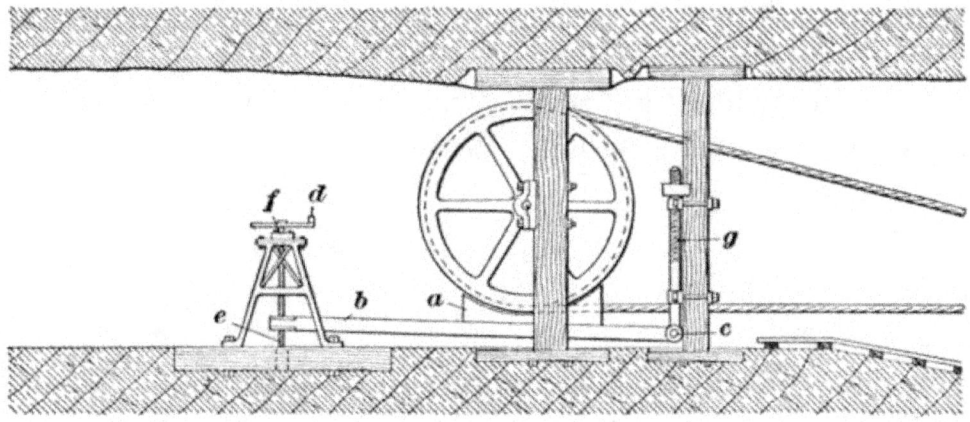

Fig. 9

brake. The brake a is controlled by lever b, whose fulcrum is at c. The handle d works a screw e that passes through a block f just below the handle and through a screw block centered in the brake lever b. The wear of the brake can be taken up at each end of the lever; for instance, the wear at the front of the wheel can be taken up by screwing up a nut on the threaded bar g, and in the rear of the wheel by the screw e.

11. Drums for Gravity Planes.—The drum shown in Fig. 1 is the most satisfactory form for most gravity planes. It consists of a number of spider wheels keyed on a shaft. To these spiders are bolted wooden planks 2 or 3 inches thick that completely lag over the spiders and form the drum surface. The spiders at the ends of the drum are flanged so as to project a short distance above the drum surface. Near the end flange, there is another flange; and the space between this flange and the end flange forms the seat for the band brake. Sometimes, there is a brake seat at each end of the drum. The drum is mounted on substantial wooden frames of a sufficient height to permit the cars to pass under it. The frames must be thoroughly braced, and if the loads are heavy, the mud-sills should be bolted to foundations. The diameter of the drum will depend on the size of the rope and on the length of the plane. The length of the drum should be such that when the cars are about half way down the plane and not on a turnout the rope will be in the centers of the tracks. The shaft turns in journal-boxes that are bolted to the cap pieces of the timber frames. Fig. 2 shows a drum placed below the track. The housing above keeps the drum dry in rainy weather. The drum pit must always be thoroughly drained.

12. Geared Drums.—Sometimes, where there is a steep pitch and a single drum is used so that one rope winds on at the top and the other off at the bottom, and vice versa, the rope winding on at the top will be so high, if the lead is short, as to lift the cars off the track. This difficulty can be overcome by the use of two drums geared together so as to turn in opposite directions. This allows the ropes to run on and off on the same side of the drum and at the same height from the track. Both ropes can thus be kept at but a short height above the rails either by placing the drums below the tracks and letting the ropes run on and off at the top or above the track and letting the ropes run on and off at the bottom of the drums.

Fig. 10 shows the construction and method of arranging drums geared together. A and B are the drums on which the ropes wind; C are the brake wheels separated from the rope reels by a flange. The drums are mounted on shafts supplied with journals at each end, which run in journal-boxes. The wooden frames are so constructed that the gear-wheels D and E mesh and the rope drums are outside to line with the tracks, and the brake wheels are so near together that the operator can put both brakes on at the same time.

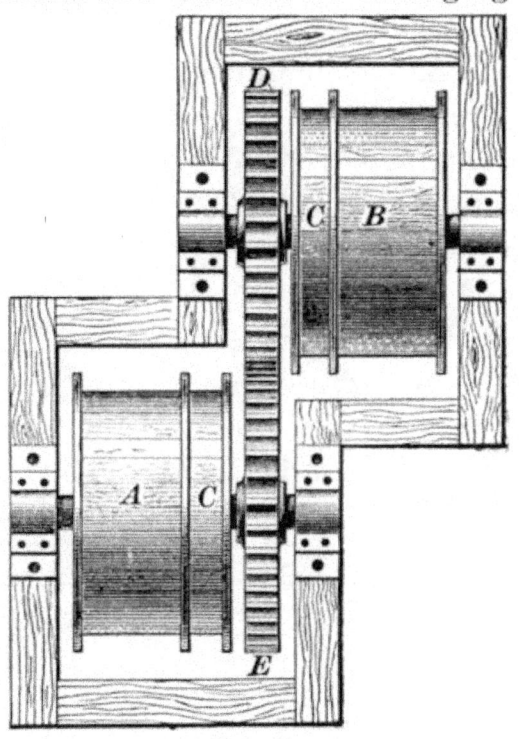

FIG. 10

3. **Head-Wheels.**—Instead of a long drum on which all or the greater part of the hoisting rope is coiled, a single wheel may be substituted around which the rope is coiled several times in order that it may not slip as the cars ascend and descend the plane. When a single wheel is used for this purpose, there is trouble on account of its tendency to travel from one side of the wheel to the other, sometimes known as **fleeting**. This side travel of the rope is illustrated in Fig. 11, which shows a wheel with a rope running on at r, and off at o, the next coil to the left running on a lesser diameter, surges over toward the coil s until at last it jams on the other side at o where the rope runs off. If the wheel is flat the surging is not continuous, and the running-on

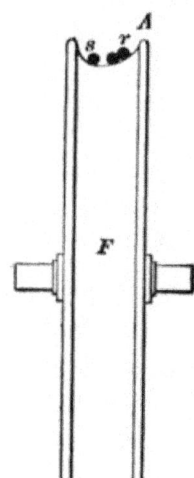

FIG. 11

rope will sometimes ride the second coil, and jumps and thuds will be more numerous than with grooved wheels; in either case there is considerable wear and tear. The rope running on crowds over toward the flange A and produces the surging mentioned. Sometimes, the intermediate coil slips, and then a longer interval elapses before the coil r presses against the flange; but when it does, another surge and jerk takes place. It has been proposed to run a single wheel with a spiral groove; this, however, is unsatisfactory, since the groove prevents the rope from surging and it will soon crowd off the wheel, or ride adjacent coils.

14. Multiple Grooved Wheels.—The surging of the rope may be obviated by the use of two grooved wheels instead of a single wheel. The rope passes around these wheels as shown in Fig. 12, only one-half the circumference

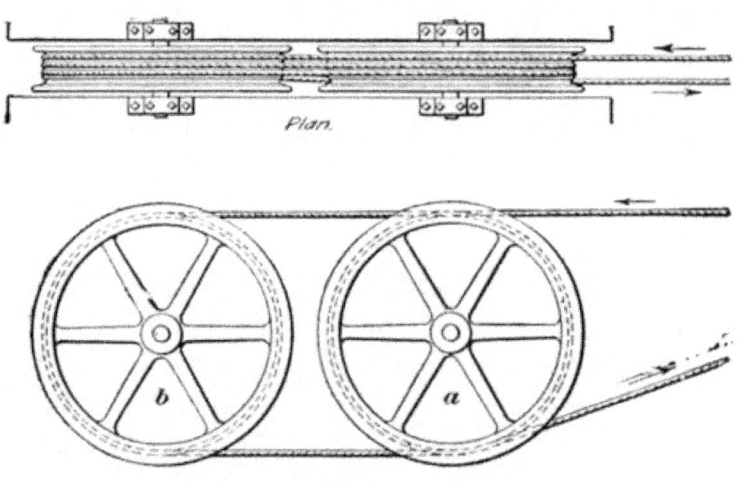

Fig. 12

of each pulley being in contact with the rope. The wheels have four grooves each. The rope passes over the wheel a, and, making a half turn about b, passes under a into the second groove, then back to b over the latter to the third groove of a back to b and finally off from b. By this arrangement, there is ordinarily no slipping of the rope. While a rope does not grip the wheels quite as firmly as when it entirely encircles a single wheel, this can be easily compensated for by one or two extra laps on the wheels. The ropes are sometimes passed from one wheel to another in the form of

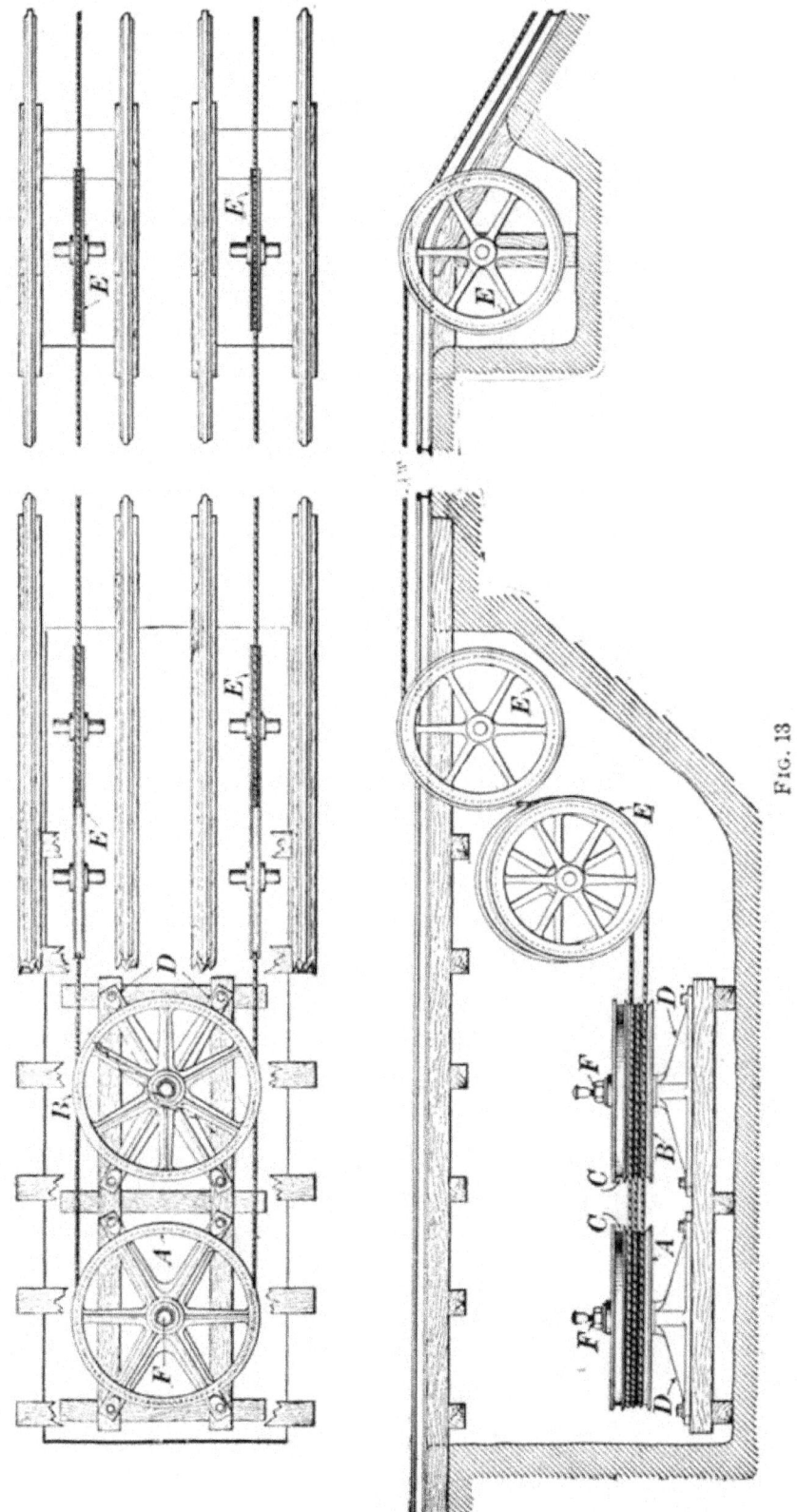

Fig. 13

a figure 8, but the reverse stresses on the rope thus bent are very hard on the rope and it is much better to arrange the rope as shown in Fig. 12.

15. Fig. 13 shows two grooved drum wheels A and B placed horizontally instead of vertically. These wheels are provided with brake wheels C and are mounted on shaft pedestals D. The wheels are of such a diameter that the plane ropes will center in the tracks. In order to bring the rope low enough to properly run on and off these wheels, deflecting sheaves E are employed, and similar sheaves are placed at the knuckle to center the rope. The cost of grooved wheels is less than that of drums; and while they require less rope and less space than drums, the rope wears quickly; and when the brake is applied, there is a tendency for the rope to bind and make starting hard. Grooved wheels also give trouble in wet weather, as the rope will then slip at times. The varying tension in the different parts of the rope passing about several wheels in this way produces unequal wear of the grooves and consequently excessive stress on the rope that quickly destroys it and forms a serious objection to such drums or sheaves.

16. Differential Drum.—A very good device for overcoming the excessive stress produced on a wire rope running about two drums is a **differential drum,** arranged as shown in Fig. 14. The grooved surface of this drum is composed of a number of independent, grooved steel rings a that are placed loosely between the flanges, there being as many rings as there are grooves desired. The surfaces of contact between the rings and the rim b must be such that the frictional adhesion between the two is less than that between the rope and the rings, but sufficient to transmit the desired amount of power. If there is any unequal wear in the rings, they will adjust

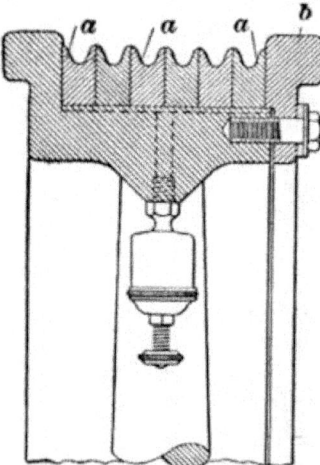

Fig. 14

themselves to compensate for this difference by slipping sufficiently to maintain a uniform tension in the different parts of the rope. The rope and the ring on which it lies thus become practically one piece of the machine, and instead of the rope slipping in the groove and thus being worn, the rings move and the wear is between the rings and the rim of the drum. The pressure is thus distributed over a greater area and the wear is correspondingly less. The rings can be replaced as they become too deeply worn. Differential rings of this kind are said to increase the life of a rope from 15 to 50 per cent.

17. Barneys.—When cars are attached to haulage ropes by couplings, as is frequently the case on inclined planes, they are subject to strains that rack them; again, there is considerable risk from accident, such as the cars running over the knuckle or angle at the top of the plane before the rope is hooked on. To these disadvantages may be added the trouble of attaching and detaching the cars and the wear of the ropes at the ends due to kinking when unhooked from the cars. To obviate these disadvantages as much as possible, small cars l, Fig. 1, called **barneys**, for pushing the cars are used to replace hooks. The barney ordinarily runs on a narrow track placed between the car rails until it reaches the bottom of the plane, when it runs into a barney pit and permits the cars to pass over it. The barneys are attached securely to the plane ropes and holds the cars as shown in Fig. 1 during their journey up and down the plane. The use of the barney is safer than attaching cars directly to the rope, since there is no danger of the cars running away unless the rope breaks. The cost of its instalment adds to the first cost of a plane, but its saving in wear and tear on the cars and rope will offset this expense. In order to do away with the expense of laying a barney track inside the car track, barneys are used with adjustable grooved wheels that require only a short piece of track at the bottom of the incline to guide them into the barney pit.

18. The **improved barney** shown in Fig. 15 is constructed as follows: The body or frame can be either iron or timber. The top, sides, and ends are secured to one or more central webs or bearing plates. Each wheel has two grooves or guideways a, a', as shown in Fig. 15 (b), these guideways always being adapted to fit over the rails. The wheels are self-oiling, having a chamber b and a cap piece with a hole in it through which the oil is introduced. The wheel slides on an axle d between the central web e and

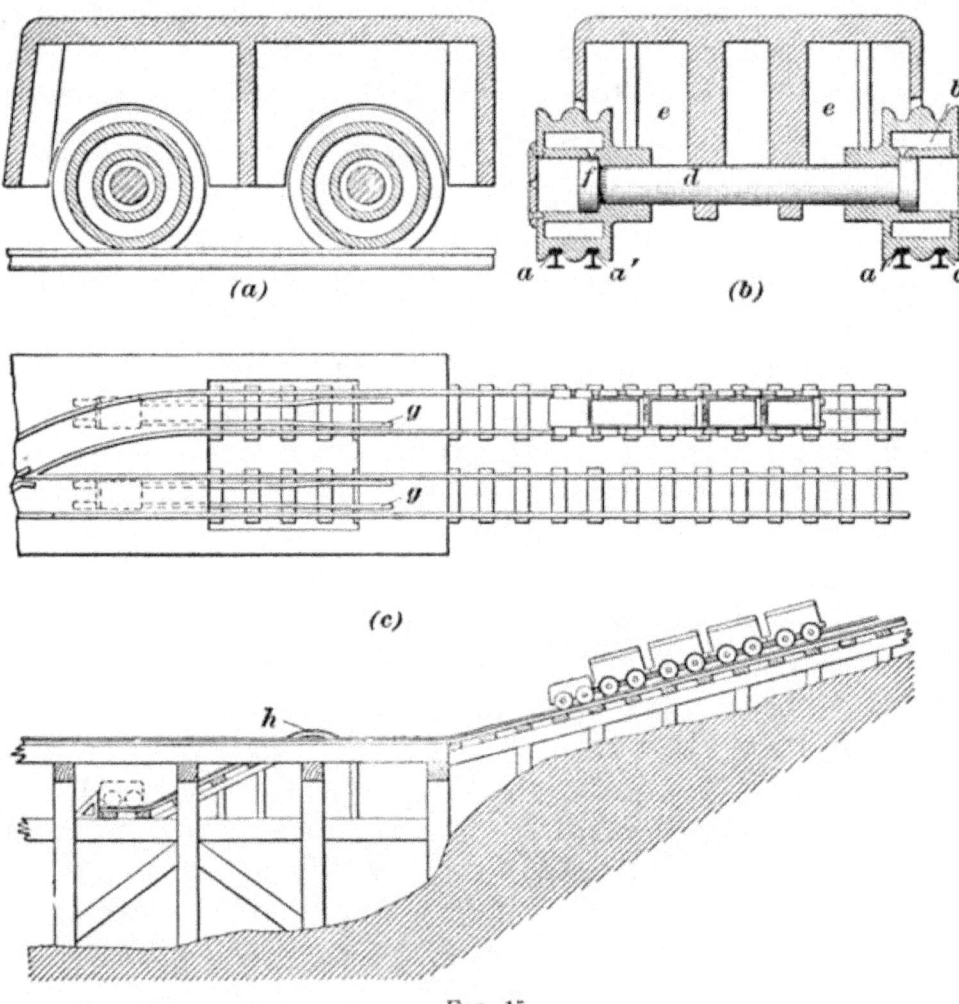

Fig. 15

collar f, which is at the end of the axle and acts as a stop for the wheels, and fits loosely enough to give the oil easy access to the oil cavity.

The method of operating the barney is shown in Fig. 15 (c). The main haulage way is a single track, but at the bottom of the hoisting plant there is a supplementary track *g*. On the plane, the barney runs on the car track, the outer grooves *a* engaging the rails. When the barney reaches the supplementary track, the inner groove *a'* of the wheel engages with this track, which rises gradually, as shown at the point *h*. When the wheels have raised so as to clear the outer rail, they slide inwards on their axles as the track *g* gradually decreases in gauge until they become flush with the outside of the barney, when the latter drops into the barney pit as shown. The weight of the barney must be sufficient to prevent its being lifted from the track by the weight of the cars pushing against it when it reaches the bottom.

The great advantage of this barney is evident, as it does not require an extra track throughout the length of the hoisting plane, but only for a distance of 30 feet at the bottom of the plane.

19. Anchoring Gravity Plane Tracks.—For the successful working of gravity planes, the track should be laid on a

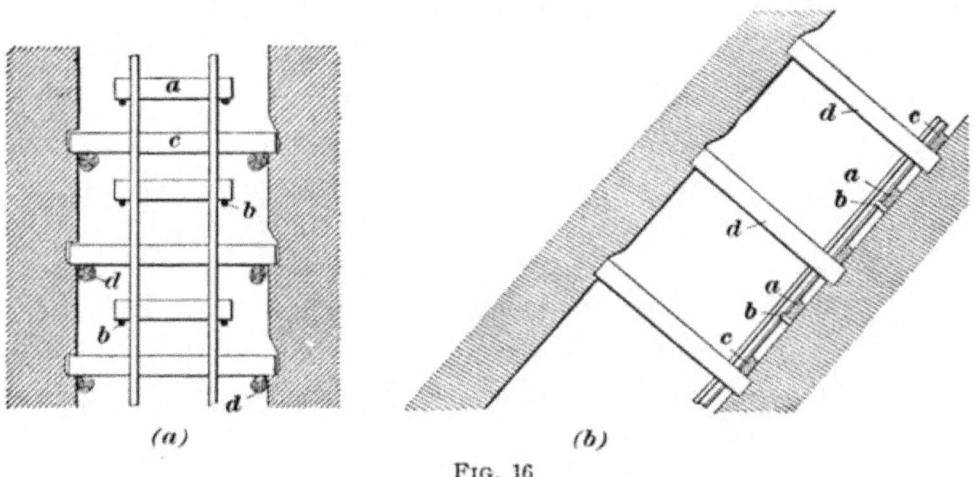

Fig. 16

uniformly graded, solid roadbed with the ties thoroughly tamped and anchored, because the movement of the cars will have a tendency to move the ties down the plane. On the surface, in addition to the movement of the cars, the variation in temperature will cause expansion and contraction of the rails and move them down the plane, and unless the ties are well tamped and supported they will also be moved, to the detriment of the plane. Fig. 16 shows a method of laying a track on an incline to prevent the track from slipping down the slope. The short ties a are held in place by pins b driven into the bottom. The long ties c rest in notches cut in the sides and also against the posts d set into notches in the top and bottom.

20. Automatic Switches.—Every gravity plane is provided with switches that are thrown by the cars passing over them either up or down the plane, so that the loaded and empty trips will automatically take the proper tracks. Two kinds of these automatic switches are in use, one, called a *spring switch*, that requires a spring to close it, and one that is so placed that it is moved to its proper position by the car. Both switches are automatic and are shown in Fig. 17.

21. The **spring switch** has a tongue or movable point a, Fig. 17 (a), attached to a spring b that holds it close against rail c as shown. When a car comes along track A, the flange of the wheel causes the movable point a to take the position shown by the dotted lines; however, as soon as the car has passed, the spring forces the point back to its original position.

22. The switch in Fig. 17 (b) has two switch points d and e that are pivoted to the ends of the lead rails. These points are held rigidly by a bridle f in order to control the movement of the points. When the points are in the position shown, a car coming from track B will take track D, and a car coming from track A will take track B; a car coming from track C, however, will throw the switch de to the dotted position, so that a car coming from track *A* will take track *C*. As the loaded trips on gravity planes take alternate tracks, the usefulness of the switch *d e* becomes evident. The guard-rails *g* assist the cars in running over and off the switches and at the same time prevent their jumping the track.

23. In those gravity planes in which three rails are used from the head of the slope to the parting and only two from the end of the parting, an automatic switch similar to Fig. 17 (*b*) must be placed at the junction where the two rails unite with the parting. Fig. 18 shows an automatic switch that may be used at such a place. It does not differ materially from the switch *d e*, Fig. 17 (*b*); two switch points *a* and *b* connected by an iron switch rod or bridle are pivoted at *c* and *d*, respectively, in such a manner that they may move freely from side to side. With the switch points in the position shown, the empty cars coming along track *e* will be pulled up the slope on track *f*, while the descending loaded cars coming down the slope

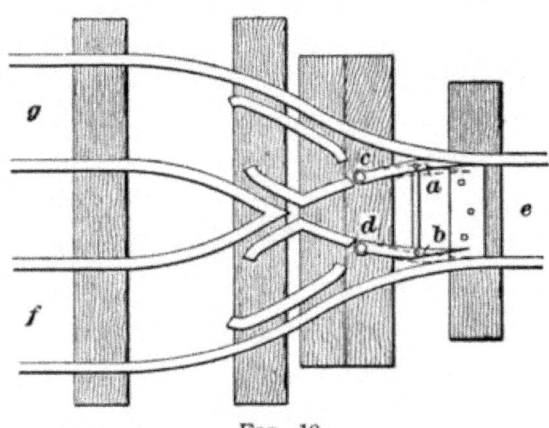

Fig. 18

on track *g* will shift the switch points over to the position shown in dotted lines, and go along track *e*. At the next trip, the empty cars coming along track *e* will be pulled up the slope on track *g*, while the descending loaded cars coming down the slope on track *f* again shift the switch points to their original position. The foregoing operation is then repeated, the empty cars being pulled alternately up the tracks *f* and *g*.

24. Derailing, or Safety, Switches.—The only reliable method of derailing a runaway car is by means of the automatic safety switch that is always set to run the cars off the track, except when the operator holds it open for the cars to pass down the incline. The switch is opened by the ascending cars and closed automatically after they have passed. A switch of this kind is shown in Fig. 19. The two switch points *a* and *b* are fastened to a chain *c*, one end of which is connected to one leg of the bell-crank *d*, having a weight *w* hung to the other leg, which causes the switch to close automatically, the points always taking the position shown

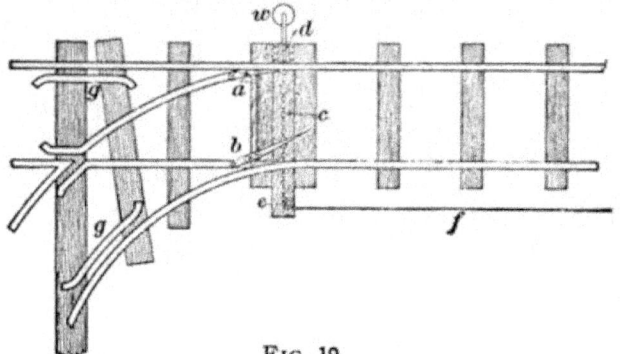

Fig. 19

in the figure. The other end of the chain *c*, after being led around the pulley *e*, which is fastened to the cross-tie, is attached to a wire *f* that connects with a lever at the head of the plane. This arrangement admits of the empty cars being run up the slope, since the wheels force over the points *a* and *b* to the position shown in dotted lines. When the loaded cars come down the slope, the points *a* and *b* must be pulled over by the wire *f* to the position shown by the dotted lines, so that the cars may pass. In case the hoisting rope breaks at any point above the switch, the wire *f* is not pulled, since the switch points are in a position to derail or side-track the cars.

25. Safety Blocks.—A safety block should be placed at the head of a slope or plane near the knuckle to prevent cars from descending the plane before they are properly attached to the rope.

A good form of safety block is shown in Fig. 20, in which *A* and *B* are two timbers pivoted at *C* and *D*, respectively. The end of each timber is iron-bound. Directly in the center between these timbers is fastened an iron plate *E* having a slot in it through which the vertical part *F* of the rod *G* may be moved back and forth. The timbers *A* and *B* are connected by two wrought-iron links *H* and *I*, which form a togglejoint. The ends of these links, meeting in the center, are fastened to the end of the rod *F* projecting up through the slot in the plate *E*. *J, J* are wrought-iron levers placed outside of the track close to the rails. These levers are pivoted at K and are connected by a rod L, at the center of which the rod G is so fastened that when either of the levers is moved the other moves with it. With the timbers in the position shown, a train of loaded cars coming along track M takes track N, and goes along until the wheel on the inner rail of the front car strikes the timber B. After the rope has been attached to the cars, the runner shifts either of the levers J to the position J', shown by the dotted lines. In doing this, the rod G is pulled to the left, the vertical part F of it slides to the left, also, in the slot of the plate E, and causes the ends of

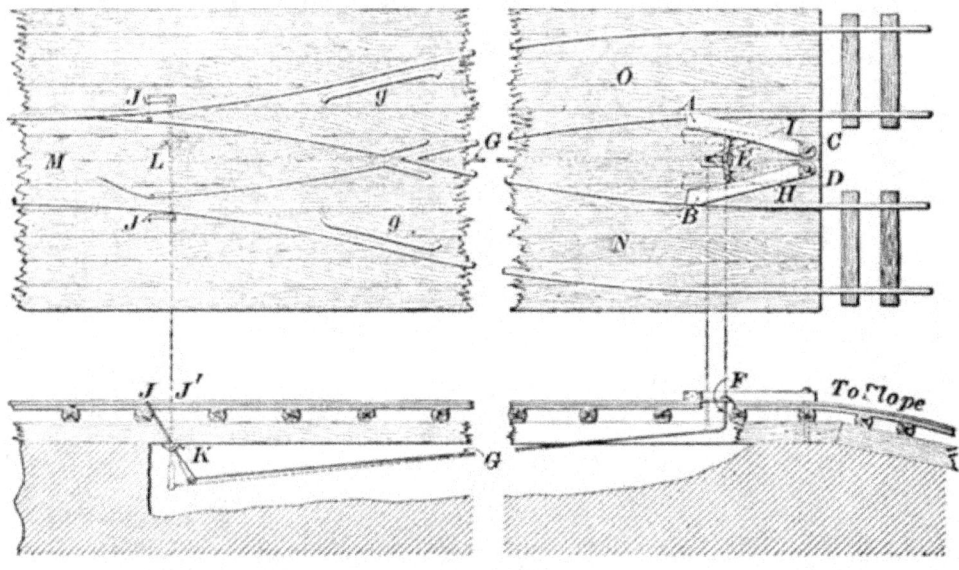

Fig. 20

the links H and I fastened to it to go with it, thereby causing the blocks A and B to take the position shown by the dotted lines. The tracks being thus freed, the cars may be let down the slope. The empty cars coming up the plane on track O pass along until the first car reaches the lever J on the outside of this track, which is now in the position J' shown by the dotted lines, and moves it to its original position J, in passing to track M, thereby again placing the timbers A and B over the tracks. On the next trip, the loaded cars coming along track M take track O, and are prevented from descending the slope by the timbers being placed over the inner rails by the last train of empty cars. This operation is repeated every trip, the empty cars coming up the plane and automatically placing the timbers over the rails, to prevent the loaded ones from running away before they are fastened to the rope.

26. Fig. 21 shows a safety block that consists of a heavy wrought-iron bar a firmly keyed on a shaft b held in position by bearings c that are bolted to suitable supports. The top of the front end a' of the bar is inclined, and is caused by the weight w to project up in the center of the track to such a height that it will strike the axle of the cars, this height

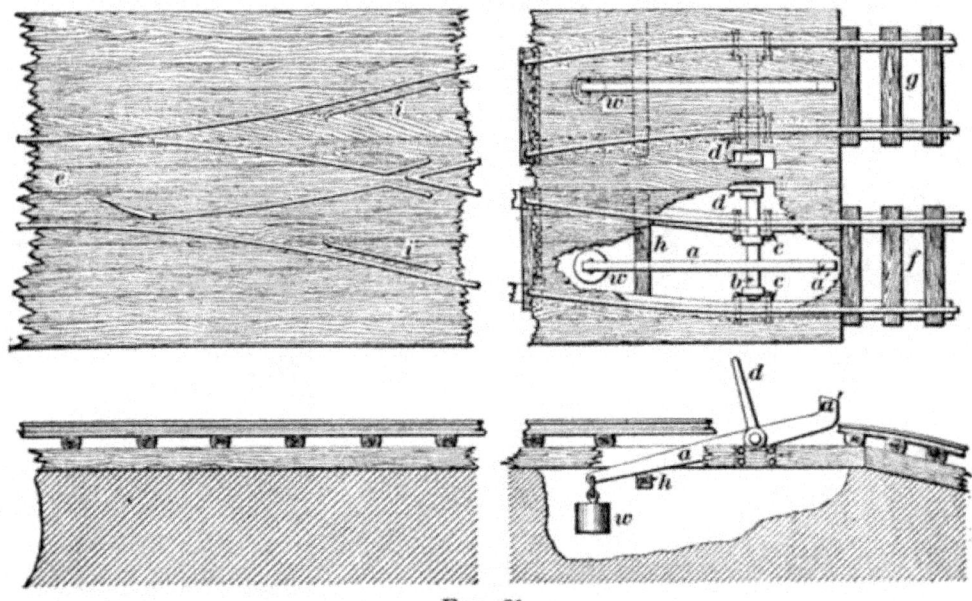

Fig. 21

being governed by the timber h. At one end of the shaft b is keyed a lever d (placed at one side of the track) by which the block is operated. One of the blocks is placed in the center of each track. With the tongues of the switch in the position shown, the loaded cars coming along track e take track f and run along it until the front axle of the first car strikes the projecting part a' of the bar a. After the rope has been attached to the cars, the lever d is pulled toward the plane, which causes the projecting part a' to swing down, when the cars may be let down the slope. The projecting part a' should be held down until all the cars have passed over it; the lever d is then released, and the part a' is again brought up to its proper height by the weight w. The trip of empty cars coming up the slope on track g finds the block in the position shown. The axles of the cars strike the inclined projection a' and force it down, allowing the trips to pass over the track c. After the cars have passed over it, the projection a' is again brought up by the weight w to its original position, as shown. A like arrangement is put on track g and the two are operated alternately. When this block is used, it is impossible for the cars arriving at the head of the slope to run down the plane against the will of the operator, since they are always in the proper position to prevent the cars from passing. The guard-rails i are to prevent the cars from being derailed at the frog.

27. Fig. 22 shows a safety block that may be used on

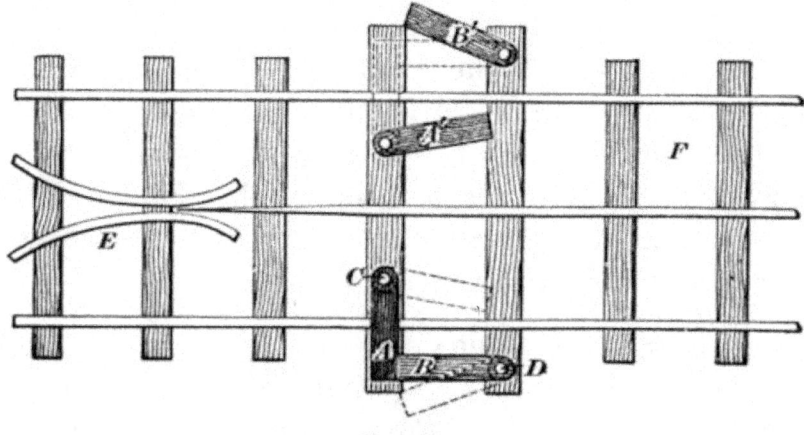

FIG. 22

gravity planes where light loads are run. This consists of two iron-bound timbers A and B, pivoted at C and D, respectively, in such a manner that the timber A can be moved over the top of the rail. One of these blocks is used for each track. With the timbers in the position shown, loaded cars coming along track E will be prevented from descending the plane. After the rope has been fastened to the cars, the timber B is moved to one side, so as to allow A to take the position shown by the dotted lines. The empty cars coming up the plane on track F find the timbers in the position shown on that track, and pass along, after which the timber A' is swung over the track, and is locked by the timber B', as shown by the dotted lines. The next trip of loaded cars coming down track F finds the track closed. After the rope has been fastened to the cars, the timber B' is moved over and the cars are let down the slope, the timber A' being moved by the wheels to its original position. The empty cars coming up the slope on track E find the timbers A and B in the position shown by the dotted lines, and pass along, after which the timbers A and B are again placed on the track by the car runner. This operation of locking and unlocking the blocks is repeated on each track alternately.

28. Knuckle Sheaves.—On a sharp knuckle, the track rollers are placed as close together as possible, or very frequently vertical sheaves are used instead of rollers, as shown at E, Fig. 13. The bending stresses in the rope are greatly reduced by using a sheave of large diameter, but the weight of the large sheave causes it to revolve for a long time after the rope has passed over it, so that it may be still revolving when the rope returns, going in the opposite direction. The friction between a rope and a sheave moving in opposite directions wears the rope considerably.

29. Compensating Sheaves.—To overcome the difficulty from revolving sheaves, the arrangement of compensating sheaves shown in Fig. 23 may be used. The two

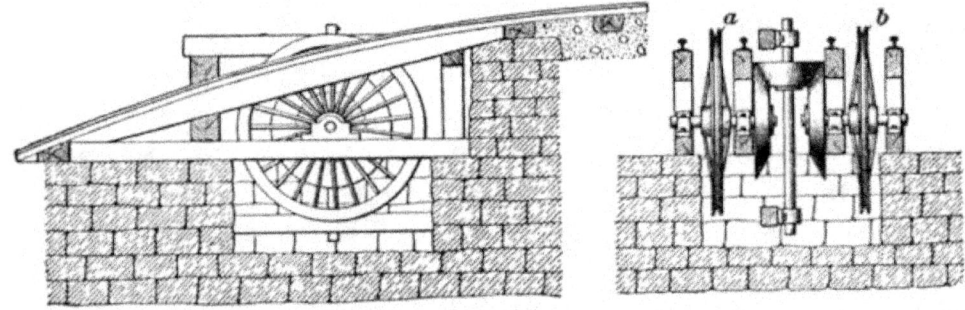

Fig. 23

sheaves a and b are connected by friction wheels, as shown. As either sheave is moved, the other is set in motion in the opposite direction, so that when the rope, traveling at a high rate of speed, strikes a sheave, it finds it already revolving in the proper direction and the heavy friction, that, with ordinary sheaves would be developed in setting them in motion, is thus avoided.

30. The **drag bar,** or **dog,** shown in Fig. 24 is a provision made for safety during the ascent of heavy trips on a plane. If the rope breaks, this bar prevents the cars from running back. The drag bar shown is simply a strong iron bar m hooked to the rear of the last car in the trip. Its mode of operation is as follows: As the trip runs up the incline, the bar trails over the ground. Should the rope or one of the

FIG. 24

coupling links break, the bar sticks into the ground and prevents the trip from running back. The drag is removed and thrown on a car when a trip is lowered down the plane in order to prevent its dragging over the ties and rollers.

31. Hooking-On Chain.—On inclined planes, the rope is usually attached to the car by means of a chain from 15 to 20 feet in length, socketed or clamped to the rope and provided with a clevis or hook. This chain saves the rope from injury to which the end of a haulage rope is always subject in attaching and detaching. The rope wears more rapidly at the end than in other places, and as it wears short pieces are cut off. When the rope shows signs of wear throughout its length, it may be changed end for end, that is to say the chain is cut off and this end made fast to the drum; the chain is then attached to the former drum end of the rope.

32. Inclination of Gravity Planes.—Wherever possible, gravity planes should be so constructed that the brakeman at the top of the plane may be able to see the cars at all points on the plane and thus control their speed. Where the pitch is considerable and the length is short, only one car is run at a time; but with a long plane and a similar pitch, it may require two or more cars to lift the long heavy rope. In either case, a powerful brake is needed, because when the trip has passed the parting the ropes are no longer in balance and the weight of the descending rope adds increasingly to the gravity pull on the descending side. When the inclination of the plane is comparatively small, long trips are a necessity; and even then it may be necessary to provide an increased fall at the top of the incline and a reduced grade at the foot of the incline in order to obtain sufficient gravity pull for the descending load to perform its work.

To operate a gravity plane successfully, the inclination must be sufficient for trips of reasonable size to be run, so as to handle the desired output, and there must be a sufficient number of loaded cars in a down trip on a given incline to overcome the resistance due to the weight of the rope and the empty cars and the friction of the system.

CALCULATIONS FOR GRAVITY PLANES

33. In considering the action of gravity planes, the following facts should be understood: The moving force is the weight of material in the car. The weights of the ascending and descending cars balance each other and the forces opposing motion are the gravity resistance due to the weight of the rope that must be pulled up the plane, and the frictional resistance due to the moving parts of the system, that is, to the rope, drum, rollers, sheaves, cars, and material.

Let W_l = weight of loaded car, in pounds;
W_e = weight of empty car, in pounds;
W_r = weight of rope, in pounds ($= wl$);
w = weight of hauling rope, in pounds per foot;
l = length of plane, in feet;
X = angle of inclination of plane;
f = coefficient of friction (usually taken as $\frac{1}{40}$);
N = required number of loaded cars in trip to produce equilibrium of two sides of plane.

The force required to overcome the gravity resistance due to the weight of the rope is equal to $W_r \sin X$.

The force required to overcome the frictional resistance due to the rope, rollers, sheaves, and drums is equal to $f W_r (\cos X)$.

Let P_1 denote the total pull on the rope required to overcome the weight and friction of the rope, then

$$P_1 = W_r \sin X + f W_r \cos X$$
$$= (\sin X + f \cos X) W_r \qquad (1)$$

Then, if $f = \frac{1}{40}$,

$$P_1 = (40 \sin X + \cos X) \frac{W_r}{40} \qquad (2)$$

EXAMPLE. 1.—What pull is required to move a rope 2,000 feet long and weighing 4,000 pounds up a plane having an inclination of 8°, assuming a coefficient of friction of $\frac{1}{40}$?

SOLUTION.—Substituting in formula **2**,
$$P_1 = (40 \times .13917 + .99027) \frac{4000}{40} = 655.71 \; (656 \text{ lb.}). \text{ Ans.}$$

Let P_2 denote the frictional resistance of the cars as given by the formula,

$$P_2 = N f (W_l + W_e) \cos X \qquad (3)$$

As the cars balance each other on the plane, except for their frictional resistance, as found by formula **3**, the force that produces motion down the plane is due to the load in the cars; the gravity pull of this load denoted by P is given by the formula,

$$P = N (W_l - W_e) \sin X \qquad (4)$$

The force P must overcome the frictional resistance of the cars, P_2, and the resistance of the rope, P_1; but in order to determine the number of loaded cars that are required to produce motion on the plane, it is necessary first to determine the condition of equilibrium shown by the expression,

$$P = P_1 + P_2$$

then, from formulas **1, 3,** and **4,** by substitution,

$$N (W_l - W_e) \sin X = (\sin X + f \cos X) W_r + N f (W_l + W_e) \cos X$$

$$N = \frac{(\sin X + f \cos X) W_r}{(W_l - W_e) \sin X - f (W_l + W_e) \cos X}$$

$$N = \frac{(\sin X + f \cos X) W_r}{(\sin X - f \cos X) W_l - (\sin X + f \cos X) W_e} \qquad (5)$$

If $f = \frac{1}{40}$ the formula becomes,

$$N = \frac{(40 \sin X + \cos X) W_r}{(40 \sin X - \cos X) W_l - (40 \sin X + \cos X) W_e} \qquad (6)$$

In order that a plane may be self-acting, the number of cars must be taken as greater than that found by formula **6**.

EXAMPLE 2.—A gravity plane has an inclination of $8°$; it is 2,000 feet long, the rope weighs 4,000 pounds, a loaded car weighs 3,000 pounds, and an empty car weighs 1,800 pounds; what number of cars must be in the trip to start it, assuming a coefficient of friction of $\frac{1}{40}$?

SOLUTION.—Substituting these values in formula **6**,

$$N = \frac{(40 \times .13917 + .99027)\,4,000}{(40 \times .13917 - .99027)\,3,000 - (40 \times .13917 + .99027)\,1,800} = 13.6$$

The load would thus be balanced and at least 14 cars would be required to produce motion on the plane. Ans.

EXAMPLE 3.—The grade of an incline is $3\frac{3}{4}°$, the length of the road is 2,000 feet, the weight of the rope is 4,000 pounds, the weight of a loaded car is 4,000 pounds, and that of an empty car is 1,700 pounds; how many cars must there be in a trip for this incline to be self-acting, assuming that $f = \frac{1}{40}$?

SOLUTION.—Substituting in formula **6**,

$$N = \frac{(40 \times .06540 + .99786)\,4,000}{(40 \times .06540 - .99786)\,4,000 - (40 \times .06540 + .99786)\,1,700}$$

$= 43.9$ cars, at least 44 cars would be required. Ans.

The result of example 3 shows the conditions given in the problem to be impracticable, for if a car is only 8 feet long, center to center of couplings, a trip of 44 cars would be 352 feet long and there is not room inside a mine especially, to handle so many cars at once even were it practicable otherwise to do so.

34. Calculations for Jig Planes.—In considering a jig plane where a balance truck is used, formula **6** of Art. **33**, may show the plane to be self-acting as far as the descent of the loaded cars and the ascent of the balance truck is concerned, but the balance truck may not prove heavy enough to return the empty cars to the top of the plane; hence, to find out if a jig plane will be self-acting with a given number of cars and a balance truck of a given weight, it is necessary to calculate, by means of formula **1** or **2** of Art. **33**, and formulas derived from formulas **3** and **4** of Art. **33**, the amount of pull available to move the system in either direction. If there is a gravity pull in the required direction in excess of that necessary to hold the system in equilibrium, the plane will, theoretically, be self-acting. In making the calculations for jig planes, formulas 3 and 4 of Art. **33**, cannot be used directly on account of the fact that the balance truck takes the place of the empty cars when the loaded cars are descending and takes the place of the loaded cars when the empties are ascending and does not weigh the same as either.

For descending loaded cars and ascending balance truck, let

W_l = weight of a loaded car, in pounds;
W_e = weight of an empty car, in pounds;
W_t = weight of balance truck, in pounds;
N = number of empty or loaded cars;
f = coefficient of friction;

then formula **3** of Art **33**, becomes

$$P_2' = f(NW_l + W_t) \cos X \quad (1)$$

and formula **4** of Art. **33**, becomes

$$P' = (NW_l - W_t) \sin X \quad (2)$$

For descending balance truck and ascending empty cars, formula **3** of Art. **33**, becomes

$$P_2' = f(W_t + NW_e) \cos X \quad (3)$$

and formula **4** of Art. **33**, becomes

$$P' = (W_t - NW_e) \sin X \quad (4)$$

EXAMPLE.—A jig plane has an inclination of 20° and a length of 200 feet; a loaded car weighs 3,600 pounds, an empty car 1,400 pounds, the balance weight 4,500 pounds, and the rope 1.2 pounds per foot. Show that this plane is self-acting when two cars are used and the coefficient of friction is taken as $\frac{1}{40}$.

SOLUTION.—When the loaded cars are at the top of the plane and the balance truck is at the bottom, the resistance due to the weight and friction of the rope is found by formula **2** of Art. **33**,

$$P_1 = (40 \sin X + \cos X)\frac{W_r}{40}$$

$P_1 = (40 \times .34202 + .93969) \times \frac{240}{40} = 87.72$ lb.

The frictional resistance of the cars and balance truck is found by formula **1** of Art. **34**,

$$P_2' = f(NW_l + W_t) \cos X$$

$P_2' = \frac{1}{40} \times (2 \times 3,600 + 4,500) \times .93969 = 274.86$ lb.

the total resistance is then $87.72 + 274.86 = 362.58$ lb.

The gravity pull that tends to produce motion is found by formula **2** of Art. **34**,

$$P' = (NW_l - W_t) \sin X$$

$P' = (2 \times 3,600 - 4,500) \times .34202 = 923.45$ lb.

The gravity pull of the loaded cars exceeds the resistance by $923.45 - 362.58 = 560.87$ lb., which is more than ample to cause the plane to act in this direction.

With the balance truck at the top of the plane and the empty cars at the bottom: The resistance due to the weight and friction of the rope is the same as before, $P_1 = 87.72$ lb. The frictional resistance of the cars and balance truck is found by formula **3** of Art. **34**,

$$P_2' = f(W_t + N W_e) \cos X$$
$$P_2' = \tfrac{1}{40} \times (4{,}500 + 2 \times 1{,}400) \times .93969 = 171.49 \text{ lb.}$$

the total resistance is then $87.72 + 171.49$ lb. $= 259.21$ lb.

The gravity pull that tends to produce motion is found by formula **4** of Art. **34**,

$$P' = (W_t - N W_e) \sin X$$
$$P' = (4{,}500 - 2 \times 1{,}400) \times .34202 = 581.43 \text{ lb.}$$

The gravity pull of the balance truck exceeds that of the empty cars and rope by $581.43 - 259.21 = 322.22$ lb., which is also ample to produce motion, hence it is evident that this plane is self-acting. Ans.

In order to determine the number and weight of the cars and the weight of the balance truck that may be necessary for a self-acting jig plane, it is first necessary to assume them and then make calculations similar to the preceding to find out if the conditions are suitable; if they are not, the proper weights may be found by repeated assumptions and trial calculations. Under some conditions, it may be necessary to alter the grades of the top or bottom of the plane.

35. Varying Grades.—In Fig. 25 T, the dotted line shows a uniform grade whose angle is equal to BAC. It is possible that this uniform grade may be a complete failure, because the gravity pull of the descending load may not be sufficient to lift the long heavy rope and overcome the friction due to the drum, rollers, rope, and cars, and yet the same incline might be made a complete success by so altering the grade as to give it an increased fall at the top and a reduced fall at the bottom; by doing this, the gravity pull of the loaded trip is increased and the resistance due to the weight of the empty trip and the rope is decreased, thus giving a greater available force for overcoming the inertia of the system at rest. For example, by altering the uniform grade AB to the varying grade $AEDB$, the gravity pull of the load is increased until the loaded trip has reached the point *D* and the empty trip has reached the point *E*, from which positions they will have sufficient momentum to carry them to the passing point. Then, after the trains pass, as the rope attached to the loaded cars continues to lengthen and as the rope attached to the empty cars continues to shorten, the rope attached to the loaded cars will assist the gravity pull of the load. Although the trains have acquired such a high velocity as to run the empty cars up the steep incline from *D* to *B* and the full cars along the level from *E* to *A* by their momentum, the velocity of the empty train, on reaching *B*, and the velocity of the full train, on reaching *A*, may be so slow as not to require brake power to stop the trains.

In Fig. 25 *S* is shown a case where the fall cannot be increased immediately from the top, but the grade is made such that when the loaded trips start from the head of the plane *H* the empty trip leaving *F* has considerable distance

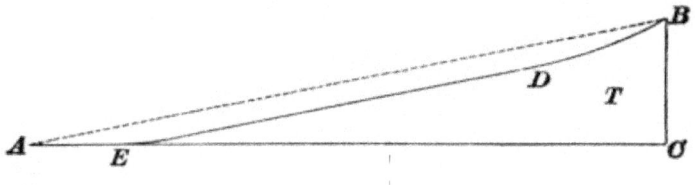

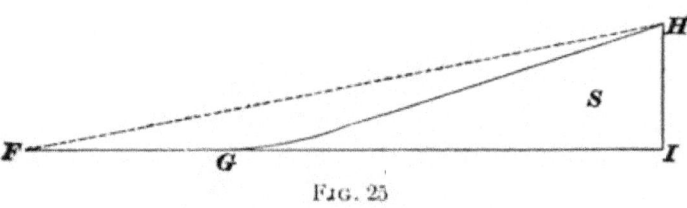

Fig. 25

to run on a level track. The speed thus becomes sufficiently accelerated before the empties arrive at the foot of the plane *G* to give sufficient momentum to carry the loaded cars to the end of the run on the level track from *G* to *F*, and to run the empties up to and over the top *H*.

EXAMPLES FOR PRACTICE

1. What is the total pull or resistance of the rope due to friction on a gravity plane 2,100 feet long and having a pitch of 12°, if the rope weighs 2 pounds per foot of length and the coefficient of friction is $\frac{1}{20}$? Ans. 1,078.6 lb.

2. A gravity plane has a grade of 6°, the length is 1,000 feet, the weight of the rope is 2,000 pounds, a loaded car weighs 3,000 pounds, and an empty car 1,000 pounds; if the coefficient of friction is $\frac{1}{40}$, how many cars must there be in a trip for this plane to be self-acting? Ans. 2.3, or practically 3 cars

3. A gravity plane has a grade of 12° and a length of 2,000 feet, the weight of the rope is 4,000 pounds, the weight of a loaded car is 4,200 pounds, and that of an empty car is 2,000 pounds; if the coefficient of friction is $\frac{1}{40}$, how many cars are required in a trip in order that this plane will be self-acting? Ans. 3.04, or practically 4 cars

4. A jig plane has an inclination of 18° and a length of 250 feet; a loaded car weighs 4,000 pounds and an empty car 1,600 pounds, the balance truck 4,800 pounds, and the rope 375 pounds. Show that this plane is self-acting when two cars are used in a trip and the coefficient of friction is $\frac{1}{40}$.

ENGINE PLANES

36. An **engine plane** is an inclined plane over which the loaded cars are raised and the empty cars lowered by means of a stationary engine. The principles of engine-plane haulage and hoisting in an inclined shaft are identical, and formulas for hoisting in inclined shafts apply equally to engine-plane haulage. The equipment is different to conform to the different conditions. The inclination of the plane must be sufficient for the cars to move down the plane by gravity and pull the haulage rope after them. Engine planes are used largely underground for raising the coal from levels below the bottom of the main shaft or drift level.

37. General Arrangement.—The engine may be located either at the top or at the bottom of the plane and there may be a single track, a double track, or three rails and a turnout.

The drum on an engine used for a haulage plane is loose on the engine shaft, being operated by a clutch so that it is not necessary for the engine to be reversed to lower a trip.

Fig. 26 A shows a plane with an engine located at the top near the hoisting shaft. The loaded cars are hauled up the grade from b to a by the engine and the empty cars run back by gravity. Fig. 26 B shows a plane on which the loaded cars run by gravity down the grade dc to the shaft and the empty cars are hauled up by an engine. In Fig. 26 C,

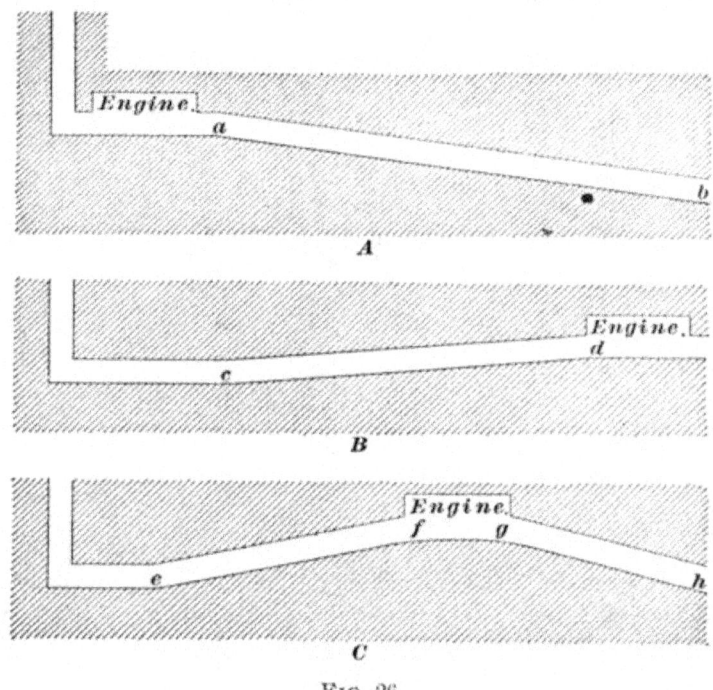

Fig. 26

the loaded cars are hauled up the plane from h to g by the engine and then run by gravity from f to e, while the empty cars are hauled from e to f and run by gravity from g to h.

38. Fig. 27 shows an engine plane with an engine located at the top to haul the cars up the plane; this is a common arrangement for underground planes and also where the material is hauled to the surface through a slope opening. The empty cars in running down the plane must have a gravity pull sufficient to unwind the rope from **the drum and pull it down the plane after them.**

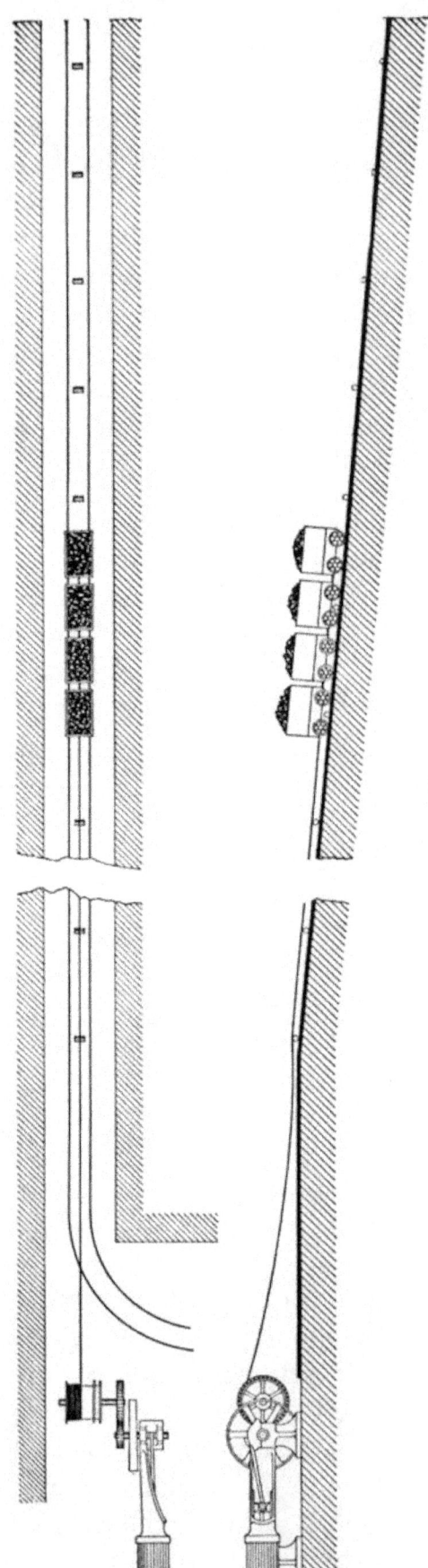

Fig. 27

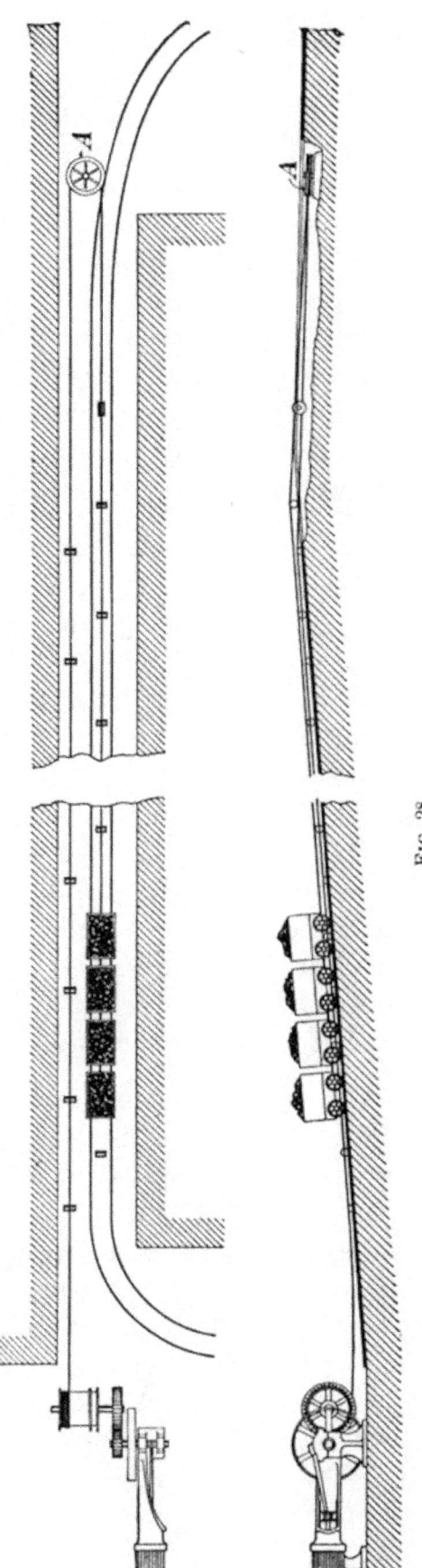

Fig. 28

Fig. 28 shows an engine plane with the engine located at the foot of the plane. The rope and drum are lined up outside of the track with a horizontal sheave A at the top of the plane, about which the rope passes, and is then hitched to the end of a car or trip of cars. By this arrangement, the hoisting rope does not come in contact with the cars on the plane. To make this haulage system work satisfactorily, the inclination of the plane must be greater than where the engine is located at the top in order to overcome the extra frictional resistance due to the extra length of rope and to the sheave wheel A at the top of the plane.

39. Fig. 29 shows a plan and elevation of an engine plane used extensively on the surface for stocking mineral or for dumping waste material. The engine in this case is situated where most convenient, but not at the top or foot of the plane. The plant is arranged so that the car A is hoisted or lowered by the aid of a barney M. When the car gets over the knuckle, it runs slowly by gravity or due to its momentum and is stopped by the head-man, although sometimes there are wooden blocks placed on the rails to stop the car at the dumping place. The barney pit N is shown both in plan and elevation. It will be noticed that there is a comparatively level track P over the barney pit and extending to the foot of the plane, so that the car will always be ahead of the barney when the engine starts.

40. Fig. 30 shows an engine plane arranged to hoist trips from several levels. In a large mine, it is frequently necessary to have several levels working at the same time, and the plane must therefore be arranged to take cars from any one of these levels as desired. If a single track is used on the level haulage way, as shown, a parting is made near the slope for holding both empty and loaded cars. By means of deflecting sheaves a and proper switches, the empty cars are dropped into the upper track b which is generally arranged so that the cars will run back for a short distance from the parting by gravity after the rope has been unhooked. The loaded track can sometimes be built at a lower elevation so

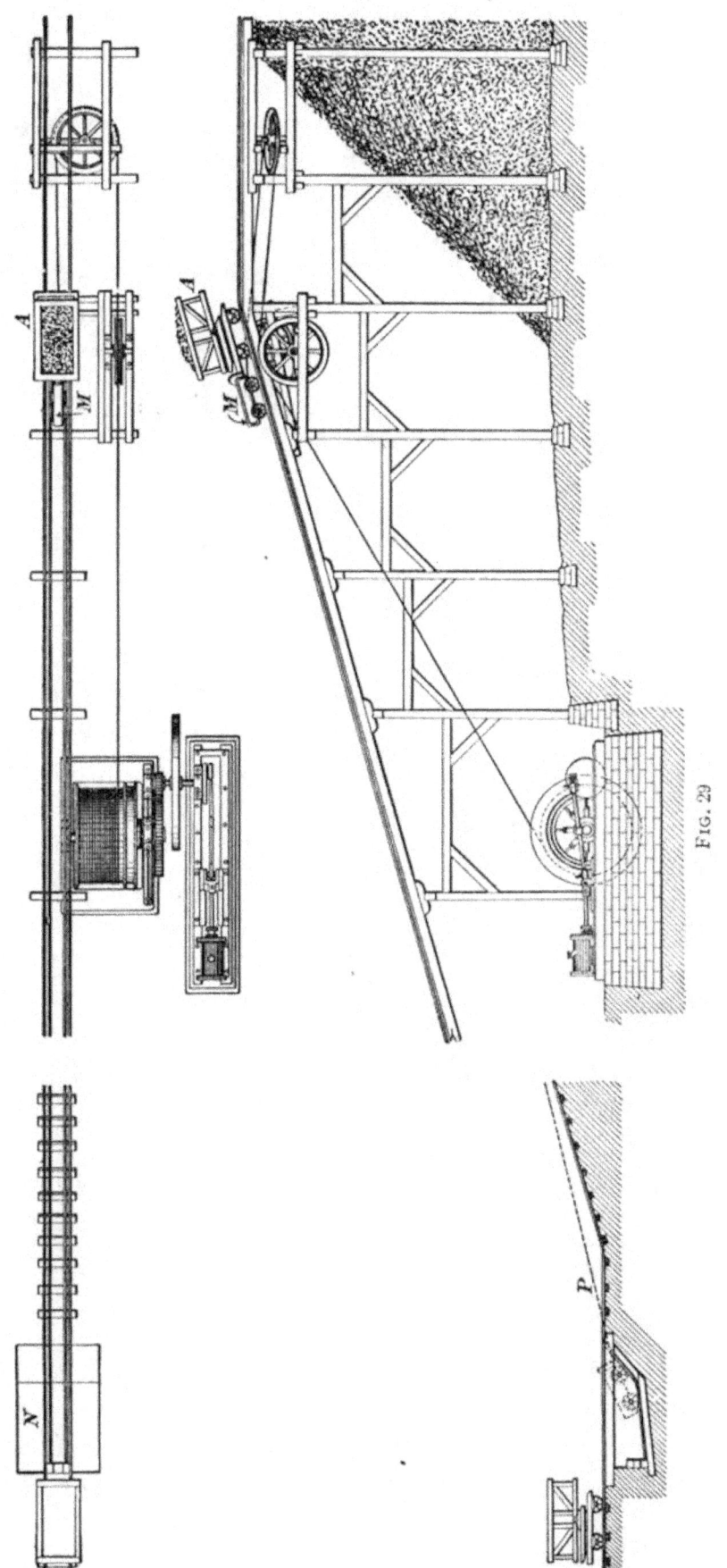

Fig. 29

that the cars can be run by gravity from a point some distance back from the parting to the point where the rope is attached. After the rope has been unhooked from the empty trip c, it is hooked to the loaded trip d and is placed around other guide sheaves as shown, and the loaded trip is then hoisted up the slope.

41. In engine-plane haulage, it is important that trains of reasonable length be run, since it requires, practically, the same time to haul and lower one car as it does a number of cars. The number of cars in a train is determined by the conditions at the mine—such as local haulage, engine power, size of rope, length of partings, etc. The grade should not be less than 3 per cent. to attain an average speed of 10 miles an hour when running back. A train of empty cars in good working order, and running on a good track will acquire a good speed on a pitch of 2.25 per cent.; and a train of full cars under the same conditions will run at a good speed on a pitch of 2 per cent.; but for all-around good work, a pitch of 3 per cent. is the most reliable, and, therefore, should be the minimum.

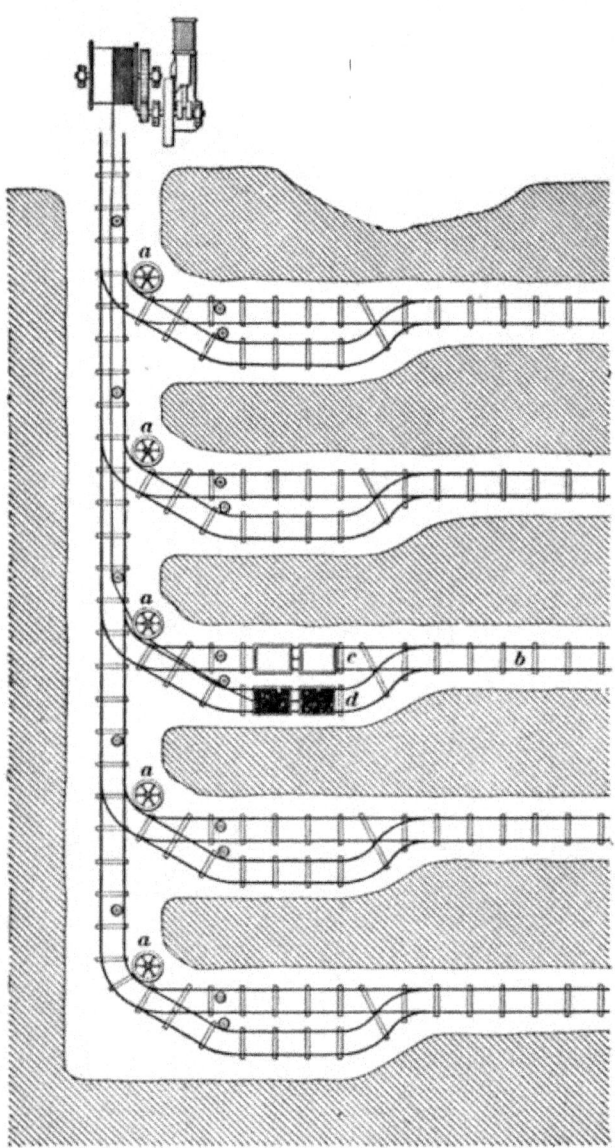

Fig. 30

42. Ropes Operated Through Bore Holes.—The disadvantages arising from piping steam or air into mines and the trouble of dealing with the exhaust steam from an engine situated inside a mine have led to the locating on the surface of the engine for a plane or other rope-haulage system and adoption of bore holes as a means of conducting ropes into mines that are worked through shaft or slope openings. In some cases, the ropes may be placed in the air or pump compartment of the main shaft; in other cases, it may be advisable to drill holes purposely for these ropes directly over the point where the rope is desired inside. When holes are drilled from the surface, surveys are made, and the holes located with reference to the haulage roads near the shaft bottom. The holes, which are usually from 5 to 6 inches in diameter, are sometimes cased with iron pipe if there is water or loose rock in the holes; if, however, the rock is solid and there is no water, a casing is not needed. Diamond drills are sometimes employed for boring the holes, but owing to the fact that the holes are small and difficult to bore vertically, the common practice is to use a percussive drill for the purpose.

43. Surface Sheave Frames.—In some localities, it

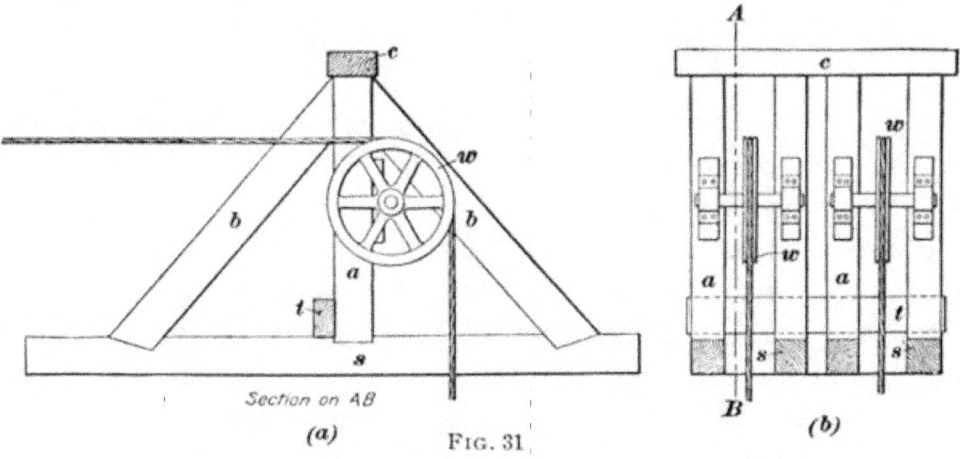

Fig. 31

may be possible to bore the holes so that tipple or drum frames may be used as supports for the sheaves over which the rope passes into the mine. In the majority of cases, it is probable that frames must be constructed purposely for these sheaves. Fig. 31 (*a*) shows a side sectional elevation of such a frame, on the line *A B* of Fig. 31 (*b*), which is an end elevation with the braces *b* removed. The posts *a* are mortised into mud-sills *s* and plate *c*, and are prevented from moving laterally by the tie *t*. Each post is tied by front and back braces *b* to the long sills *s* in order to give the frame stability.

The sheaves *w* are of a size recommended by wire-rope makers for the given diameter of rope. The construction of this frame is such that the sheave wheels can be readily centered over a bore hole.

44. Mine Sheave Frames.—Fig. 32 shows one method of supporting a sheave frame underground at the point where the rope enters the workings.

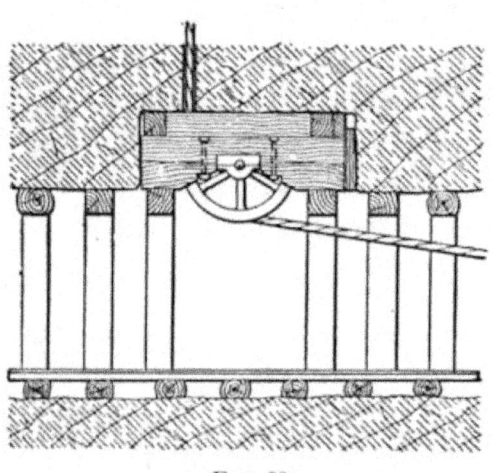

Fig. 32

A hitch is cut in the roof rock deep enough to permit the insertion of blocking and sheaves of the proper size. The blocking is supported on frames and is wedged tight after the sheave has been so set that the rope groove centers with the bore hole. Although this is a rather expensive method of arrangement, it is made necessary when the bore hole comes directly over a road on which men or animals travel, and it is not thought advisable to place the sheaves in the floor.

The objections to placing a sheave in the floor are: There is a tendency to lift the sheave necessitating very strong and exact timbering. If the sheave is not placed at one side of the track, the cars cannot pass the rope. If it is unnecessary for the cars to pass the rope, some simpler and less expensive plan should be adopted as shown in Fig. 33.

These sheaves and timbers should be watched and cared for, since there is great strain on the sheaves and the timbers are likely to rot. The sheaves should have their journals

CALCULATION FOR AN ENGINE PLANE

45. The method of calculating the tension or pull on the hauling rope of an engine plane depends on whether the plane is a single or double track and whether the engine

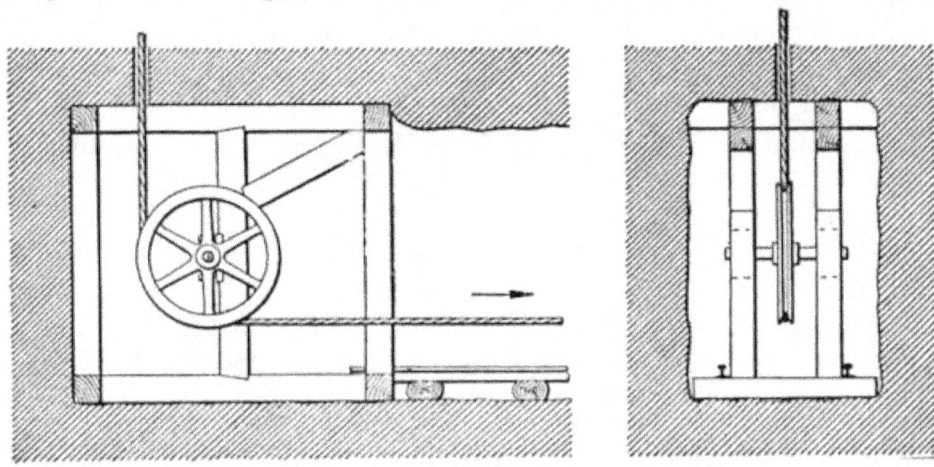

Fig. 33

is located at the top or bottom of the plane, and whether hoisting is done in balance.

46. Single-Track Plane.—If the engine is at the head of the plane, the pull on a single-track plane is equal to the gravity resistance plus the frictional resistance of the loaded trip and the rope at the bottom of the plane.

If L_w = working load, or pull necessary to raise the loaded trip, in pounds;

W_l = weight of loaded car, in pounds;

W_e = weight of empty car, in pounds;

W_r = total weight of rope, in pounds;

N = number of cars in trip;

X = angle of inclination of plane;

f = coefficient of friction;

then,

$$L_w = (NW_l + W_r)\sin X + f(NW_l + W_r)\cos X \quad (1)$$

or if $f = \tfrac{1}{40}$,

$$L_w = (NW_l + W_r)\sin X + (NW_l + W_r)\frac{\cos X}{40} \quad (2)$$

well oiled, and the timbers should be given antiseptic washes from time to time in order to prevent the fungus growth that rots mine timbers.

If the engine is at the foot of the plane, the gravity pull of the ascending rope is balanced by the gravity pull of the descending rope and the friction pull of the rope is twice as great as in the first case, or, in other words, the rope is twice as long. Formula **1** then becomes,

$$L_w = N W_l \sin X + f(N W_l + W_r) \cos X \qquad (3)$$

or if $f = \tfrac{1}{40}$,

$$L_w = N W_l \sin X + (N W_l + W_r) \frac{\cos X}{40} \qquad (4)$$

EXAMPLE 1.—What pull is required to haul up a single-track inclined plane, a loaded trip of ten mine cars weighing 1,000 pounds each and carrying a load of 2,000 pounds each, if the inclination of the plane is 16° and its length 500 yards, the rope weighing 2 pounds per foot and the engine is at the head of the plane?

SOLUTION.—Since the engine is at the head of the plane, this example should be solved by means of formula **2**, in which $N W_l = 10 \times (1{,}000 + 2{,}000) = 30{,}000$ lb., $W_r = 3 \times 500 \times 2 = 3{,}000$ lb., $N W_l + W_r = 33{,}000$ lb., $\sin 16° = .27564$, $\cos 16° = .96126$. Substituting in the formula, $L_w = (N W_l + W_r) \sin X + (N W_l + W_r) \dfrac{\cos X}{40}$.

$$33{,}000 \times .27564 + 33{,}000 \times \frac{.96126}{40} = 9{,}889+ \text{ lb. Ans.}$$

EXAMPLE 2.—Assuming the same conditions as in example 1, except that the engine is at the foot of the plane, what pull will be required to raise the loaded trip?

SOLUTION.—In this case, the rope is 1,000 yd., or 3,000 ft., long; hence, $W_r = 2 \times 3{,}000 = 6{,}000$ lb. Then substituting these values in formula **4**,

$$L_w = 30{,}000 \times .27564 + (30{,}000 + 6{,}000) \times \frac{.96126}{40} = 9{,}134+ \text{ lb.}$$
Ans.

47. Double-Track Plane.—If the engine is at the head of the plane and both the hoisting and lowering drums are clutched to the engine shaft, in other words, if the engine is working in balance, the load that the engine must hoist is the difference between gravity pulls of ascending and descending trips and the rope, plus the friction pull of both trips and one rope, since only one rope is on the plane at a time.

The total gravity pull of the system is $[N(W_l - W_c) + W_r] \sin X$, and the friction pull is $f[N(W_l + W_c) + W_r] \cos X$; then,

$$L_w = [N(W_l - W_c) + W_r] \sin X + f[N(W_l + W_c) + W_r] \cos X$$

EXAMPLE.—Assume the same conditions as in example 1 of Art **46**, except that the plane is double-tracked and that there are as many empty cars descending as there are loaded cars ascending the plane.

SOLUTION.—The tension on the rope is the same in this case as in a single-track plane, but the load that must be hoisted by the engine is not the same, since the cars in the two trips balance. $W_l = 3,000$, $W_e = 1,000$, $W_r = 3 \times 500 \times 2 = 3,000$, $N = 10$. Substituting in the formula,

$$L_w = [N(W_l - W_e) + W_r] \sin X + f[N(W_l + W_e) + W_r] \cos X,$$
$$L_w = [10 \times (3,000 - 1,000) + 3,000] \times .27564 + \tfrac{1}{40} \times [10 \times 3,000 + 1,000) + 3,000] \times .96126 = 7,373+ \text{ lb. Ans.}$$

An engine plane is, however, seldom run in balance, but when two tracks are used each track is operated by an independent drum that runs loosely on the shaft as a trip descends; hence, the calculation for a double track is the same as for single track unless it is stated that the engine hoists in balance.

EXAMPLES FOR PRACTICE

1. What pull is required for an engine at the top of the plane to haul a loaded trip of thirteen mine cars weighing 1,800 pounds each, carrying a load of 2,200 pounds each, up a single-track engine plane 2,000 feet long and having an inclination of 18°, if the rope weighs 4,000 pounds and the coefficient of friction is $\tfrac{1}{40}$?   Ans. 18,636.8 lb.

2. What pull is required for an engine at the foot of a plane to haul a loaded trip of eight mine cars weighing 1,400 pounds each, carrying a load of 2,000 pounds each, up a single-track engine plane 2,200 feet long and having an inclination of 20°, if the rope weighs 1.8 pounds per foot and the coefficient of friction is $\tfrac{1}{40}$?   Ans. 10,127.9 lb.

3. Assume the same conditions as in example 1, except that there is a double track and the number of descending cars is equal to the number of ascending cars; what will be the load on the engine?
   Ans. 11,962.18 lb.

A KEY

TO ALL THE

QUESTIONS AND EXAMPLES

INCLUDED IN THE

EXAMINATION QUESTIONS.

It will be noticed that the Keys have been given the same section numbers as occur on the headlines of Examination Questions to which they refer. All article references refer to the Instruction Paper bearing the same section number as the Key in which it occurs, unless the title of some other Instruction Paper is given in connection with the article number.

To be of the greatest benefit, the Keys should be used sparingly. They should be used much in the same manner as a pupil would go to a teacher for instruction with regard to answering some example he was unable to solve. If used in this manner, the Keys will be of great help and assistance to the student, and will be a source of encouragement to him in studying the various papers composing the Course.

the handling, it is often dumped in the mine into a large skip or gunboat (Art. **17**) to be hoisted; in this case, the mine cars are not taken to the surface. Skips or gunboats are generally made self-dumping. See Art. **19**.

(6) See Art. **26**.

(7) See Art. **27**.

(8) The direction of the resultant in this case is vertical. The load on the sheave or the amount of this resultant is $\sqrt{5^2 + 5^2} = 7.07$ T. See Fig. 28, Art. **29**.

(9) The resultant here will make an angle of $30 \div 2 = 15°$ with the vertical, and the amount of this resultant for a pull on the rope of 10 T. would be $2(10 \times \cos 15°) = 2(10 \times .9659) = 19.318$ T.

(10) Whenever there are tensile stresses produced in any members of the head-frame, a steel head-frame is better adapted to resist the stresses than a wooden one, because steel is adapted to withstand both compressive and tensile strains. See Art. **32**.

(11) Springs are sometimes placed under the journal boxes or in the cage connection of the hoisting rope. See Art. **35**.

(12) See Art. **37**.

(13) See Art. **39**.

(14) See Art. **41**.

(15) See Art. **42**.

HAULAGE
(PART 1)

(1) The size of the mine car is determined by the size of the mine passage through which the cars must pass when such passages are restricted. If the passageways are not restricted in size, the size of the car is determined by the ease with which they may be handled and loaded. See Art. **2**.

(2) (*a*) See rule, Art. **3**.
    (*b*) Area of $a\,b\,c\,d$ and topping
$$(10 + 5)\,(5 \times 12) = 15 \times 60 = 900 \text{ sq. in}$$
$$\text{Area of } b\,c\,f\,e \left(\frac{4+5}{2}\right) 12 \times 12 = 648 \text{ sq. in.}$$
$$\text{Area of } e\,f\,h\,g\; 10\,(4 \times 12) = 480 \text{ sq. in.}$$
$$\overline{2{,}028 \text{ sq. in.}}$$

The length 7 ft. 6 in. is 90 in., hence $\dfrac{2028 \times 90}{1728} = 105.6$ cu. ft.
$\dfrac{105.6}{40} = 2.64$ T. Ans.

(3) $8 \times 3\tfrac{1}{2} \times 3 = 84$ cu. ft.; $84 \div 40 = 2.1$ T. Ans.

(4) See Arts. **8** and **9**.

(5) See Art. **17**.

(6) (*a*) and (*b*) See Art. **24**.

(7) (*a*) and (*b*) See Art. **25**.

(8) (*a*) See Art. **27**.
    (*b*) and (*c*) See Art. **29**.

(9) See Art. **38**.

(10) (*a*) and (*b*) See Art. **52**.

(11) (*a*) and (*b*) See Art. **41**.
    (*c*) See Art. **42**.

(12) (*a*) See Art. **47**.

(*b*) The grade resistance varies directly with **the sine of the angle** of inclination, and the greater the sine of the angle of inclination, the greater is the grade resistance. See Art. **47**.

(13) It may be calculated by means of the formula given in Art. **47**, or it may be estimated as so many pounds of resistance per ton of weight hauled as explained in Art. **50**.

(14) See Art. **52**.

(15) (*a*) See Art. **54**.
      (*b*) See Art. **55**.

(16) (*a*) See Art. **57**.
      (*b*) See Art. **58**.

(17) See Art. **60**.

(18) See Arts. **63** and **64**.

(19) See Arts. **66** and **67**.

HAULAGE
(PART 2)

(1) See Art. **2**.

(2) (*a*) and (*b*) See Art. **5**.

(3) (*a*) See Art. **10**.
(*b*) See Art. **11**.

(4) Substituting the values in the formula in Art. **14**,
$$T = \frac{D^2 \times L \times .85\, p}{d}$$
$$T = \frac{100 \times 14 \times .85 \times 150}{30} = 5{,}950 \text{ lb. Ans.}$$

(5) The drawbar pull on a level may be considered the same as the tractive force or 5,950 lb. The resistance due to the grade is 20 lb. for each per cent. of grade. Since the locomotive weighs 16,000 lb. or 8 T., this resistance is $8 \times 20 \times 3 = 480$ lb. This resistance is deducted from the drawbar pull on a level; $5{,}950 - 480 = 5{,}470$ lb. Ans.

(6) The locomotive weighs 16,500 lb. or 8.25 T. Applying the rule in Art. **18**,
$$\frac{5{,}560}{8.25 \times 10} = 67.394 \text{ T.}; \quad 67.394 - 8.25 = 59.144 \text{ T. Ans.}$$

(7) See Art. **20**.

(8) Applying the formula in Art. **22**, $R = \dfrac{BW}{5\,r}$,
$$R = \frac{4 \times 14{,}000}{5 \times 60} = 186.7 \text{ lb. Ans.}$$

(9) Applying the formula in Art. **25**, $F = DW(C + R + G)$. In this case, $D = 5{,}280$; $W = 4{,}100$ lb. $= 2.05$ T.; $C = .01 \times 2{,}000 = 20$ lb. per T.; $R = 0$; G is $+$ and $= 20 \times 2 = 40$ lb. per T. Hence,
$$F = 5{,}280 \times 2.05\,(20 + 40) = 649{,}440 \text{ ft.-lb. Ans.}$$

(10) See Art. **27**.

(11) The volume of free air required is
$$\frac{(800 - 140) \times 160}{14.7} = 7{,}183.7 \text{ cu. ft.}$$

Now, substituting in the formula in Art. **31**, $C = \dfrac{nc}{t}$,
$$C = \frac{1 \times 7{,}183.7}{30} = 239.46 \text{ cu. ft. per min.} \quad \textbf{Ans.}$$

(12) (*a*) See Art. **36**.
(*b*) See Art. **37**.

(13) Substituting in formula **2**, Art. **38**, $V = \dfrac{v(p - p')}{P - p}$, in which $v = 150$ cu. ft.; $P = 950$ lb. per sq. in.; $p = 800$ lb. per sq. in.; $p' = 130$ lb. per sq. in.,
$$V = \frac{150 \times (800 - 130)}{950 - 800} = 670 \text{ cu. ft.} \quad \textbf{Ans.}$$

(14) The volume of 1 lineal ft. of $5\tfrac{1}{2}$-in. pipe is $\dfrac{5.5^2 \times .7854 \times 12}{1{,}728}$ = .16499 cu. ft. The number of lineal feet required will be $670 \div .16499$ = 4,061 ft. **Ans.**

(15) See Art. **52**.

(16) See Art. **54**.

HAULAGE
(PART 3)

(1) (*a*) Electricity can be more readily transmitted long distances than any other kind of power. See Art. **1**.

(*b*) It cannot be used in any place where explosive gas or mixtures of gas and coal dust may accumulate, owing to the risk of the gas being ignited by electric arcs, or sparks. See Art. **1**.

(2) See Art. **3**.

(3) See Art. **5**.

(4) See Art. **10**.

(5) (*a*) Round, figure **8**, and grooved. The grooved shape has the preference for mine locomotives. See Art. **12**.

(*b*) A large wire has a correspondingly large current-carrying capacity beside giving greater stiffness, so that the trolley wheel is more likely to follow it and is less likely to spark. See Art. **13**.

(6) Substituting the given values in formula of Art. **15**,
$$A = \frac{14 \times 4{,}000 \times 100 \times 100}{250 \times 10} = 224{,}000 \text{ cir. mils}$$

(7) Each branch line should have installed near its junction with the main line a separate fused switch. It is better, however, to have separate feeder lines for the haulage line and for the other machinery operated by electricity. See Art. **17**.

(8) It consists of an inverted cast-iron cup or hood into which is molded some insulating material that holds firmly in place a stud pin, to the lower end of which is attached the trolley hanger. This insulating cup is screwed to the protruding end of an expanding bolt that is inserted and held firmly in a hole prepared for it in the roof. The trolley wire is clamped between two parts or jaws attached to the projecting stud pin. See Art. **21**.

(9) An automatic switch, constructed of a rocker-arm so arranged that a locomotive, when going in, automatically closes connections to the sections it is approaching, and when returning, automatically

opens them again, is installed at the entrances to branch lines and sometimes at points along the main line. See Art. **24**.

(10) A rail bond is an electric conductor usually made of copper and used to connect the ends of consecutive rails in a track, so as to reduce the track resistance by making the return path for the current continuous. See Arts. **25** and **26**.

(11) (*a*) See Art. **28**.

(*b*) In order to reduce to a minimum the shock and jar caused by the unevenness of the road, when the locomotive is in motion. See Art. **28**.

(12) (*a*) Because the ordinary series-parallel controller is so arranged that the motors are connected in series when starting and with a very heavy load, such as mine locomotives usually have to start with, if one pair of wheels begin to slip their friction with the rails will be small and they will spin very rapidly so that the motor connected to them will generate a high counter electromotive force, thus limiting the current through both motors and thus limiting the torque of the other one whose wheels are not sliding. See Art. **39**.

(*b*) See Art. **43**.

(13) (*a*) Reversing the motors when they are connected in parallel and running at high speed with the controller handle in the off-position.

(*b*) An equalizer connection is provided on the power drum so that when the controller handle is at off-position the two series-fields are in parallel. See Art. **47**.

(14) See Art. **48**.

(15) By means of the sprocket wheels, which run in the combined traction and third rail. See Art. **59**.

(16) A gathering locomotive is one used to distribute empties and gather loaded cars from the faces of rooms, where it would be difficult for the regular haulage locomotives to enter. They are usually lighter and more compact than the haulage locomotives and are provided with a reel of flexible cable for use in entering rooms where there is no trolley wire. See Art. **61**.

(17) The weight must be great enough to keep the wheels from slipping until the motors are overloaded to the danger point, and the motors must be large enough to take advantage of the traction afforded by the full weight of the locomotive. See Art. **69**.

(18) Total weight of trip $15 \times 6,000 = 90,000$ lb. Total resistance to be overcome $2 + 2 = 4$ per cent. for cars and $2 + 1 = 3$ per cent. for the locomotive. $90,000 \times .04 = 3,600$ lb. drawbar pull. $3,600 \times 6 = 21,000$ lb. weight on drivers. $21,600 \times .03 = 648$ lb. tractive power required to move the locomotive up the grade. $3,600 + 648 = 4,248$ lb.

total tractive power. The maximum starting effort of a 10-T. (20,000-lb.) locomotive on dry rails with sand, per Table IV, is 5,000 lb.; therefore, a 10-T. locomotive will do the work with a safe margin. The curve for a 10-T. locomotive, Fig. 29, shows that it is capable of hauling 45 T. (90,000 lb.) up a **2-per-cent.** grade and this is the weight of the trip.

(19) See Arts. **82** and **83.**

(20) See Art. **89.**

HAULAGE
(PART 4)

(1) See Arts. **3, 40,** and **52.**

(2) See Arts. **5** and **52.**

(3) (*a*) See Art. **6.**
 (*b*) See Art. **55.**

(4) See Art. **6.**

(5) Since the hold or the grip of the rope increases directly as the square of the number of turns or coils around the drum, the ratio of the grips is $2^2 : 4^2 = 4 : 16$, or $1 : 4$; therefore, the rope with four turns on the drum has four times the grip that the rope with two turns has. See Art. **8.**

(6) See Art. **9.**

(7) See Art. **11.**

(8) See Art. **22.**

(9) See Art. **28.**

(10) See Art. **29.**

(11) See Art. **29.**

(12) See Art. **30.**

(13) (*a*) The pull on the rope is determined by means of formula **4**, Art. **37**, $L_w = f\left[\dfrac{lo}{v}\left(\dfrac{W_l + W_e}{W_l - W_e}\right) + W_r\right]$, in which $f = \dfrac{1}{40}$, $l = 4{,}720$ ft., $o = \dfrac{976 \times 2{,}240}{10 \times 60} = 3{,}643.7$ lb. per min., $v = \dfrac{2.5 \times 5{,}280}{60} = 220$ ft. per min., $W_l = 4{,}000$ lb., $W_e = 1{,}200$ lb., $W_r = 2 \times 1.5 \times 4{,}720 = 14{,}160$ lb. Substituting these values in the formula,

$$L_w = \frac{1}{40} \times \left[\frac{4{,}720 \times 3{,}643.7}{220} \times \left(\frac{4{,}000 + 1{,}200}{4{,}000 - 1{,}200}\right) + 14{,}160\right] = 3{,}983.5 \text{ lb.}$$
Ans.

(b) The horsepower of the engine is found by substituting the preceding values in formula **5**, Art. **37**, $HP = \dfrac{f}{33,000}\left[lo\left(\dfrac{W_l + W_e}{W_l - W_e}\right) + vW_r\right]$

$$HP = \dfrac{1}{40 \times 33,000} \times \left[4,720 \times 3,643.7 \times \left(\dfrac{4,000 + 1,200}{4,000 - 1,200}\right) + 220 \times 14,160\right]$$
$$= 26.56 \text{ H. P. Ans.}$$

(14) From Art. **39**, the size of the rope would be $\sqrt{\dfrac{3}{5}} \times \dfrac{5}{3} = 1.29$ in. Ans.

(15) (a) See Art. **39**.
    (b) See Art. **73**.

(16) (a) Substituting in formula **3**, Art. **37**, $N = \dfrac{lo}{v(W_l - W_e)}$, in which $l = 5,230$ ft., $o = \dfrac{2,500 \times 2,000}{8 \times 60} = 10,416.6$ lb. per min., $v = \dfrac{2 \times 5,280}{60} = 176$ ft. per min., $W_l = 4,000$ lb., $W_e = 1,600$ lb.

$$N = \dfrac{5,230 \times 10,416.6}{176 \, (4,000 - 1,600)} = 128.9, \text{ or } 129 \text{ cars. Ans.}$$

(b) The distance apart of the cars is the length of the haulage road divided by the number of cars on the rope at one time, or $\dfrac{5,230}{129} = 40.5$ ft. Ans.

(17) See Art. **54**.

(18) See Art. **55**.

(19) (a) To find pull on the rope or the load on the engine when the main rope is pulling, it must be remembered that 7,000 ft. of main rope and 7,000 ft. of tail-rope are out when the trip starts from the gathering station; substituting the values, $f = \dfrac{1}{40}$, $N = 20$, $W_l = 4,500$, $W_r = (.6 + .7) \times 7,000 = 9,100$ lb., in formula **1**, Art. **69**, $L_m = f(NW_l + W_r)$,

$$L_m = \dfrac{1}{40} \times (20 \times 4,500 + 9,100) = 2,477.5 \text{ lb. Ans.}$$

When starting to pull back the empty trip, 14,000 ft. of tail-rope is out and no main rope. Substituting the values as before, except that $W_e = 1,500$ lb. and $W_r = .6 \times 14,000 = 8,400$ lb., in formula **2**, Art. **69**, $L_t = f(NW_e + W_r)$

$$L_t = \dfrac{1}{40} \times (20 \times 1,500 + 8,400) = 960 \text{ lb. Ans.}$$

(b) The maximum load $L_w = L_m = 2,477.5$ lb., and $v = \dfrac{5,280 \times 10}{60} = 880$ ft. per min. Substituting these values in the formula in Art. **70**, $HP = \dfrac{L_w v}{33,000}$,

$$HP = \dfrac{2,477.5 \times 880}{33,000} = 66.1 \text{ H. P. Ans.}$$

(20) See Art. **74**.

HAULAGE
(PART 5)

(1) See Art. **1**.

(2) See Art. **3**.

(3) See Art. **3**.

(4) See Art. **4**.

(5) See Art. **7**.

(6) See Art. **8**.

(7) See Art. **12**.

(8) See Art. **14**.

(9) See Art. **16**.

(10) See Art. **17**.

(11) See Art. **19**.

(12) See Art. **20**.

(13) See Art. **24**.

(14) (*a*) and (*b*) See Art. **25**.

(15) See Art. **29**.

(16) This example may be solved by means of formula **6** of Art. **33**. $W_l = 2{,}700$ lb., $W_e = 1{,}500$ lb., $W_r = 3{,}600$ lb., $X = 10°$, $f = \frac{1}{40}$, $\sin 10° = .17365$, $\cos 10° = .98481$. Substituting in the formula,

$$N = \frac{(40 \sin X + \cos X) W_r}{(40 \sin X - \cos X) W_l - (40 \sin X + \cos X) W_e}$$

$$N = \frac{(40 \times .17365 + .98481) \times 3{,}600}{(40 \times .17365 - .98481) \times 2{,}700 - (40 \times .17365 + .98481) \times 1{,}500}$$

$= 6.8$, or practically 7 cars. Ans.

(17) This example may be solved by means of the formulas given in Art. **34**, in which, $W_l = 3{,}800$, $W_e = 1{,}600$, $W_t = 4{,}500$, $W_r = 450$, $N = 2$, $f = \frac{1}{40}$, $X = 22°$, $\sin 22° = .37461$, $\cos 22° = .92718$. With

descending loaded cars and ascending balance truck, the resistance due to the weight and friction of the rope is found from formula **2** of Art. **33,**

$$P_1 = (40 \sin X + \cos X) \frac{W_r}{40}$$

$$P_1 = (40 \times .37461 + .92718) \frac{450}{40} = 17.90 \text{ lb.}$$

The frictional resistance of the cars and balance truck is found from formula **1** of Art. **34,**

$$P_2' = f(NW_l + W_t) \cos X$$

$$P_2' = \frac{1}{40}(2 \times 3,800 + 4,500) \times .92718 = 280.47 \text{ lb.}$$

The total resistance is $P_1 + P_2' = 17.90 + 280.47 = 298.37$ lb. The gravity pull of the load that tends to produce motion is found from formula **2** of Art. **34,**

$$P' = (NW_l - W_t) \sin X$$

$$P' = (2 \times 3,800 - 4,500) \times .37461 = 1,161.29 \text{ lb.}$$

Hence, there is an excess of gravity pull of $1,161.29 - 298.37 = 862.92$ lb., which is ample to operate the plane in this direction.

With descending balance truck and ascending empty cars, the resistance due to the weight and friction of the rope is the same as before, $P_1 = 17.90$ lb. The frictional resistance of the balance truck and empty cars is found by formula **3**, Art. **34,**

$$P_2' = f(W_t + NW_e) \cos X$$

$$P_2' = \tfrac{1}{40}(4,500 + 2 \times 1,600) \times .92718 = 178.48 \text{ lb.}$$

The total resistance is $P_1 + P_2' = 17.90 + 178.48 = 196.38$ lb. The gravity pull of the load that tends to produce motion is found from formula **4** of Art. **34,**

$$P' = (W_t - NW_e) \sin X$$

$$P' = (4,500 - 2 \times 1,600) \times .37461 = 486.99 \text{ lb.}$$

Hence, there is an excess of gravity pull of $486.99 - 196.38 = 290.61$ lb., which is ample to operate the plane in this direction; therefore, the plane will be self-acting.

(18) See Art. **36.**

(19) See Art. **41.**

(20) This example may be solved by formula **2** of Art. **46,** in which $W_l = 4,000$, $W_e = 1,800$, $W_r = 900 \times 2.4 = 2,160$ lb., $N = 15$, $X = 25°$, $f = \tfrac{1}{40}$, $\sin 25° = .42262$, $\cos 25° = .90631$. Substituting these values in the formula,

$$L_w = (NW_l + W_r) \sin X + (NW_l + W_r) \frac{\cos X}{40}$$

$$L_w = (15 \times 4,000 + 2,160) \times .42262 + (15 \times 4,000 + 2,160) \times \frac{.90631}{40}$$

$$= 27,678.5 \text{ lb. Ans.}$$

(21) This example may be solved by means of the formula in Art. **47**, in which the values are the same as given in example 20. Substituting these values in the formula,

$L_w = [N(W_l - W_e) + W_r] \sin X + f[N(W_l + W_e) + W_r] \cos X$

$L_w = [15 \times (4{,}000 - 1{,}800) + 2{,}160] \times .42262 + \tfrac{1}{40} \times [15 \times (4{,}000 + 1{,}800) + 2{,}160] \times .90631 = 16{,}879.47$ lb. Ans.

INDEX

Note.—All items in this index refer first to the section and then to the page of the section. Thus, "Advantages of bucket drainage, §60, p40," means that advantages of bucket drainage will be found on page 40 of section 60.

A

A type of head-frame, §53, p36.
Abandoned mine workings, Drainage of, §60, p7.
   workings, Tapping and draining, §60, p18.
Adaptability of whims, §50, p5.
Adhesive power of a locomotive, §55, p11.
Adit drainage, §60, pp4, 12.
Advantages of bucket drainage, §60, p40.
   of Koepe system, §52, p23.
   of locomotive haulage, §55, p1.
Air, compressed, Reheating, §55, p41.
   compressors, §55, p23.
   compressors, Power for, §55, p23.
   in a siphon, §60, p35.
   -power reverse, §50, p24.
   -tank tenders, Locomotives with, §55, p34.
Alternating-current hoists, §50, p38.
Anchoring gravity-plane tracks, §58, p17.
Animal haulage, §54, p47.
   versus mechanical haulage, §54, p51.
   versus steam-locomotive haulage, §55, p5.
Animals, Purchasing, §54, p51.
Anthracite mine cars, §54, p12.
Appliances, Hoisting, §51, p1; §52, p1.
Areas, robbed, Drainage of, §60, p7.
Armature, §56, p26.
Arrangement of tracks, §58, p2.
Automatic bucket dumps, §60, p42.
   dumping cages, §53, p12.
   reducing valve, §55, p40.
   switches, §58, p18.
   trolley switch, §56, p16.
Auxiliary tanks, §55, p39.
Axle bearings, §56, p22.
Axles and wheels, §56, p21.
   Mine-car, §54, pp27, 32.

B

Balance or tension cars, §57, p7.
Balanced slide-valve engine with plain drum, §50, p17.
   throttle valves, §50, p21.
Baldwin-Westinghouse locomotive, §56, p19.
Bar, Drag, §58, p25.
Barneys, §58, p15.
Barrier pillars, §60, p21.
Bearings, §50, p15.
   Axle, §56, p22.
Bedplates, Specifications for, §51, p53.
Beekman friction clutch, §52, p30.
Bells, Electric, §53, p49.
Bending stress, §51, p25.
Bent friction clutches, §52, p29.
Bituminous mine car, §54, p18.
Block brakes, §52, p32.
Blocks, Safety, §58, p20.
Blowing off pipes, §55, p26.
Bolts, Foundation, §51, p49.
Bond, Double-loop protected, §56, p18.
Bonds, Protected, §56, p17.
   Rail, §56, p17.
   Unprotected, §56, p17.
Bonnet, §53, p7.
Bore holes, Ropes operated through, §58, p39.
Brake, Crank, §52, p43.
   Differential, §52, p38.
   Post, §52, p32.
   shoes, §56, p23.
   Strap, §52, p34.
Brakes, §52, p32; §56, p23.
   Block, §52, p32.
   Mine-car, §54, p34.
   Power, §52, p42.

Brakes—(Continued)
   Power for, §52, p39.
Branch lines, §56, p11.
Brick dams, §60, pp27, 28.
Bucket drainage, §60, pp39, 40.
   drainage, Advantages of, §60, p40.
   dumps, Automatic, §60, p42.
Buckets, §53, p23.
Bushing, Car-wheel, §54, p30.
Butterfly valves, §60, p46.

C

Cable shields, §51, p11.
Cage chairs, §53, p30.
   guides, §53, p25.
   rests, §53, p27.
   Skip, §53, p23.
Cages, §53, p6.
   Automatic dumping, §53, p12.
   for vertical shafts, §53, p6.
   Multiple-deck, §53, p11.
Calculating capacity of coal cars, §54, p4.
   compressor capacity, §55, p24.
   size of wires, §56, p9.
Calculation for an engine plane, §58, p41.
Calculations for gravity planes, §58, p26.
   for jig planes, §58, p28.
Capacity of air compressors, §55, p24.
   of coal cars, Calculating, §54, p4.
   of locomotives, Hauling, §55, p15.
   of mine cars, §54, p3.
   of motors, §56, p52.
   of pipes, Storage, §55, p28.
   Stationary storage, §55, p27.
Car and rope couplings, §57, p64.
   Bituminous mine, §54, p18.
   couplings, §54, p35.
   Diamond, or Hollenback, §54, p12.
   locks, §53, p23.
   resistance, §54, p39.
   Scoop-box, §54, p10.
   Size of, §54, p1.
   Steel coal, §54, p23.
   -wheel bushing, §54, p30.
   wheels, Water-tank, §60, p47.
Care of locomotives, §56, p67.
   of mules, §54, p52.
   of rope rollers, §57, p28.
Carriage, Slope, §53, p15.
Carrying shafts and rollers, §53, p5.
   sheaves, §57, p24.
Cars, Anthracite mine, §54, p12.
   Balance or tension, §57, p7.
   Calculating capacity of coal, §54, p4.
   Capacity of mine, §54, p3.
   Coal, §54, p12.
   Gathering, with mules, §54, p48.

Cars—(Continued)
   Grip, §57, p21.
   in motion, Frictional resistance to, §54, p38.
   Mine, §54, p1.
   Oil for mine, §54, p33.
   Ore, §54, p6.
   Rotary dump, §54, p9.
Cast-iron rollers, §57, p25.
   -iron sheave, §53, p1.
   -steel ropes, §51, p2.
Catch, Heavy-steel, §53, p10.
   Light-steel, §53, p11.
Catches, Safety, §53, p7.
Chain, Hooking-on, §58, p25.
Chairs, Cage, §53, p30.
   Landing, §53, p27.
Charging compressed-air locomotives, §55, p26.
   Stationary storage system of, §55, p26.
   valves, §55, p28.
Choice of haulage systems, §54, p43.
Circuit-breakers, Use of, §56, p65.
Clevis hitchings, §54, p35.
   hook, §54, p36.
Clutch, Beekman friction, §52, p30.
   Friction, §52, p29.
   Jaw, §52, p28.
Clutches, §52, p28.
   Bent friction, §52, p29.
Coal cars, §54, p12.
   cars, Calculating capacity of, §54, p4.
   cars, Steel, §54, p23.
Coil, Field, §56, p26.
Column indicators, §52, p1.
Combination two- and three-rail inclines, §58, p3.
Combined endless-rope and gravity plane, §57, p13.
Commutators, §56, p26.
Comparative sizes of locomotives, §56, p51.
Comparison of cylindrical and conical drums, §52, p15.
   of endless- and tail-rope haulage, §57, p50.
Compensating sheaves, §57, p9; §58, p24.
Compressed-air gathering locomotive, §55, p35.
   -air hoisting engines, §50, p29.
   -air locomotive haulage, §55, p22.
   -air locomotives, §55, p30.
   -air locomotives, Charging, §55, p26.
   -air locomotives, Direct system of charging, §55, p26.
   -air locomotives, Tractive power of, §55, p43.
   air, Reheating, §55, p41.
   air, Use of, expansively, §55, p42.

INDEX

Compressor capacity, Calculating, §55, p24.
    service and pipe line, §55, p30.
Compressors, Air, §55, p23.
Conditions governing use of first-motion engines, §50, p15.
Conductors, §53, p25.
Conical and cylindrical drums, Comparison of, §52, p15.
    drums, §52, p9.
    drums, Hoisting with, §52, p11.
Connecting-rods, Specifications for, §51, p53.
Connections of mine locomotives, §56, p31.
Construction, Material for, §51, p46.
    of steam mine locomotive, §55, p5.
Continuous-motion engines, §50, p11.
Contour drainage map, §60, p5.
Controller, §56, p26.
    Use of, §56, p63.
Controllers, Series-parallel, §56, p27.
Core, §51, p3.
Corliss hoisting engine, §50, p19.
Cornish horse whim or gin, §50, p3.
Cost of electric haulage, §56, p69.
    of loading mine cars, §54, p46.
    of mule haulage, §54, p50.
    of operating wire-rope haulage systems, §57, p51.
    of water-hoisting plants, §60, p51.
Couplings, §57, p64.
    Car, §54, p35.
    Car and rope, §57, p64.
    Flanged, §55, p25.
    Rope, §57, p64.
    Union, §55, p25.
Crab winch, §50, p2.
Crank-brake, §52, p43.
    -shaft, Specifications for, §51, p53.
Crooked-entry haulage, §57, p45.
Cross-compound hoisting gin, §50, p6.
    -entry drainage, §60, p5.
Crossheads, Specifications for, §51, p53.
Crucible-steel ropes, §51, p2.
Current and voltage, §56, p2.
Curves, §57, p29.
    Grade reduction of, §55, p17.
    Resistance due to, §55, p16.
    Resistance on, §55, p18.
    Reverse, §57, p32.
Cut gears, §50, p14.
Cylinder pressure, Reducing, §55, p40.
    Reverse, §56, p29.
Cylinders, Specifications for, §51, p52.
Cylindrical and conical drums, Comparison of §52, p15.
    dams, §60, p32.
    drum, Hoisting with, §52, p9.
    drums, §52, p7.

D

Dams, Brick, §60, pp27, 28.
    Cylindrical, §60, p32.
    Location of, §60, p22.
    Mine, §60, p22.
    mine, Thickness of, §60, p31.
    Shaft, §60, p30.
    Spherical, §60, p32.
    Stone, §60, p29.
    Straight, §60, p32.
    Water pressure on, §60, p23.
    wooden, Flat, §60, p26.
    wooden, Wedge-shaped, §60, p24.
Deposits, Flat, above water level, §60, p4.
    Flat, below water level, §60, p8.
    Inclined, above water level, §60, p12.
    Inclined, below water level, §60, p15.
Depth of hoist, §51, p34.
Derailing, or safety, switches, §58, p19.
Detaching hooks, §53, p46; §57, p65.
Determination of size required for a given service, §56, p59.
Dial indicators, §52, p3.
Diameter of drum or sheave, §51, p41.
    of sheave, §53, p5.
Diamond, or Hollenback, cars, §54, p12.
Differential brake, §52, p38.
    drum, §58, p14.
    lever, §52, p41.
Dimensions of electric mine locomotives, §56, p52.
    of mine locomotives, §55, p9.
Direct-acting hoisting engines, §50, p15.
    -current hoist, §50, p34.
    system of charging compressed-air locomotives, §55, p26.
Disadvantages of Koepe system, §52, p23.
Discharge of a siphon, §60, p38.
District haulage, §57, p13.
    tail-rope haulage, §57, p42.
Ditches on levels, §60, p15.
Diverter, §56, p33.
Dogs, Landing, §53, p27.
Double clevis hitching, §54, p35.
    -loop protected bond, §56, p18.
    -seat throttle, §50, p23.
    -track plane, §58, p42.
    -tracked incline, §58, p2.
Drag bar, §58, p25.
Drainage, Adit, §60, pp4, 12.
    bucket, §60, p40.
    Cross-entry, §60, p5.
    in stope mining, §60, p17.
    map, Contour, §60, p5.
    of robbed areas, §60, p7.
    of temporarily abandoned mine workings, **§60, p7.**

INDEX

Drainage—(Continued)
  of working places, §60, p17.
  Room, §60, p17.
  Swamp, §60, p5.
  tunnels, §60, p13.
  Underground, §60, p4.
Draining, §60, p1.
  abandoned workings, §60, p18.
  off water, §60, p19.
Drawbar pull, §55, p10; §56, p52.
  pull from electrical input, §56, p54.
  pull on a grade, §55, p14; §56, p56.
Drawbars, §54, p24.
Drum, cylindrical, Hoisting with, §52, p9.
  Differential, §58, p14.
  or sheave, Diameter of, §51, p41.
  Shell of, §52, p7.
Drums and reels, §52, p6.
  Comparison of cylindrical and conical, §52, p15.
  Conical, §52, p9.
  conical, Hoisting with, §52, p11.
  Cylindrical, §52, p7.
  Endless-rope, §57, p7.
  for gravity planes, §58, p10.
  Geared, §58, p10.
  hoisting, Specifications for, §51, p54.
Dump cars, Rotary, §54, p9.
Dumping cages, Automatic, §53, p12.
  skips, Method of, §53, p19.
Dumps, Automatic bucket, §60, p42.

E

Eccentrics, Specifications for, §51, p54.
Effort, Maximum starting, §55, p11.
Ehrenfeld haulage plant, §56, p3.
Electric bells, §53, p49.
  generators, §56, p3.
  haulage, Cost of, §56, p69.
  hoists, §50, p32.
  locomotive haulage, §56, p1.
  mine haulage, Wiring for, §56, p7.
  mine locomotives, §56, p19.
  mine locomotives, Dimensions of, §56, p52.
  mine locomotives, Operation of, §56, p61.
  mine locomotives, Rails for, §56, p7.
Electrical input, Drawbar pull from, §56, p54.
End-dump tanks, §60, p44.
Endless- and tail-rope haulage, Comparison of, §57, p50.
  -rope, Combined, and gravity plane, §57, p13.
  -rope drums, §57, p7.
  -rope engines, §57, p4.
  -rope haulage, §57, p2.
  -rope haulage calculations, §57, p33.
  -rope haulage on inclines, §57, p34.

Endless—(Continued)
  -rope system, Operation of, §57, p36.
  -rope system, Power for driving, §57, p4.
  tail-rope system, §57, p51.
Engine, Balanced slide-valve, with plain drum, §50, p17.
  Geared hoisting, §50, p7.
  Hoisting, §50, p1.
  hoisting, Corliss, §50, p19.
  hoisting, Parts of, §50, p20.
  hoisting, Second-motion, §50, p7.
  hoisting, Size of, §51, p33.
  Horsepower of, §57, p73.
  Load on, §51, p36; §57, p71.
  Moment of, §51, p38.
  plane, Calculations for, §58, p41.
  planes, §58, p32.
  reversing, Specifications of, §51, p54.
  Uniform load on, §57, p4.
  Winding, §50, p1.
  Work of, §51, p37.
Engines, Continuous-motion, §50, p11.
  Endless-rope, §57, p4.
  first-motion, Conditions governing, §50, p15.
  haulage, Tail-rope, §57, p52.
  hoisting, Compressed-air, §50, p29.
  hoisting, Direct-acting, §50, p15.
  hoisting, Gasoline, §50, p30.
  hoisting, Kinds of, §50, p6.
  hoisting, Overwinding device for, §50, p27.
  hoisting, Stationary, §50, p8.
  Oil, §50, p30.

F

Fans, §53, p26.
  Hydrostatic, §53, p28.
  Pneumatic, §53, p30.
Fastenings, Rope, §52, p27.
Field coils, §56, p26.
Fillers, Rope, §51, p11.
First-motion engines, Conditions governing, §50, p15.
  -motion hoisting engines, §50, p15.
Fitting, Rope-end, §51, p11.
Flanged couplings, §55, p25.
Flat deposits above water level, §60, p4.
  deposits below water level, §60, p8.
  rope and reel, Calculating size of, §52, p18.
  ropes, §51, pp7, 31.
  ropes, Attaching, §51, p14.
  -rope reels, §52, p17.
  wooden dams, §60, p26.
Flattened strand ropes, §51, p4.
Flow of water into mines, §60, p2.
Force, Tractive, §55, p10.
Foundation bolts, §51, p49.
  Hoisting-engine, §51, p44.

INDEX

Foundations, Permanent, §51, p46.
    Temporary, §51, p46.
Four-wheel, Connected, steam mine locomotives, §55, p7.
Frame, Locomotive, §56, p21.
Frames, Mine sheave, §58, p40.
    Surface sheave, §58, p39.
Friction clutch, §52, p29.
    clutch, Beekman, §52, p30.
    in pipes, §55, p26.
    Shaft, §51, p26.
Frictional resistance, §54, p37.
Frogs, Trolley, §56, p16.

G

Gasoline hoisting engines, §50, p30.
Gathering locomotive, General Electric, §56, p45.
    locomotive, Goodman, §56, p50.
    locomotive, Jeffrey, §56, p48.
    locomotives, §56, p45.
    locomotives, Compressed-air, §55, p35.
Gear-cases and gears, §56, p22.
    -teeth, Herring bone, §50, p14.
    -teeth, Strengthening, §50, p14.
Geared drums, §58, p10.
    hoisting engine, §50, p7.
Gears and gear-cases, §56, p22.
    Cut, §50, p14.
    Machine-molded, §50, p14.
    Wooden-tooth, §50, p13.
General Electric gathering locomotive, §56, p45.
    Electric locomotive, §56, p35.
Generators, Electric, §56, p3.
Gilberton water shaft, §60, p49.
Gin, Cornish horse, §50, p1.
    hoisting, Cross-compound, §50, p6.
Gong signal, Pneumatic, §53, p50.
Goodman gathering locomotive, §56, p50.
    single-motor locomotive, §56, p39.
Grade, Drawbar pull on, §55, p14; §56, p56.
    reduction on curves, §55, p17.
    resistance, §54, p39.
Grades, Varying, §58, p30.
Gravity plane, §57, p13.
    plane, Three-rail, §58, p3.
    -plane tracks, Anchoring, §58, p17.
    planes, §58, p1.
    planes, Calculations for, §58, p26.
    planes, Drums for, §58, p10.
    planes, Inclination of, §58, p25.
Grip and tension, §57, p7.
    cars, §57, p21.
Grips, §57, p18.
Grooved wheels, §58, p12.
Grounds, Shocks caused by, §56, p62.

Guides, Cage, §53, p25.
    Rope, §57, p32.
Gunboats, or skips, §53, p17.

H

Hammer-and-plate signal, §53, p48.
Hand power, §50, p1.
Handling of mine water, §60, p1.
Hanger, Petticoat, §56, p13.
Haulage, Animal, §54, p47.
    Animal versus mechanical, §54, p51.
    calculations, §57, p33.
    calculations, Tail-rope, §57, p71.
    Compressed-air locomotive, §55, p22.
    Cost of electric, §56, p69.
    Crooked-entry, §57, p45.
    District, §57, p13.
    District tail-rope, §57, p42.
    Electric locomotive, §56, p1.
    Endless-rope, §57, p2.
    Endless-rope, on inclines, §57, p34.
    engines, Tail-rope, §57, p52.
    Locomotive, §55, p1.
    locomotive, Advantages of, §55, p1.
    locomotives, §56, p35.
    Mule, §54, p47.
    on an incline, Tail-rope, §57, p74.
    plant at Ehrenfeld, §56, p3.
    Resistance to mine-car, §54, p37.
    Rope, §57, p1; §58, p1.
    ropes, §51, p1.
    Steam-locomotive, §55, p4.
    system, Overhead endless-rope, §57, p22.
    systems, Choice of, §54, p43.
    systems, wire-rope, Cost of operating, §57 p51.
    Tail-rope, §57, p37.
    Wiring for electric, §56, p7.
Hauling capacity of locomotives, §55, p15.
Head-frame, A type of, §53, p36.
    -frame, Enclosing a, §53, p43.
    -frame, Specifications of, §53, p44.
    -frame, Square type of, §53, pp38, 40.
    -frame with inclined brace, §53, p40.
    -frame without inclined brace, §53, p38.
    -frames, §53, p30.
    -frames, Types of, §53, p36.
    -shaft, §53, p42.
    -sheave, Specifications for, §51, p56.
    wheels, §58, p11.
Headlights, §56, p24.
Herring-bone gear-teeth, §50, p14.
Hitching, Double clevis, §54, p35.
Hitchings, Clevis, §54, p35.
Hoist, Depth of, §51, p34.
    Direct-current, §50, p34.
Hoisting and haulage ropes, §51, p29.

INDEX

Hoisting—(Continued)
  appliances, §51, p1; §52, p1.
  drums, Specifications for, §51, p54.
  engine, §50, p1.
  engine, Corliss, §50, p19.
  -engine foundation, §51, p44.
  engine, Geared, §50, p7.
  engine, Parts of, §50, p20.
  engine, Second-motion, §50, p7.
  engine, Size of, §51, p33.
  -engine specifications, §51, p51.
  engines, Compressed-air, §50, p29.
  engines, Direct-acting, §50, p15.
  engines, Gasoline, §50, p30.
  engines, Kinds of, §50, p6.
  engines, Overwinding device for, §50, p27.
  engines, Stationary, §50, p8.
  gin, Cross-compound, §50, p6.
  Inclined-shaft, §53, p14.
  indicators, §52, p1.
  Koepe system of, §52, p21.
  machinery, §50, p1.
  Modified Whiting system of, §52, p26.
  ropes, §51, p1.
  Slope, §53, p14.
  Speed of, §51, p35.
  Tail-rope, §51, p37.
  versus pumping water, §60, p53.
  Water, §60, p39.
  water on slope, Speed of, §60, p47.
  water on slopes, §60, p46.
  with conical drums, §52, p11.
  with cylindrical drum, §52, p9.
Hoists, Alternating-current, §50, p38.
  Electric, §50, p32.
  Portable, §50, p8.
  Steam-power, §50, p6.
Hollenback car, §54, p12.
Hook, Clevis, §54, p36.
Hooking-on chain, §58, p25.
Hooks, Detaching, §53, p46; §57, p65.
Horizontal brick dams, §60, p28.
  sheaves, §57, p58.
Horse whim or gin, Cornish, §50, p3.
Horsepower, §50, p3.
  of engine, §57, p73.
Hydrostatic fans, §53, p28.

I

Improved barney, §58, p15.
Inclination of gravity planes, §58, p25.
Incline, Double-tracked, §58, p2.
  Tail-rope haulage on, §57, p74.
Inclined deposits above water level, §60, p12.
  deposits below water level, §60, p15.
  -shaft hoisting, §53, p14.

Inclines, Combination two- and three-rail, §58, p3.
  Endless-rope haulage on, §57, p34.
  Three-rail, §58, p3.
Index wheel, §56, p29.
Indicator, Specifications for, §51, p55.
Indicators, §57, p57.
  Column, §52, p1.
  Dial, §52, p3.
  Hoisting, §52, p1.
  Special, §52, p4.
  Types of, §52, p1.
Input, electrical, Drawbar pull from, §56, p54.
Inspection of ropes, §51, p9.
Iron-ore cars, §54, p8.
  -wire ropes, §51, p2.
Jaw clutch, §52, p28.
Jeddo tunnel, §60, p14.
Jeffrey gathering locomotive, §56, p48.
  six-wheel locomotive, §56, p38.
Jig planes, §58, p8.
  planes, Calculations for, §58, p28.

J

Joint, Moran flexible, §55, p29.
Journal-box, Roller, §57, p26.

K

Keeping surface water from mine, §60, p1.
Keeps, §53, p26.
Kinds of hoisting engines, §50, p6.
Knock-off links, or detaching hooks, §57, p65.
Knuckle sheaves, §58, p24.
Koepe system, Advantages of, §52, p23.
  system of hoisting, §52, p21.

L

Landing chairs, §53, p27.
  dogs, §53, p27.
  fans or keeps, §53, p26.
Lang-lay ropes, §51, p7.
Lay of a rope, §51, p6.
Level, Water, §60, p4.
  water, Flat deposits above, §60, p4.
  water, Flat deposits below, §60, p8.
  water, Inclined deposits above, §60, p12.
  water, Inclined deposits below, §60, p15.
Levels, Ditches on, §60, p15.
Lever, Differential, §52, p41.
Lines, Branch, §56, p11.
  Pipe, §55, p24.
Links, Knock-off, §57, p65.
  reversing, Specifications for, §51, p54.
Load, Moment of, §51, p38.
  on engine, §51, p36; §57, p71.
  on engine, Uniform, §57, p4.
  Working, §51, p25.

INDEX

Loading mine cars, Cost of, §54, p46.
  skips, Method of §53, p18.
Location of dams, §60, p22.
  of sheaves, §57, p61.
  of wires, §56, p12.
Locked-wire rope, §51, p6.
Locks, Car, §53, p23.
Locomotive, Adhesive power of, §55, p11.
  Baldwin-Westinghouse, §56, p19.
  frame, §56, p21.
  gathering, General Electric, §56, p45.
  gathering, Goodman, §56, p50.
  gathering, Jeffrey, §56, p48.
  General Electric, §56, p35.
  Goodman single-motor, §56, p39.
  haulage, §55, p1.
  haulage, Compressed-air, §55, p22.
  haulage, Electric, §56, p1.
  Jeffrey six-wheel, §56, p38.
  mine, Connections of, §56, p31.
  Morgan-Gardner, §56, p38.
  Morgan-Gardner combined third- and rack-rail, §56, p41.
  Overcylindered, §55, p12.
  Power of, §55, p10.
  Special geared, §55, p34.
  Speed of, §55, p18.
  storage tanks, §55, p38.
  storage tanks, Size of, §55, p39.
  Undercylindered, §55, p12.
  wiring, §56, p34.
  Work of, §55, p20.
Locomotives, Care of, §56, p67.
  Comparative size of, §56, p51.
  Compressed-air, §55, p30.
  compressed-air, Charging, §55, p26.
  compressed-air, Direct system of charging, §55, p26.
  Compressed-air gathering, §55, p35.
  Construction of steam-mine, §55, p5.
  electric, Speed of, §56, p64.
  Gathering, §56, p45.
  Haulage, §56, p35.
  Hauling capacity of, §55, p15.
  mine, Dimensions of, §55, p9.
  mine, Dimensions of electric, §56, p52.
  mine, Electric, §56, p19.
  mine, Four-wheel, connected, steam, §55, p7.
  mine, Power for, §55, p3.
  mine, Rails for electric, §56, p7.
  mine, Six-wheel, connected, steam, §55, p8.
  Mine tracks for, §55, p2.
  Operation of electric mine, §56, p61.
  Surplus power for, §55, p22.
  Tractive power of compressed-air, §55, p43.
  Weight of, and capacity of motors, §56, p52.
  with air-tank tenders, §55, p34.

Lodgment, Shaft, §60, p9.
Lubrication of ropes, §51, p10.
Lubricators and oil cups, Specifications for, §51, p56.

M

Machine-molded gears, §50, p14.
Machinery, Hoisting, §50, p1.
Man power for transportation, §54, p44.
Map, Contour drainage, §60, p5.
Material for construction, §51, p46.
  Specifications for, §53, p44.
  used in making wire ropes, §51, p2.
Maximum starting effort, §55, p11.
Mechanical versus animal haulage, §54, p51
Method of dumping skips, §53, p19.
  of loading skips, §53, p18.
  of operating tail-rope system, §57, p41.
Methods of splicing ropes, §51, p14.
  of supporting wires, §56, p12.
Mine-car axle, §54, p32.
  car, Bituminous, §54, p18.
  -car brakes, §54, p34.
  -car details, §54, p24.
  -car haulage, Resistance to, §54, p37.
  -car wheels and axles, §54, p27.
  -car wheels, Comparison of, §54, p27.
  Carrying wires into, §56, p7.
  cars, §54, p1.
  cars, Anthracite, §54, p12.
  cars, Capacity of, §54, p3.
  cars, Cost of loading, §54, p46.
  cars, Oil for, §54, p33.
  dams, §60, p22.
  dams, Thickness of, §60, p31.
  haulage, electric, Wiring for, §56, p7.
  -haulage systems, §54, p43.
  Keeping surface water from, §60, p1.
  locomotive, Connections of, §56, p31.
  locomotives, Dimensions of, §55, p9.
  locomotives, Dimensions of electric, §56, p52.
  locomotives, Electric, §56, p19.
  locomotives, electric, Operation of, §56, p61.
  locomotives, electric, Rails for, §56, p7.
  locomotives, Four-wheel, connected, steam, §55, p7.
  locomotives, Power for, §55, p3.
  locomotives, Six-wheel, connected, steam, §55, p8.
  locomotives, Tracks for, §55, p2.
  sheave frames, §58, p40.
  stables, §54, p53.
  water, Handling of, §60, p1.
  workings, abandoned, Drainage of, §60, p7.
Mines, Flow of water into, §60, p2.
Mining hoist, Direct-current, §50, p34.
  stope, Drainage in, §60, p17.

Moment of engine, §51, p38.
  of load, §51, p38.
Moran flexible joint, §55, p29.
Morgan-Gardner combined third- and rack-rail locomotive, §56, p41.
  -Gardner locomotive, §56, p38.
Motormen, §56, p61.
Motorneers, §56, p61.
Motors, §56, p24.
  Capacity of, §56, p52.
Mule haulage, §54, p47.
  haulage, Cost of, §54, p50.
  haulage in mine, Speed of, §54, p49.
  haulage, Safe grade for, §54, p50.
  teams, §54, p48.
Mules, Care of, §54, p52.
  Feeding, §54, p52.
Multiple-deck cage, §53, p11.
  -grooved wheels, §58, p12.

O

Objections to bucket drainage, 60, p40.
Oil cups, Specifications for, §51, p56.
  engines, §50, p30.
  for mine cars, §54, p33.
  -saving device, §54, p33.
Operating tail-rope, §57, p52.
Operation of electric mine locomotives, §56, p61.
  of endless-rope system, §57, p36.
  of tanks, §60, p47.
Ore cars, §54, p6.
  cars, Iron, §54, p8.
Overcylindered locomotives, §55, p12.
Overhead endless-rope haulage system, §57, p22.
Overloads, §56, p58.
Overwinding device for hoisting engines, §50, p27.

P

Painting, Specifications for, §51, p56; §53, p45.
Parts of a hoisting engine, §50, p20.
Permanent foundations, §51, p46.
Petticoat hanger, §56, p13.
Pillars, Barrier, §60, p21.
Pipe line and compressor service, §55, p30.
  lines, §55, p24.
Pipes, Blowing off, §55, p26.
  Friction in, §55, p26.
  steam, Specifications for, §51, p55.
  Storage capacity of, §55, p28.
Piston rods, Specifications for, §51, p52.
Pistons, Specifications for, §51, p52.
Plane, Double-track, §58, p42.
  engine, Calculation for, §58, p41.
  gravity, Three-rail, §58, p3.

Plane—(Continued)
  Self-acting, §58, p1.
  Single-track, §58, p41.
Planes, Engine, §58, p32.
  Gravity, §58, p1.
  gravity, Calculations for, §58, p26.
  gravity, Drums for, §58, p10.
  gravity, Inclination of, §58, p25.
  Jig, §58, p8.
  jig, Calculations for, §58, p28.
Plants, water-hoisting, Cost of, §60, p51.
Plow-steel ropes, §51, p2.
Pneumatic fans, §53, p30.
  gong signal, §53, p50.
Portable hoists, §50, p8.
Post brake, §52, p32.
Power brakes, §52, p42.
  for air compressors, §55, p23.
  for brakes, §52, p39.
  for driving endless-rope system, §57, p4.
  for mine locomotives, §55, p3.
  for operating tail-rope, §57, p52.
  Hand, §50, p1.
  of a locomotive, §55, p10.
  reverse, §50, p24.
  Tractive, §55, p10.
Pressure on dams, §60, p23.
  Reducing cylinder, §55, p40.
  Working, §55, p40.
Protected bonds, §56, p17.
Pull, Drawbar, §55, p10; §56, p52.
  Drawbar, on a grade, §55, p14; §56, 56.
Pumping water versus hoisting, §60, p53.

R

Rail bonds, §56, p17.
Rails for electric mine locomotives, §56, p7.
Reducing cylinder pressure, §55, p40.
  valve, Automatic, §55, p40.
Reel and flat rope, Calculating size of, §52 p18.
Reels and drums, §52, p6.
  Flat-rope, §52, p17.
Regular lay rope, §51, p6.
Relation between diameter of wire and diameter of rope, §51, p4.
Relief valve, Safety, §50, p23.
  valves, §50, p23.
Reheating compressed air, §55, p41.
Resistance, Car, §54, p39.
  due to curves, §55, p16.
  Frictional, §54, p37.
  Grade, §54, p39.
  on curves, §55, p18.
  to cars in motion, §54, p38.
  to mine-car haulage, §54, p37.
  Track, §54, p39.

INDEX

Resistances, Starting, §56, p33.
Rests, Cage, §53, p27.
Return sheave, §57, p55.
Reverse, Air-power, §50, p24.
    curves, §57, p32.
    cylinder, §56, p29.
    Power, §50, p24.
    Steam- or air-power, §50, p24.
Reversing engines, Specifications for, §51, p54.
    links, Specifications for, §51, p54.
Robbed areas, Drainage of, §60, p7.
Rocker-arms, Specifications for, §51, p54.
Roller journal-box, §57, p26.
Rollers and carrying shafts, §52, p5.
    Cast-iron, §57, p25.
    Roof rope, §57, p28.
    rope, Care of, §57, p28.
    rope, Spacing, §57, p29.
    Spacing of, §57, p40.
    Track, §57, p24.
    Wooden, §57, p24.
Roof rope rollers, §57, p28.
Room drainage, §60, p17.
Rope and wire, Relation between diameters of, §51, p4.
    couplings, §57, p64.
    -end fitting, §51, p11.
    fastenings, §52, p27.
    fillers, §51, p11.
    guides, §57, p32.
    haulage, §57, p1; §58, p1.
    Lay of a, §51, p6.
    Locked-wire, §51, p6.
    Regular-lay, §51, p6.
    rollers, Care of, §57, p28.
    rollers, Spacing, §57, p29.
    sections, §51, p3.
    Size required, §51, p29.
    six-strand, wire, Splicing a thimble into, §51, p13.
    socket, §51, p11.
    stress, §51, p25.
    thimble, §51, p12.
    travel, §51, p35.
    wheels, §52, p21.
Ropes, §51, p1.
    Cast-steel, §51, p2.
    Crucible-steel, §51, p2.
    Flat, §51, pp7, 31.
    flat, Attaching, §51, p14.
    Flattened-strand, §51, p4.
    Haulage, §51, p1.
    Hoisting, §51, p1.
    Inspection of, §51, p9.
    Iron-wire, §51, p2.
    Lang-lay, §51, p7.

Ropes—(Continued)
    Lubrication of, §51, p10.
    operated through bore holes, §58, p39.
    Plow-steel, §51, p2.
    Round, §51, p2.
    Round hoisting and haulage, §51, p29.
    Round-strand, §51, p2.
    Taper, §51, p8.
    Transmission, §51, p1.
    Universal-lay, §51, p7.
    Vegetable-fiber, §51, p1.
    Wire, §51, p2.
    wire, Materials used in making, §51, p2.
    wire, Splicing, §51, p14.
    wire, Strength of, §51, p25.
    wire, Varieties of, §51, p2.
    wire, Wear of, §51, p18.
Rotary dump cars, §54, p9.
Round hoisting and haulage ropes, Table, §51, p29.
    ropes, §51, p2.
    -strand ropes, §51, p2.

S

Safety blocks, §58, p20.
    catches, §53, p7.
    relief valve, §50, p23.
    switches, §58, p19.
Sand boxes, §56, p23.
Schedule, §56, p67.
Scoop-box car, §54, p10.
Second-motion hoisting engine, §50, p7.
Sections, Rope, §51, p3.
Self-acting plane, §58, p1.
    -oiling wheels, §54, p31.
Series-parallel controllers, §56, p27.
Shaft dams, §60, p30.
    friction, §51, p26.
    Gilberton water, §60, p49.
    lodgment, §60, p9.
    tumbling, Specifications for, §51, p54.
Shafts and rollers, §53, p5.
    vertical, Cages for, §53, p6.
Shape of trolley wires, §56, p8.
Sheave, Cast-iron, §53, p1.
    Diameter of, §53, p5.
    frames, Mine, §58, p40.
    frames, Surface, §58, p39.
    head, Specifications for, §51, p56.
    or drum, Diameter of, §51, p41.
    Return, §57, p55.
    Shell of a drum, §52, p7.
Sheaves, §53, p1.
    Carrying, §57, p24.
    Compensating, §57, p9; **§58, p24.**
    Horizontal, §57, p58.
    Knuckle, §58, p24.

INDEX

Sheaves—(Continued)
   Location of, §57, p61.
   Tail or return, §57, p58.
   Vertical, §57, p60.
   Wood-lined, §53, p4.
Shocks caused by grounds, §56, p62.
Signal, Hammer and plate, §53, p48.
   Pneumatic gong, §53, p50.
Signaling, §53, p48; §57, p56.
Single-motor locomotive, Goodman, §56, p39.
   -track plane, §58, p41.
Siphon, Air in a, §60, p35.
   Discharge of, §60, p33.
Siphons, §60, p33.
Six-wheel, connected, steam, mine locomotives, §55, p8.
   -wheel locomotive, Jeffrey, §56, p38.
Size of car, §54, p1.
   of flat rope and reel, Calculating, §52, p18.
   of hoisting engine, §51, p33.
   of locomotive storage tanks, §55, p39.
   of rope required, §51, p29.
   of trains, §57, p76.
   of trolley wire, §56, p8.
   of wires, Calculating, §56, p9.
   required for a given service, §56, p59.
Skip cage, §53, p23.
Skips, Method of dumping, §53, p19.
   Method of loading, §53, p18.
   or gunboats, §53, p17.
Slide-valve engine, Balanced, with plain drum, §50, p17.
Slipping of wheels, §56, p62.
   point, §55, p11.
Slope carriage, §53, p15.
   or inclined-shaft hoisting, §53, p14.
Slopes, Hoisting water on, §60, p46.
   Speed of hoisting water on, §60, p47.
Socket, Rope, §51, p11.
Spacing of rollers, §57, p40.
   rope rollers, §57, p29.
Speaking tubes, §53, p50.
Special geared locomotive, §55, p34.
   indicators, §52, p4.
Specifications, Hoisting-engine, §51, p51.
   for head-frames, §53, p44.
Speed of a locomotive, §55, p18.
   of electric locomotives, §56, p64.
   of hoisting, §51, p35.
   of hoisting water on slope, §60, p47.
   of mule haulage in mine, §54, p49.
   of trains, §57, p75.
Spherical dams, §60, p32.
Splicing a thimble into a six-strand wire rope, §51, p13.
Spring loaded valve, §50, p24.
   switch, §58, p18.

Stables, Mine, §54, p53.
   Surface, §54, p53.
Star wheel, §56, p29.
Starting effort, §54, p38.
   effort, Maximum, §55, p11.
   resistances, §56, p33.
Stationary hoisting engines, §50, p8.
   storage capacity, §55, p27.
Storage system of charging, §55, p26.
Steam locomotive versus animal haulage, §55, p5.
   locomotive haulage, §55, p4.
   mine locomotives, Construction of, §55, p5.
   or air-power reverse, §50, p24.
   pipes, Specifications for, §51, p55.
   -power hoists, §50, p6.
Steel coal cars, §54, p23.
Stone dams, §60, p29.
Stop-valve, The, §55, p40.
Stope mining, Drainage in, §60, p17.
Storage capacity of pipes, §55, p28.
   system of charging, §55, p26.
   tanks, Locomotive, §55, p38.
   tanks, locomotive, Size of, §55, p39.
Straight dams, §60, p32.
Strap brake, §52, p34.
Strength of wire ropes, §51, p25.
Strengthening gear-teeth, §50, p14.
Stress, Bending, §51, p25.
   Rope, §51, p25.
Sumps, §60, p8.
Supporting wires, Methods of, §56, p12.
Surface sheave frames, §58, p39.
   stables, §54, p53.
   tail-rope system of haulage, §57, p55.
   water, Keeping, from mine, §60, p1.
Surplus power for locomotives, §55, p22.
Swamp drainage, §60, p5.
Switch, Automatic trolley, §56, p16.
   Spring, §58, p18.
Switches, Automatic, §58, p18.
   Derailing, §58, p19.
   Safety, §58, p19.
Systems of mine haulage, §54, **p43**.
   Rope haulage, §57, p1.

T

Tail or return sheaves, §57, p58.
   -rope and endless haulage, Comparison of, §57, p50.
   -rope haulage, §57, p37.
   -rope haulage calculations, §57, p71.
   -rope haulage, District, §57, p42.
   -rope haulage engines, §57, p52.
   -rope haulage on an incline, §57, p74.
   -rope haulage, Third-rail, §57, p67.
   -rope hoisting, §51, p37.

INDEX

Tail—(Continued)
   -rope, Power for operating, §57, p52.
   -rope system, Endless, §57, p51.
   -rope system, Method of operating, §57, p41.
   -rope system of haulage, Surface, §57, p55.
Tanks and water buckets, §60, p41.
   Auxiliary, §55, p39.
   End-dump, §60, p44.
   Locomotive storage, §55, p38.
   Operation of, §60, p47.
Taper ropes, §51, p8.
Tapping and draining abandoned workings, §60, p18.
   water from below, §60, p20.
Teeth, gear, Strengthening, §50, p14.
Telephones, §53, p50.
Temporarily abandoned mine workings, Drainage of, §60, p7.
Tenders, air-tank, Locomotives with, §55, p34.
Tension and grip, §57, p7.
Thickness of mine dams, §60, p31.
Thimble, Rope, §51, p12.
Third-rail tail-rope haulage, §57, p67.
Three-rail gravity plane, §58, p3.
   -rail inclines, §58, p3.
Throttle, Double-seat, §50, p23.
   valve, Balanced, §50, p21.
   valves, §50, p20.
Torque, §56, p10.
Track resistance, §54, p39.
   rollers, §57, p24.
Tracks, Arrangement of, §58, p2.
   for mine locomotives, §55, p2.
   gravity plane, Anchoring, §58, p17.
Tractive force, §55, p10.
   power, §55, p10.
   power of compressed-air locomotives, §55, p43.
Trains, Size of, §57, p76.
   Speed of, §57, p75.
Tramming, §54, p45.
Transmission ropes, §51, p1.
Transportation, Man power for, §54, p44.
Trolley frogs, §56, p16.
   switch, Automatic, §56, p16.
   wire, Shape of, §56, p8.
   wire, Size of, §56, p8.
Tubes, Speaking, §53, p50.
Tumbling shaft, Specifications for, §51, p54.
Tunnel, Jeddo, §60, p14.
Tunnels, Drainage, §60, p13.
Twisted-link coupling, §54, p36.
Two- and three-rail inclines in combination, §58, p3.
Types of head-frames, §53, p36.
   of indicators, §52, p1.

U

Undercylindered locomotive, §55, p12.
Underground drainage, §60, p4.
Uniform load on engine, Advantages of, §57, p4.
Union couplings, §55, p25.
Universal-lay ropes, §51, p7.
Unprotected bonds, §56, p17.
Unwatering, §60, p1.
Use of circuit-breakers, §56, p65.
   of compressed air expansively, §55, p42.
   of controller, §56, p63.
   of first-motion engines, Conditions governing, §50, p15.
   of mine dams, §60, p22.

V

Valve, Automatic reducing, §55, p40.
   Balanced throttle, §50, p21.
   Safety relief, §50, p23.
   Spring loaded, §50, p24.
   The stop, §55, p40.
Valves, Butterfly, §60, p46.
   Charging, §55, p28.
   Relief, §50, p23.
   Throttle, §50, p20.
   Water-bucket, §60, p41.
Variation in amount of water, §60, p3.
Varieties of wire ropes, §51, p2.
Varying grades, §58, p30.
Vegetable-fiber ropes, §51, p1.
Vertical sheaves, §57, p60.
Voltage and current, §56, p2.

W

Water, §60, p43.
   -bucket valves, §60, p41.
   buckets, §60, p43.
   buckets and tanks, §60, p41.
   Draining off, §60, p19.
   Flow of, into mines, §60, p2.
   hoisting, §60, p39.
   -hoisting plants, Cost of, §60, p51.
   level, §60, p4.
   level, Flat deposits above, §60, p4.
   level, Flat deposits below, §60, p8.
   level, Inclined deposits above, §60, p12.
   level, Inclined deposits below, §60, p15.
   mine, Handling of, §60, p1.
   pressure on dams, §60, p23.
   shaft, Gilberton, §60, p49.
   -tank car wheels, §60, p47.
   Tapping, from below, §60, p20.
   Variation in amount of, §60, p3.
Wear of wire ropes, §51, p8.
Wedge-shaped wooden dams, §60, p24.
Weight of locomotive and capacity of motors, §56, p52.

Wheel, Index, §56, p29.
    Star, §56, p29.
Wheelbarrow transportation, §54, p44.
Wheels and axles, §56, p21.
    car, Water-tank, §60, p47.
    Comparison of mine-car, §54, p27.
    Head, §58, p11.
    Mine-car, §54, p27.
    Multiple-grooved, §58, p12.
    Slipping of, §56, p62.
    Rope, §52, p21.
    Self-oiling, §54, p31.
Whim, Cornish horse, §50, p3.
Whims, Adaptability of, §50, p5.
Whiting system of hoisting, §52, p23.
    system of hoisting, Modified, §52, p26.
Winch, Crab, §50, p2.
Winding engine, §50, p1.
Windlass, §50, p1.
Wings, §53, p27.
Wire and rope, Relation between diameters of, §51, p4.
    -rope haulage systems, Cost of operating, §57, p51.
    rope, six-strand, Splicing a thimble into, §51, p13.
    ropes, §51, p2.
    ropes, Materials used in making, §51, p2.

Wire—(Continued)
    ropes, Splicing, §51, p14.
    ropes, Strength of, §51, p25.
    ropes, Wear of, §51, p8.
    ropes, Varieties of, §51, p2.
    trolley, Shape of, §56, p8.
    trolley, Size of, §56, p8.
Wires, Calculating size of, §56, p9.
    in mine, §56, p7.
    Location of, §56, p12.
    Methods of supporting, §56, p12.
Wiring for electric mine haulage, §56, p7.
    Locomotive, §56, p34.
Wood-lined sheaves, §53, p4.
Wooden dams, Flat, §60, p26.
    dams, Wedge-shaped, §60, p24.
    ore cars, §54, p6.
    rollers, §57, p24.
    -tooth gears, §50, p13.
Work of a locomotive, §55, p20.
    of engine, §51, p37.
Working load, §51, p25.
    places, Drainage of, §60, p17.
    pressure, §55, p40.
Workings, abandoned mine, Drainage of, §60, p7.
Workmanship, Specifications for, §53, p44.

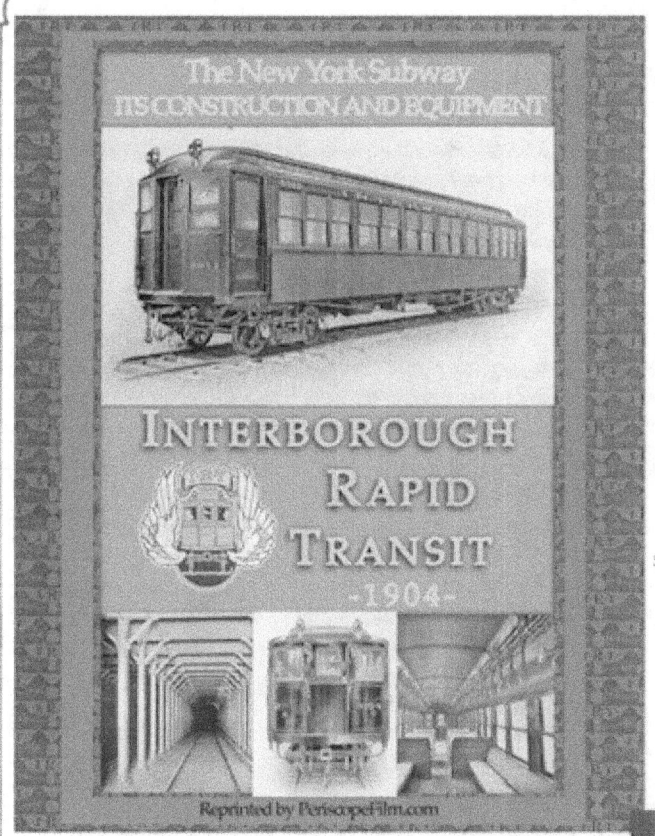

On October 27, 1904, the Interborough Rapid Transit Company opened the first subway in New York City. Running between City Hall and 145th Street at Broadway, the line was greeted with enthusiasm and, in some circles, trepidation. Created under the supervision of Chief Engineer S.L.F. Deyo, the arrival of the IRT foreshadowed the end of the "elevated" transit era on the island of Manhattan. The subway proved such a success that the IRT Co. soon achieved a monopoly on New York public transit. In 1940 the IRT and its rival the BMT were taken over by the City of New York. Today, the IRT subway lines still exist, primarily in Manhattan where they are operated as the "A Division" of the subway. Reprinted here is a special book created by the IRT, recounting the design and construction of the fledgling subway system. Originally created in 1904, it presents the IRT story with a flourish, and with numerous fascinating illustrations and rare photographs.

Originally written in the late 1900's and then periodically revised, A History of the Baldwin Locomotive Works chronicles the origins and growth of one of America's greatest industrial-era corporations. Founded in the early 1830's by Philadelphia jeweler Matthais Baldwin, the company built a huge number of steam locomotives before ceasing production in 1949. These included the 4-4-0 American type, 2-8-2 Mikado and 2-8-0 Consolidation. Hit hard by the loss of the steam engine market, Baldwin soldiered on for a brief while, producing electric and diesel engines. General Electric's dominance of the market proved too much, and Baldwin finally closed its doors in 1956. By that time over 70,500 Baldwin locomotives had been produced. This high quality reprint of the official company history dates from 1920. The book has been slightly reformatted, but care has been taken to preserve the integrity of the text.

NOW AVAILABLE AT
WWW.PERISCOPEFILM.COM

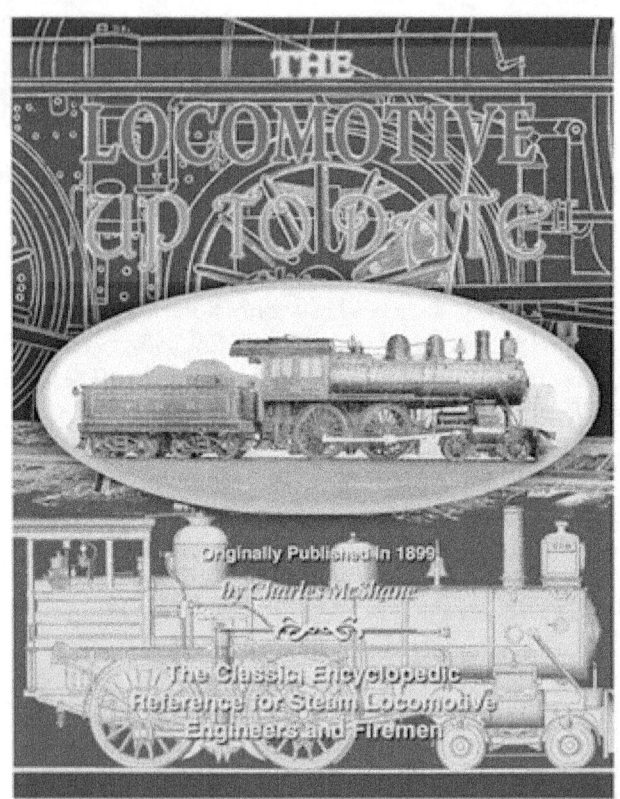

When it was originally published in 1899, **The Locomotive Up to Date** was hailed as "...the most definitive work ever published concerning the mechanism that has transformed the American nation: the steam locomotive." Filled with over 700 pages of text, diagrams and photos, this remains one of the most important railroading books ever written. From steam valves to sanders, trucks to side rods, it's a treasure trove of information, explaining in easy-to-understand language how the most sophisticated machines of the 19th Century were operated and maintained. This new edition is an exact duplicate of the original. Reformatted as an easy-to-read 8.5x11 volume, it's delightful for railroad enthusiasts of all ages.

Originally printed in 1898 and then periodically revised, **The Motorman...and His Duties** served as the definitive training text for a generation of streetcar operators. A must-have for the trolley or train enthusiast, it is also an important source of information for museum staff and docents. Lavishly illustrated with numerous photos and black and white line drawings, this affordable reprint contains all of the original text. Includes chapters on trolley car types and equipment, troubleshooting, brakes, controllers, electricity and principles, electric traction, multi-car control and has a convenient glossary in the back. If you've ever operated a trolley car, or just had an electric train set, this is a terrific book for your shelf!

ALSO NOW AVAILABLE FROM PERISCOPEFILM.COM!

The technology of mining is the subject of this fascinating book, and two companion volumes, all of which were originally published in 1907. Mining: Haulage details the railways that operate in the underground world of the mine. The book contains over 300 pages of text, numerous illustrations, and a set of examination questions for the mining sciences student. It contains chapters about steam locomotives, electric locomotives and wiring, and cable railway systems and the principles behind them. It also examines compressed air, gravity and rope, and animal haulage. This historic book has been reprinted in its entirety. It's a treat for anyone who ever worked underground, or for anyone who ever wondered, "How does that work?"

331 Pages, 8.5x11, softbound

Mining: Hoisting details the elevators, hoists and component machinery used to lift miners, supplies and ore. It contains over 200 pages of text, numerous illustrations and a set of examination questions for the mining sciences student. The book examines electric, steam and hand-powered hoists and explains the principles behind them in detail. It also delves into the control and signaling systems used to ensure safe and reliable operation.

232 Pages, 8.5x11, softbound

ALSO NOW AVAILABLE FROM PERISCOPEFILM.COM!

©2008-2010 Periscope Film LLC

All Rights Reserved

www.PeriscopeFilm.com

ISBN #978-1-935700-13-5

www.ingramcontent.com/pod-product-compliance
Lightning Source LLC
Chambersburg PA
CBHW082107230426
43671CB00015B/2628